W9-CRL-155

Dawn of the Dinosaurs

Life in the Triassic

NICHOLAS FRASER

ILLUSTRATED BY
DOUGLAS HENDERSON

INDIANA UNIVERSITY PRESS
Bloomington and Indianapolis

This book is a publication of Indiana University Press
601 North Morton Street Bloomington, IN 47404-3797 USA

http://iupress.indiana.edu
Telephone orders   800-842-6796
Fax orders   812-855-7931
Orders by e-mail   iuporder@indiana.edu

© 2006 by Nicholas Fraser and Douglas Henderson
All rights reserved
HIDDEN TREASURE
Words and Music by STEVE WINWOOD and JIM CAPALDI
STEVE WINWOOD and JIM CAPALDI
1972 F.S.MUSIC LTD. and ISLAND MUSIC LTD.
All Rights for F.S.MUSIC LTD. Administered by WARNER-TAMERLANE PUBLISHING
CORP.
All Rights Reserved
Lyrics Reprint with the Permission of Alfred Publishing Co., Inc.

No part of this book may be reproduced or utilized in any form or by any means, electronic or
mechanical, including photocopying and recording, or by any information storage and retrieval
system, without permission in writing from the publisher. The Association of American University
Presses' Resolution on Permissions constitutes the only exception to this prohibition.
The paper used in this publication meets the minimum requirements of American National
Standard for Information Sciences—Permanence of Paper for Printed Library Materials, ANSI
Z39.48-1984.

Printed in China

Library of Congress Cataloging-in-Publication Data

Fraser, Nicholas C.
    Dawn of the dinosaurs : life in the Triassic / Nicholas Fraser ; illustrated by Douglas
Henderson.
        p. cm.—(Life of the past)
    Includes bibliographical references and index.
    ISBN 0-253-34652-5 (cloth : alk. paper)
    1. Paleontology—Triassic.   2. Paleoecology—Triassic.   3. Dinosaurs.   I. Title.   II. Series.
    QE732.F73 2006
    560'.1762—dc22

                                                                2005035986

1 2 3 4 5 11 10 09 08 07 06

For Bev Halstead

Take a walk down by the river
There's a lot that you can learn
If you've got a mind that's open, if you've got a heart that yearns

Message in the deep, from a strange eternal sleep
That is waiting there for you
Like hidden treasure

STEVIE WINWOOD/JIM CAPALDI, "Hidden Treasure"

contents

preface

An informal group of zoologists and geologists used to meet in a Cambridge, England, pub on a regular basis under the unassuming title of PALS. As you might expect from such a name, it was a friendly gathering and it was devoted to examining "Paleontology as a Life Science"—discussing how the animals we see today represented by fossils actually lived. This is the theme that we have endeavored to capture in this book. Paleontology does not have to be a science as drab as certain muds that contain some of its main subject matter, although admittedly when it is hot and humid and a plague of flies has been sent to eat you alive, it can be difficult to appreciate the finer points of fieldwork. On a field trip to the Tomahawk locality in the Richmond Basin of Virginia (see chapter 11), a good friend and colleague remarked, "What do you hope to find in this mudhole?" It is our aim to share with you the jewels of such mudholes. Certainly the site in question, which is little more than a tick-infested drainage ditch at the side of a dirt road, did not inspire much confidence. Yet the fossil wonders that it yielded tell an incredible story of a rich and varied world in the tropics 220 million years ago. Such unpretentious Triassic localities are not uncommon: an abandoned quarry with rusting machinery is home to one of the most spectacular assemblages of fossil insects in the world. On the other hand, one or two of the localities are surrounded by some of today's most awe-inspiring landscapes—the magnificent alpine scenery of northern Italy and the Chinle exposures in the Petrified Forest National Park, to name just two examples.

Each fossil wrestled from the ancient mud represents part of a once-living organism that interacted with a host of others in a vibrant, living world. Although soft parts are not commonly preserved, and details of such things as color and behavior are at best sketchy, new discoveries and new techniques, coupled with inferences based on modern ecosystems, can often provide us with a complete image of life in the past. Rather than examining individual fossils in isolation from their world, in this book we look at the Triassic period in traditional natural history terms. In particular, we have focused on those areas that reveal wonders and a diversity of life that were on a par with the national parks and World Heritage sites of today. Although the terrestrial vertebrates are the main focus of the book, we fully recognize that these animals did not live in isolation. The vast swaths of plant life and myriads of invertebrates that lived at the same time molded the environment, and we have therefore endeavored to include details of these wherever possible.

This book is certainly not intended to be a comprehensive study of all terrestrial life in the Triassic. We freely admit that there are significant omissions. But we have tried to pick a selection of regions and fossil deposits that give the overall flavor of Triassic life. We make no apology for the fact that the book is skewed toward the Upper Triassic. We consider this to be a pivotal time in the history of life on earth because it documents the foundation of our modern-day terrestrial ecosystems.

acknowledgments

Over the years it has been a pleasure to make so many good friends through a mutual interest in paleontology. Numerous colleagues have generously shared their ideas and willingly discussed all topics relating to life in the Mesozoic. I particularly want to express my appreciation to Hans Sues, Paul Olsen, Olivier Rieppel, Kevin Padian, and Mike Benton for so many long discussions, often extending into the night over several beers; my mind—if not my physical health—has certainly been enriched by these five gentlemen! Dave Grimaldi has patiently taught me about fossil insects, but I regret to say that I still have a long way to go. Without Brian Axsmith's assistance and keen eye, I would almost certainly have been responsible for robbing paleobotany of several significant specimens. I have benefited from so many friends who have provided advice and encouragement. They include Eric Buffetaut, Bob Carroll, Sankar Chatterjee, Arthur Cruickshank, Susan Evans, Jacques Gauthier, Andy Heckert, Adrian Hunt, Spencer Lucas, Zhexi Luo, Andrew Milner, Stefania Nosotti, Silvio Renesto, Andrei Sennikov, Bob Sullivan, Dave Norman, Robert Reisz, Rainer Schoch, Mary Schweitzer, Dave Unwin, Dave Weishampel, Dave Whiteside, Rupert Wild, and Xiao-chun Wu.

Gordon Walkden has been a valued mentor. He nurtured me through the finer points of carbonate sedimentology and introduced me to the wonderful world of vertebrate paleontology. He also taught me how to get out of a quarry when you were locked in late at night!

I shall always be grateful to Bev Halstead. His excitement and enthusiasm for paleontology, politics, and partying were infectious. I shall certainly never forget the grueling six-hour-long doctoral viva that he put me through, followed by an excursion into the drinks cabinet of the Geology Department of Aberdeen University! Our science lost a great advocate on his untimely death.

Other influences on my career who, sadly, have passed away include Alick Walker, Allen Charig, and Pamela Robinson. Zak Erzinclioglu was a wonderful inspiration who gave me pause for thought on innumerable occasions. He was a man who understood the richness of the natural world and took great pleasure in investigating its complexity and diversity. I hope this book encourages the reader to follow in Zak's footsteps and take delight in the hidden treasures of the natural world that surround us all.

The crew at Cambridge—Ken Joysey, Adrian Friday, Sue Gay, Henry Gee, Per Ahlberg, Janet Harker, Enid McRobbie, Christine McKie, and Jenny Clack—provided a wonderful atmosphere for scientific debate. That wonderful English institution of afternoon tea was a source of much inspiration for me in the zoology department.

Many people have assisted my field expeditions into the Triassic, but I especially want to recognize Brian Axsmith, Alton Dooley, Christa Hampton, Phil Huber, Pete Le-Tourneau, Paul Olsen, Julian MacCarthy, and Gordon Walkden for their labors of love.

John Roberts was a tremendous support for my work in the Virginia Solite Quarry. While manager of the Solite Quarry, C. H. Gover always welcomed me and my field crew with a grin. We would then sweat pounds by toiling in the furnace that is the typical July day in southern Virginia, until our heads began to swim. A sincere thank you to all the quarry managers who have unfailingly trusted me next to the most dangerous faces in their quarries both in Britain and the United States.

My scientific colleagues associated with the Virginia Museum of Natural History have been a wonderful support. A big thank you to Jim Beard, Alton Dooley, Richard Hoffman, Nancy Moncrief, Elizabeth Moore, Clayton Ray, Bill Shear, Lauck Ward, and Judy Winston.

I have benefited from visits to many museums worldwide and appreciate all the assistance that I have received from curators and collections managers. The list is too long to detail here, but I especially want to thank the staffs of the Natural History Museum, London; Museo Civico di Scienze Naturali "Enrico Caffi," Bergamo; Museo Civico di Storia Naturale, Milano; Staatliches Museum für Naturkunde, Stuttgart; Paläontologisches Institut und Museum der Universität, Zurich; the Paleontological Institute, Moscow; Institute of Palaeontology and Palaeoanthropology, Beijing, New Mexico Museum of Natural History and Science, the Field Museum, University of California Museum of Paleontology, Berkeley; Museum of Comparative Zoology, Harvard; and the National Museum of Natural History, Washington, D.C.

Without the guidance of Jim Farlow, Bob Sloan, Dawn Ollila, and Miki Bird, this book would never have seen the light of day. Jim and Bob patiently ironed out all my writing foibles. Karen Hellekson meticulously corrected all the inconsistencies in my final text.

Images of the insects from the Vosges (figs. 4.1, 4.2, and 4.3) and of *Monilipartus tenuis* (fig. 4.4) are courtesy of Francois Papier, and those on modern caves (figs. 10.2 and 10.3) are courtesy of Mike Simms. The photograph of *Icarosaurus* (fig. 11.6) is courtesy of Bob Sullivan. Anna Paganoni and Heinz Fürrer provided me unlimited access to the collections in Bergamo and Zurich and allowed me to photograph the specimens.

The words to "Hidden Treasure" are reproduced courtesy of Universal Music Publishing Group and Alfred Publishing Company, Incorporated.

The financial support of the National Science Foundation and National Geographic on various occasions has certainly had an impact on this book.

Finally, but by no means least, I want to thank Chris, Hannah, and Amy for accepting my lengthy periods away from home to pursue my passion, and for understanding my lack of attention for so many evenings as I sat at my computer with my head buried in books. You certainly deserve "presents for girls!"

introduction

For many, the end of the Triassic period is notable for marking the beginning of the Age of Dinosaurs. For others, it is also a world of truly fantastic creatures. But it is much more than that: the Triassic signaled the first steps toward the establishment of modern terrestrial ecosystems. It was a melting pot of the ancient and modern.

The Triassic can be viewed as a drama in four acts. Set against a backdrop of a world with a single large continent and no polar icecaps, the curtain opens on a stage that has been decimated by the Permian extinctions. By Act Two, center stage is dominated by a terrestrial fauna of archosaurs and mammal-like reptiles. The climax of the play comes first with an explosion of bizarre life-forms toward the end of the Triassic, followed by a major extinction event. In the last act, we see the origin of practically all the major groups of modern tetrapods, plus, of course, the dinosaurs and pterosaurs.

It could become all too easy to fantasize at the sight of some of the amazing fossils from the Triassic and consequently come to look at that world as one far removed from our own. We must remember that in the Triassic, just like today, animals needed to breathe oxygen, and their waste matter needed to be disposed of and recycled. We must therefore try to strike a balance between images of the world as we know it and some bizarre planet inhabited by weird mutants.

Background to the Triassic

The Triassic is the first period of the Mesozoic era, and it is so named because of its tripartite division in the rocks of Germany. Here the marine Muschelkalk Group is sandwiched between two terrestrial groups: the Bundsandstein below and the Keuper above. The Muschelkalk Group is made up of dolomites, limestones, marls, and evaporites that were deposited in an epicontinental basin that covered much of Germany, but also parts of Denmark, Poland, and the North Sea. The carbonates include marine faunas that connected from time to time with the faunas of the southern European (Teythyan) basins.

The beginning of the period is marked by what has been called the "Mother of all Extinctions," the terminal event of the Permian. There is some debate as to whether the curtain dropped on the Triassic stage in the form of another mass extinction, triggered perhaps by the impact of a large bolide or the result of excessive volcanism (see chapter 12).

As Padian (1986) remarks, the Triassic was an extraordinary time in vertebrate history. Just like the rocks of the Germanic Basin, the terrestrial tetrapods can be split into three main divisions: the groups that survived the end Permian event; the groups that are unique to the Triassic; and the groups that appear at the end of the Triassic but attain their peak diversities in later times. It should be emphasized, though, that the temporal distributions of these three somewhat arbitrary groups do not coincide with the three divisions of the sediments in the Germanic Basin. The Early Triassic world was inhabited by holdovers from the Paleozoic, with the most predominant forms being the Therapsida, one of the major lineages of synapsid amniotes ("mammal-like reptiles"). The therapsids gradually diminished in importance as the Triassic progressed and finally died out in the Jurassic, but not before they had given rise to one important synapsid offshoot, the Mammalia. Other key forms in the Early Triassic terrestrial environment included the Procolophonia, a rather enigmatic group that some authors associate closely with the Chelonia (turtles and tortoises), and a variety of diapsids. Even by earliest Triassic times the differentiation into the two major diapsid lineages had occurred, and both archosauromorphs (a group that is represented today by the crocodiles and alligators) and lepidosauromorphs (the group that includes lizards and snakes) were present.

Among tetrapods, a number of groups are unique to the Triassic. Archosauromorphs such as the rhynchosaurs, phytosaurs, and aetosaurs were particularly abundant and diverse during the latter part of the Triassic. Many other groups, including the metoposaurs, plagiosaurs, rauisuchids, ornithosuchids, and traversodontids, are also confined to Triassic sediments and do not extend beyond the Triassic-Jurassic boundary. One of the mysteries that is hotly debated is just how rapidly these groups died out. Was their disappearance relatively gradual, or was a major catastrophe right at the close of the Triassic responsible for a rapid demise? It is even conceivable that both theories are partially correct, and that the gradual decline of certain tetrapods was accelerated and brought to complete closure at the very end of the Triassic by some agent of catastrophe. Whatever the reason, the third division of Triassic tetrapod life was ushered in only toward the end of the period. Although in the overall scheme of Triassic vertebrate life these new forms were perhaps not the most significant taxa, they are characterized by their explosive radiation at a later date. Thus we find in Late Triassic sediments the first mammals, turtles, crocodiles, and lissamphibians (salamanders and frogs), and of course pterosaurs and dinosaurs. By the end of the Triassic, then, essentially all the building blocks were in place for modern-day vertebrate terrestrial faunas.

Before entering the world of the Triassic and its animals and plants, we need an understanding of the rocks that entomb their remains and the distribution of these rocks worldwide. Although hints of the paleoenvironment can be gleaned from the morphology of the fossils, the sediments can also tell us much. But before any analysis of the sediments can begin, we need to negotiate a veritable minefield of terms and potential pitfalls regarding the global correlation of the rocks. Readers unfamiliar with the ins and outs of Triassic biostratigraphy will want to read appendix 1 before embarking on this journey through the Triassic.

Part One

THE END OF AN ERA
The Early Triassic

chapter one

Setting the Stage

The Paleogeography and Climate of the Triassic World

A Catastrophic Start

The coming together of the continents at the very end of the Permian to form the super-continent Pangaea has been heralded as one of the causes of the greatest mass extinction the world has ever witnessed. Undoubtedly the enormous loss of coastline that resulted from the final coalescence of the continents would have meant a huge loss of habitat for marine life. On top of all this, the eruption of the Siberian flood basalts (traps) would have had a profound effect on life worldwide. Flood basalts are the direct result of numerous closely spaced volcanic eruptions and are enormous accumulations of basaltic flows. The Siberian traps are probably the most extensive flood basalts of the entire Phanerozoic eon and cover approximately 2.5 million square kilometers. Although their age has been the subject of considerable debate, there is now a significant body of evidence suggesting that they were produced over a relatively short time (Campbell et al. 1992), and estimates of the age tend to be around, or even right at, the Permian-Triassic boundary (Baksi and Farrar 1991; Renne and Basu 1991). Some have suggested that the impact of an extraterrestrial body may also have been involved with the extinctions (Becker et al. 2001).

What precisely brought about the great mass extinction is not at all clear. It may well have been a complex sequence of factors, and there are still any number of candidates to choose from. Some authors have cited widespread anoxia (lack of oxygen); others have invoked carbon dioxide poisoning. We will need detailed studies of the groups of organisms that were affected and those that were not, together with an understanding of their physiology, to tease apart the agents of the extinctions. Only then will it be possible to determine with some measure of confidence the primary cause—or causes—of the extinction. Knoll et al. (1996) put forward a testable hypothesis: they argued that if a rise in carbon dioxide levels were primarily responsible for the Permian extinction event, then a telltale signature would be apparent. There would be marked selectivity in the extinctions of organisms because different groups of animals exhibit varying sensitivity to fluctuations in carbon dioxide levels. High levels of carbon dioxide would disrupt calcification and slow metabolism in those groups of marine organisms with heavy calcium carbonate skeletons and little physiological control over gas exchange. These groups include corals, bryozoans, and brachiopods. Animals with a noncalcium carbonate skeleton and animals without skeletons would be affected least by increases in carbon dioxide. However, organisms with moderately calcified skeletons but with well-developed physio-

Plate 1.1. The gorgonopsid (mammal-like reptile) *Lycaenops* in a Permian woodland.

logical controls over gas exchange (e.g., infaunal clams) would not be affected so severely. In fact, detailed analyses of the degree to which different groups were affected by the end Permian event support the hypothesis that raised carbon dioxide levels were directly responsible for the extinctions. Perhaps the overturning of anoxic deep oceans introduced high levels of carbon dioxide into surface environments worldwide, and in turn, this may have been the major factor for a spike of global warming that affected the terrestrial environment.

Whatever the ultimate outcome of these debates, we do know that at the beginning of the Triassic, not only was there little diversity among marine faunas, but the same was true for terrestrial faunas. The beginning of the Triassic was an empty playing field, so to speak, waiting for an explosion of evolutionary activity.

What do we know of the world's climate at the start of the Triassic? A good place to begin is with the climate at the end of the Permian. Although some studies indicate seasonal extremes of climate at high latitudes during the Late Permian, much of the fossil evidence points at much more temperate climates. For example, the *Glossopteris* flora that is so characteristic of Permian times in the southern hemisphere is widely regarded as comprising relatively cool weather plants. Yemane (1993) suggested that the presence of large lakes may have had a significant moderating effect on the Gondwanan climate. But with the approach of Triassic times these rather temperate climate conditions were not destined to last long.

The Hot and Arid Days of the Early to Middle Triassic

There is a wealth of support for the notion that the Permian-Triassic boundary marked a major change in global climate, with a warming trend and increased climatic instability (e.g., Holser and Magaritz 1987). Some of this support is drawn from climate modeling coupled with analyses of paleogeography. Throughout the entire Triassic period, all the world's continents remained joined together in the form of Pangaea (fig. 1.1). There is no doubt that a single continental landmass would have facilitated heat transfer from the equator to the poles by means of ocean currents. Unimpeded by the uneven distribution of land, ocean currents in some parts of the world would have had a direct path from the equatorial belt to the poles. Polar regions would therefore have been considerably warmer than they are now, and this would have effectively curtailed the build up of ice caps. In turn, the lack of polar ice caps might readily be taken as an indicator of a warm (greenhouse-type) climate globally.

Many older texts talk about arid conditions pervading much of the world during the Triassic. Although there are probably a variety of reasons for this, principal among them is probably the prevalence of red beds in many Triassic sequences. Certainly the concept that red beds are evidence of arid conditions could easily lead to the notion that the Triassic was a time of widespread aridity. Undoubtedly another factor is the documentation of the extensive occurrence of evaporites in Permo-Triassic sediments. For example, in parts of the European Triassic succession, there is an abundance of eolian sandstones, halite, and gypsum deposits, a combination highly indicative of hot and dry conditions. However this does not mean that such conditions were worldwide throughout the entire Triassic. Nevertheless, perhaps because Europe is where the type section for the Triassic is situated, these sequences were apparently regarded as typical of the Triassic worldwide. Whatever the reasons, many a textbook has renditions of Triassic animals wandering aimlessly across hot and barren wastelands, presumably desperately searching for any available scrap of food. As we shall see, this is by no means a completely accurate picture.

One observation that has been used as evidence of rather different climatic conditions in the Triassic when compared with today is the actual dis-

Figure 1.1. Map of the Triassic world with the great sea Tethys forming a large embayment separating Laurasia (to the north) and Gondwana (to the south). The giant ocean Panthalassa surrounds the entire Pangaean landmass.

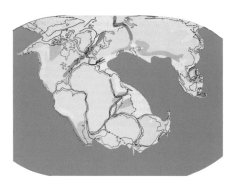

tribution of Triassic evaporites and eolian sandstones. When plotted on paleogeographic maps, they are shown to have occurred mostly in the equatorial zone. This result is rather surprising, considering that today, this belt is the world's most humid, and modern evaporites and sands typically tend to occur in the zones centered on 30° north and south—the "dry belts" of the world. To account for this apparent anomaly, it would seem that either some major differences in global climatic processes held sway during the Triassic, or our assessments of paleolatitudes and our assumptions on the positions of the continental plates are flawed. I shall come back to this concept shortly, but first let's look at some previous discussions of the global distribution of basic sedimentary beds, or facies, and their implications for climate regimes.

Tucker and Benton (1982) discussed Triassic climates and environments in general terms, with some comment on apparent regional variations. They recognized three very broad Triassic facies: (1) those dominated by mudrocks, together with coal seams and abundant plant remains, which Tucker and Benton considered to be indicative of a mild, moist temperate regime. A picture of life at times of such deposition would be one of swamps, freshwater lakes, and widespread forests; (2) fluvio-lacustrine sandstones with occasional gypsum, rare coals, some red beds, and common plant remains, which point to a warm to hot climate and variable rainfall; and (3) fluvio-lacustrine red beds with playa mudrocks, common gypsum and halite, together with eolian sands, which developed under hot, arid to semiarid conditions with generally rare or erratic rainfall.

If we examine the distribution of these broad facies, we see a picture of high latitudes

Plate 1.2. The sun rises on an arid landscape, setting the stage for another scorching day in the Late Triassic of western Europe. Although such scenes may have been common in the Triassic, Pangaea was not one huge desert, and there were also many humid and well-vegetated regions.

Plate 1.3. A group of *Aetosaurus* struggle against burial in a Triassic desert sandstorm.

(e.g., Antarctica and Australia) essentially experiencing a mild and moist temperate regime throughout the Triassic. By contrast, although a similar climate existed over southern Africa, South America, and India (mid-low latitudes) during earliest Triassic times, by the Middle Triassic, these areas apparently became much drier and hotter. Furthermore, the equatorial humid belt was poorly developed throughout the Triassic. Tucker and Benton (1982) there-fore still accept the classic picture of a trend toward hot and dry conditions in the Triassic, as exemplified by the sediments of western Europe; nevertheless, they also point out that mild and moist conditions were prevalent at the beginning of the Triassic. They also acknowledge that in some areas, there were periods of high humidity even in Late Triassic times.

Tucker and Benton's (1982) survey serves as a useful introduction, but in order to get a real feel for Triassic times, we need to go beyond the broad brushstrokes and climate models. After all, each fossil assemblage is merely a sampling of what may be unique and restricted communities. It is worth bearing in mind that even today, there is marked seasonality in some equatorial regions. For example, the Serengeti plains of Africa, just to the south of the equator, are noted for their annual dry seasons, which force the migration of vast herds of wildebeest northward. Yet just to the east, the lush forests of Ngorongoro crater are humid year round. Even more dramatic are the snows on the top of Mount Kenya just south of the equator. Differences in elevation can have a similar effect to difference in latitude—the higher the elevation and latitude, the cooler the temperature. Perhaps what this really underscores is the need to be wary of overgeneralizations. It is hard to imagine that during Triassic times similar localized and regionalized extremes in environmental conditions did not exist.

A Late Triassic Megamonsoon?

I have already alluded to the notion that some rather extraordinary climate conditions may have prevailed at certain times during the Triassic. It is interesting to note that some authors have argued in favor of strong monsoon phases during the middle and late Carnian that resulted in substantial seasonal rains over certain parts of the earth. Some of the evidence for increased rainfall comes from sedimentological studies and includes extensive fluvial sandstones that occur in otherwise thick sequences of playa-lake mudstones, indicating that periodically there was extensive runoff from the surrounding land. The sandstones occasionally contain kaolinite, which also is suggestive of humid conditions. In one shallow marine sequence in Israel, there is an extreme depletion of carbon-13 (the rarer of the two stable isotopes of carbon), and this too has been interpreted as the result of a heavy influx of freshwater runoff. Furthermore, as we shall see in chapter 10, extensive cave and underground watercourse systems were developed in limestone areas that were exposed during the Late Triassic, and this too is suggestive of high levels of runoff during the middle and late Carnian (Simms and Ruffell 1990, Simms et al. 1994). Finally, a widespread change from carbonates to clastics in marine sequences at this time could also reflect climate changes.

The development of a "megamonsoon" during Triassic times has been cited by several authors (in particular, Parrish and coworkers; e.g., Dubiel et al. 1991, Parrish 1993), and there is certainly substantial evidence from climate modeling to support this theory. Although the term *monsoon* is used today to describe the rainy seasons of much of southern and eastern Asia and of east and west Africa, it describes a more complex phenomenon.

Monsoon is a term derived from an Arabic word meaning "fixed season," and it was originally applied to a wind blowing over the Arabian Sea. Today, during the northern hemisphere summer, heating of the Eurasian landmass causes the breakdown of the normal wind systems of the midlatitudes. This is termed *sensible heating*, which refers to a change from a strongly negative net radiative flux to either a weakly negative or positive flux. What this really means is that the earth's surface absorbs more heat than radiates into space. The result is that the surface air over the Eurasian landmass warms, and so becomes less dense and rises, and then drifts south toward the colder hemisphere. Directional airflow across the equator develops because a cross-equatorial thermal (and pressure) contrast exists.

There is thus a fairly involved chain of events that goes something like this: The interior of a large continent heats up in the summer. This brings about a warming of the surface air and consequently a decrease in the air density. The less dense air rises, bringing about low pressure and thereby inducing moist air to flow onto land from the high-pressure cells lying offshore. This moist, warm air in turn rises and cools, and the result is heavy rainfall. So in the northern hemisphere summer, winds blow from the southwest across the Arabian Sea, bringing lengthy periods of rain to the Indian continent. In the

Plate 1.4. A foreboding sky heralds another massive storm in the Chinle. The dense forest lines a river laden with the fallen tree trunks from a previous flood.

northern hemisphere winter, the Eurasian landmass cools off and air pressure increases. Cold air spreads south over southern and eastern Asia bringing about dry weather—the so-called northeast monsoon.

Today the best model for Pangaea is probably Asia—a large continent lying at mid latitudes, with mountains to the south and east. In winter the mountains protect the interior from the warming effect of the Pacific and Indian Oceans, resulting in a cold interior and a strong winter monsoon. Conversely, in summer, the Himalayas protect the Indian continent from the cooler air to the north and thereby contribute to the creation of a low-pressure cell (fig. 1.2).

During the Triassic, Pangaea drifted northward through some 10 degrees of latitude (Parrish et al. 1986). During the Permian, 64 percent of the exposed landmass was south of the equator. By the Early Triassic, approximately 45 percent of the exposed landmass was

north of the equator, increasing to 51 percent by the beginning of the Jurassic. Parrish et al. suggest that this movement of the continents tended to enhance the development of monsoons. By the Late Triassic, with Pangaea centered over the equator and the continental landmass distributed approximately equally in the northern and southern hemispheres, conditions were ideal. Unlike Asia today, the cross-equatorial thermal contrast would have been even greater because the subtropical area was occupied by a large landmass (rather than a large water body, like today's southern Indian Ocean), allowing further cooling and sharper thermal contrast. Thus Parrish et al. (1986) contend that the central position of Tethys incised into the eastern part of Pangaea, the position of the Colorado Plateau, and the development of mountain ranges along the eastern part of the North American plate all served to increase the effects of the monsoon. They suggest that as the world warmed during the Permian and Triassic, the subtropical dry belts expanded before receding again during the Early Jurassic.

Demko et al. (1998) considered that as a result of this megamonsoon, even certain equatorial regions of the world experienced considerable seasonality. In a study of Late Triassic Chinle sediments in the American Southwest (see chapter 7), they felt that ferns

Figure 1.2. Diagram to show the predominant air pressure systems and wind patterns that bring about a monsoon-type climate in India today. A similar arrangement of ocean and continental landmass in the Triassic may have brought about an exaggerated monsoon climate influencing much of the Triassic world.

and other compressed plant fossils merely represented the flora growing immediately alongside streams and rivers, and that other fossils and the sediments indicated a generally arid and strongly seasonal environment in the region.

In addition to the possible development of a strong monsoon, an episode of high humidity in the Carnian may have been correlated with the mid-Carnian rifting of Pangaea (Cousminer and Manspeizer 1976). Veevers (1989) considered the final coalescence of the continents to have been immediately followed by rifting preceding the breakup of Pangaea in post-Triassic times. This major change was accompanied by increased volcanism with a concomitant rise in carbon dioxide levels that in turn resulted in the development of a greenhouse climate. The dating of this event seems to match the independently documented climate change in the mid- to late Carnian, indicating a possible link between volcanism associated with continental rifting and climate change.

Climate Shifts Caused by Rifts?

Toward the latter part of the Triassic, further shifts in climate seem to have taken place. The continents were becoming restless, and preparations were underway that would set them in motion once again. The incipient rifting and separation of the African, North American, and Eurasian plates that took place in the Late Triassic would inevitably have had some impact on global climate patterns. Indeed, this rifting caused the appearance of several large lakes, and the sediments deposited in these lakes contain some key fossils that give us major insights into life at the end of the Triassic and the evolution of modern terrestrial ecosystems—but more of that later (chapter 11). On the basis of a variety of different data, including sedimentological, paleobotanical, and modeling techniques, there seems to have been a drying out in the Colorado Plateau region and a concomitant increase in rainfall in Gondwana and parts of Laurasia.

Regional Differences—What Do the Sediments Say?

As I have already indicated, the concept of a relatively arid and hot climate for the Triassic as a whole may well have its origins in the nature of the type section in Europe. Classic continental sequences in Germany and France consist of fluviatile and eolian sandstones together with halite and gypsum deposits as well as lacustrine-playa mudstones. Tucker and Benton (1982) indicate that throughout Triassic times, much of western Europe was apparently under a hot and dry climate regime, with persistent desert conditions being widespread. Apparently they believed that the occurrence of at least superficially similar Triassic deposits in other parts of the world was sufficient to make a generalization about the global Triassic climate.

However, even as early as 1928, Roberts had begun to question certain generalizations regarding the sediments themselves. Concerning the association of red beds with arid climates, he countered that all the great deserts of the world should therefore be red—which of course they are not! With respect to the Triassic sediments of eastern North America, Roberts was particularly impressed with the abundance of plant remains, noting in particular the coals in Virginia and North Carolina, and the *Araucarioxylon*-type wood in the Danville area of Virginia. A photograph (Roberts 1928, pl. 32A) of a small creek near Otterdale, just west of Richmond, Virginia, shows large tree trunks akin to those that are so impressively preserved where they fell in Petrified Forest National Park. Alas, those in Virginia have long since been lost to the activities of collectors. Roberts also considered the numerous traces made by crustaceans and worms together with ripple marks, mud cracks, and raindrop impressions as evidence of "normal" rainfall.

As it turns out, red beds can actually occur in a variety of depositional environments, ranging from marine to desert (Ziegler and McKerrow 1975), and as Van Houten (1982) points out, all red beds are certainly not alike. With respect to alluvial and fluvial deposits (as opposed to sabkha or marine sediments), it is now thought that reddening results from alternating wet and dry climates. During the humid portion of the cycle, iron is leached from easily weathered iron-bearing minerals. The red-colored mineral hematite is then precipitated during the dry cycle. Reddening in this instance seems to be very much dependent on significant rainfall, albeit seasonal, and not a dearth. Moreover, red beds are formed along a climatic gradient, with seasonal rainfall in a mostly dry environment at one end of the spectrum to seasonal rainfall in a mostly wet environment at the other extreme.

That is not to say that we should abandon the general theme that during Early and Middle Triassic times the world was relatively warm and dry. However, there is pretty good evidence that rather humid conditions did persist, at least at times and in some parts of Pangaea, during the Triassic. We are therefore still faced with some difficult questions. How widespread were the periods of high humidity? Were these periods principally during Carnian times and associated with seasonal rains and a megamonsoon? Only by undertaking detailed studies of all Triassic sediments and compiling all the data will it be possible to begin answering these questions.

For example, in parts of western Asia, there are apparently some sequences consistent with the apparent global trend toward drier conditions from the Early Triassic through the end of the Middle Triassic, but also some notable departures. In the Cis-Ural region, the early part of the Lower Triassic (Induan and lower Olenikian) was apparently a time of relative aridity (Shiskin et al. 2000). However, the sediments of the upper Olenikian point to more humid conditions, with an increase in at least seasonal humidity. These conditions probably persisted into the Anisian (early Middle Triassic). Admittedly it is difficult to know how much of this should be attributed to the physical geography of the region at the time because many of the fossiliferous units in this area were deposited in a vast delta, and in other cases, the proximity of the coastline is apparent. Thus by Ladinian times, the available evidence, including the gray color of the sediments and very abundant plant remains (Shiskin et al. 2000), points to a much more humid climatic regime in this region, although one could argue that the coastal environments experienced a benign climate that was atypical of the region. The vast interior may well have been extremely arid, but very little record was left of such conditions.

What Do the Plants Tell Us?

Although much more work remains to be done, to date, nothing has been found in the sedimentological record to cast serious doubts on the general picture of Triassic climates that I have sketched so far. To recap, we have a relatively warm and dry start extending to Middle Triassic times, and then a return to drier conditions during the Norian (at least in some parts of the world). But we now have some evidence to support a significant departure from the classic hot and dry regime depicted by many earlier texts. There is now good evidence for significant rainfall, at least on a seasonal basis, during late Middle to early Late Triassic times. However, to accept this nice, neat story would be jumping the gun, until it can be confirmed through further study of the geological record. Although elaborate modeling and sedimentological, geochemical, and paleomagnetic data provide us with clues to the climate regime that existed, probably the best indicators are the animals and plants that were living under those conditions. Phytogeography (the distribution of different plant types) may be a particularly good indicator of paleoclimates. Terrestrial plants have the advantage of occupying a realm with a pronounced climate signal. Unlike animals, plants are sedentary, which eliminates variations attributable to migration. Unlike isotopes, they are not subject to diagenetic alteration. Unlike computer models, they represent hard data ("ground truth"). And they are abundant in many regions of the world. So in addition to the putative megamonsoon, plant fossils might tell us a great deal about the climate during the Triassic as a whole.

Ziegler et al. (1993) conducted a broad study of phytogeography and climate in the Triassic. They found that a latitudinal gradient existed for each interval, ranging from dry subtropical regions to the warm and cool temperate biomes. They believed that floras at the dry end of the spectrum could be identified by the dominance of microphyllous (small-leafed forms) conifers (e.g., *Pagiophyllum*, *Brachyphyllum*) and the regional association with evaporites. In contrast, they thought that cool temperate climates could be recognized by broad-leafed conifers (e.g., *Podozamites*?) and deciduous ginkgophytes (e.g., *Sphenobaiera*, *Baiera*). Warm temperate zones (which according to Ziegler et al. [1993] were centered around 40 degrees) have the highest diversity and contain, in addition to microphyllous conifers and other typical cool temperate forms, ferns and cycads. In their study, the boundaries of these biomes remained at relatively constant latitudes throughout the Mesozoic, perhaps suggestive of climate stasis over a very long time interval. During the Early and Middle Triassic, however, the biomes were less distinct, with warm temperate biomes extending up to 70 degrees north. Ziegler et al. felt that there were no real tropical rain forest zones. However, Ziegler et al. conducted only a very general study, and if we start to break down floras by region, the picture becomes a little more blurred and some inconsistencies start to appear.

One complicating factor is the fact that local physiographic gradients, such as mountain slopes, can mimic broader climate patterns. Even in tropical areas, which at sea level support a diverse flora adapted to warm and humid conditions, the higher the altitude, the cooler the temperature, and therefore the more conducive to the growth of temperate floras. In areas where a tropical biome might be expected, some nontropical floral assemblages could occur. Abundant cycads and ferns representing lowland floras might be mixed with ginkgophytes and broad-leafed conifers representing a flora that flourished on nearby mountains.

I shall now briefly look at the paleoflora of the southern and northern hemispheres in turn. Although I will touch on one or two of the important Triassic floral assemblages, I shall leave much of the regional data to later chapters.

Southern Hemisphere

The so-called *Dicroidium* flora is typical for the southern hemisphere and replaced the *Glossopteris* flora in the early part of the Triassic. It is worth reiterating that *Glossopteris* is widely held to have been adapted to cooler conditions than *Dicroidium*. As well as *Dicroidium* itself, the *Dicroidium* flora includes other seed ferns, spore-bearing true ferns, conifers, horsetails, cycads, and ginkgoes. These seed ferns likely would have formed forests and woods lining watercourses. Moreover, horsetails, such as *Calamites*, probably formed thickets of reeds in the shallow water margins. Ash (2001) suggests that the local climate in the immediate vicinity of modern river valleys is often little different from the landscape they pass through. Therefore, it could be argued that where these *Dicroidium* floras occur, the entire region was warm and humid and probably therefore well vegetated. At the same time, there will be more ground water available at a shallower distance below the surface in the immediate proximity of the rivers than adjacent hillsides.

In Africa during the late Scythian and Spathian (Early Triassic), the topography reportedly comprised large floodplain and lake environments. On the basis of floras, conditions are thought to have been warm, with a vegetation of temperate woodland and swampland. One particularly detailed study suggests that by Carnian times, the climate was still quite humid. Anderson et al. (1996) have carried out detailed studies of the Molteno Formation and collected from 12 different localities over a more than 2000-km² area. The Molteno itself covers approximately 25,000 km². Some of the differences the authors see between localities they attribute directly to sampling of different but adjacent

communities (coassociations). Thus they recognize floral and faunal associations for riparian forests, woodland of different types, meadows, and marshes. However, as Anderson and Anderson (1993a) also point out, "taphonomic processes are such that fossil assemblages cannot be read literally." We therefore have to be very careful in making direct comparisons of the fossil assemblage from one outcrop with that from elsewhere—different taphonomic factors may have been in action as well as sampling of different communities and ecosystems. More extensive evaporite deposits known in the Latest Triassic indicate the ushering in of warmer and drier conditions, and toward the close of the period, there is more evidence to suggest that the climate was generally semiarid.

A *Dicroidium* flora is also found in the Early Triassic of India. Interestingly, *Glossopteris* species sometimes occur in this flora, but the plants are smaller in size than older (Permian) species of the genus and generally insignificant in terms of the overall assemblage. The Parsora Formation of India also contains the lycopod *Pleuromeia*, ferns (e.g., *Cladophlebis, Sphenopteris*), pteridosperms, and cycadophytes (e.g., *Taeniopteris*). This

Plate 1.5. Triassic scene with a braided stream bed and forests clothing the bases of the mountains. Such a scene may well have been typical of some high-latitude regions of Pangaea.

Indian *Dicroidium* flora may be older than the typical *Dicroidium* floras of South Africa and Australia, and it is certainly indicative of a healthy rainfall. A well-known younger flora in India is the Maleri flora. It is notable for containing only conifers, and it does not include typical forms of either the *Dicroidium* nor *Scytophyllum* floras (see below). This is at least suggestive of relatively drier conditions. Interestingly, on the sole basis of its tetrapods, the Maleri has been established as Carnian.

Northern Hemisphere

Northern hemisphere floras were markedly different from their southern counterparts and tended to be dominated by ferns, cycadophytes (palmlike gymnosperms), and conifers. It is conceivable that this difference may be related to overall drier conditions in the north than in the south, although the data are conflicting.

At the beginning of the Triassic, gymnosperms became more abundant, and the ferns and pteridophytes that were dominant at the end of the Permian became less so. Where low-lying land became inundated, lycopods (spore-bearing plants with simple leaves) became widespread at the beginning of the Triassic. But as the Triassic progressed, the lycopod communities were eventually replaced by varied gymnosperms, presumably a direct result of the drier climate.

The North American plant record is intriguing. For example, the Chinle Formation is a floodplain environment. Some have argued that during Carnian times, hot but moist climate conditions prevailed, and that the Chinle contains strong evidence supporting the global effects of a megamonsoon. Others strongly dispute this. We shall visit this controversy in chapter 7.

Dobruskina (1994) undertook extensive studies of Laurasian Triassic floras, and it is worth looking at this study some more. Within Eurasian Triassic sediments, she was able to distinguish four stratigraphic intervals, each represented by a different plant assemblage. However, the subdivisions based on plant fossils are larger than the biostratigraphic intervals based on faunas because plant fossils have not been studied in as much detail as the faunal assemblages. But in a more general comparison with plant assemblages from elsewhere in the world, she divides the floras into three temporal sequences: (1) the Lower Triassic plus the Anisian; (2) the Ladinian through the Carnian, perhaps also including the lower Norian; and (3) Norian and Rhaetian.

1. *Lower Triassic.* The Buntsandstein floras are dominated by the conifer *Voltzia* within the western part of the area and by *Pleuromeia* in the eastern part. In some places, the two genera occur together. The *Voltzia* floras are typically found in northeastern France and along the west bank of the Rhine. Commonly, many species of *Voltzia* are found together as well as with *Albertia*, another conifer. *Yuccites* is another common taxon that is thought to be a conifer. After the conifers, the Equisetales are most abundant, with the ferns third in abundance, although only one fern species, *Anomopteris mougeotii*, is at all widely distributed. Pteridosperms, cycadophytes, and ginkgophytes may also be present, but their numbers are fairly insignificant. The *Voltzia* flora has been considered xeromorphic. The ferns contain dense overlapping pinnae, whereas those with large, widely spaced pinnae are absent. Extensive branching of the conifers might be indicative of a shrubby nature, and the relatively low diversity of Buntsandstein species may be explained by low vegetative cover in arid and saline areas. Outside western Europe, *Voltzia* floras are known in China and Kyrgyzstan. *Pleuromeia* is thought to have grown along sea- and lakeshores. Some authors have considered *Pleuromeia* to be a halophyte because it is common near to the coasts of the Tethys and Boreal Seas.

2. *Ladinian/Carnian.* The Lettenkohle is a typical Ladinian assemblage. As in the Buntsandstein, ferns, conifers, cycadophytes, and Equisetales are the principal components, but interestingly, in the Lettenkohle, there are no obvious xerophytes. Ferns have large pinnae with thin blades (mesophytes), and the pinnae of the cycadophytes do not

overlap (cycadophytes include *Pterophyllum* and *Sphenozamites*). Perhaps we are beginning to see signs of increased humidity, although the conifers do include *Voltzia*. For Eurasia as a whole, during the Ladinian/Carnian, there are only a few common taxa from one area to the next, and those of the extreme west are rather different from those occurring in the extreme east. Nevertheless, there are no clear-cut boundaries, and all the floras are, according to Dobruskina, related to one another. There are clear provincial differences in an otherwise basic floral type.

3. *Norian/Rhaetian floras.* In Laurasia, according to Dobruskina, the main standard flora for the end of the Triassic is the *Lepidopteris* flora. Cycadophytes are abundant, and typical forms include *Pterophyllum, Wielandiella, Ctenis,* and *Pseudoctenis.* Next in abundance are conifers such as *Elatocladus* and *Araucarites.* Dipteridaceous ferns are also more common. Numbers of cycadophytes, ginkgophytes, and czekanowskias have increased at the expense of the pteridosperms and sphenopsids. There is more similarity between the western and eastern floras of the Norian interval than between western and eastern floras of the preceding Ladinian/Canian interval. Thus in Eurasia, at least we see a little more homogeneity in the floras toward the close of the Triassic. There is some north-to-south variation apparent, with ginkgoes and czekanowskias decreasing from north to south and numbers of Dipteridaceae and cycadophytes increasing. This is presumably a temperature- and humidity-related phenomenon, with both increasing toward the equator.

The Big Picture

Taking the sedimentological and floral evidence together, we can say that on a global scale, there was a general trend toward increased temperature and aridity in the Early and Middle Triassic. This was followed first by an interlude of much more humid conditions, and then a final drying phase that began in the northern hemisphere. The global ubiquity of lycopods, horsetails, and ferns (all hydrophilic plants) points to at least some levels of humidity, as well as to warm conditions during long stretches of Triassic times.

At the same time, we have to be careful of overgeneralizing, remembering that there would have been marked regional variations. There are still many anomalies in the fossil record. Even on the same continents, there were many subtle differences. The climatic conditions under which the Chinle (chapter 7) was deposited were not necessarily the same as for the contemporaneous Newark Supergroup (chapter 11). This is only to be expected, on the basis of a look at the climates of individual continents today. We see great variation between, say, Southern California and New England, or southern Argentina and Ecuador. Maybe the extremes were not as well marked during Triassic times, but they would nevertheless have existed. Although the floras don't always support the concept of a hot and dry climate, particularly in the equatorial belt, we must also be mindful that such climatic conditions are not conducive to fossilization. Thus there is no guarantee that fossil assemblages are representative of widespread conditions at the time of deposition. They merely reflect a number of localized damp and humid environments. The truth is that the fossil record is too spotty to fully test hypotheses concerning paleoclimates. Vast areas of Triassic terrestrial life are almost certainly missing from our knowledge. Indeed, it is not inconceivable that many of our Triassic assemblages document the exception rather than the norm.

In the chapters that follow, we shall see evidence supporting the general scenario of a Triassic climate characterized by an extended hot and dry spell increasingly influenced by a humid megamonsoon phase. In all honesty, this is a simplistic model, but climate patterns are complex. It is therefore not surprising that we shall also encounter evidence that is inconsistent with the model.

A final thought to ponder is that today, vast swathes of the earth are covered by grasslands of various kinds, such as savannas, prairies, and steppes. But there were no flowering plants in the Triassic, and consequently no grasses. Perhaps struggling with this void, artists and paleontologists may have unwittingly enhanced the view of the Triassic world being an unforgiving place, with sparse ground cover. However, there is no reason to think that some other major group of plants did not fill the same role that grasses do today. Perhaps there were very low, shrubby conifers or even a specialized group of ferns tolerant to dry conditions that clothed the ground. We just don't know. What we do know is that there is a rich paleobotanical record for the Triassic and, in some assemblages, exceptionally dense occurrences of fossil plants.

chapter two

A Brief Phylogeny of Triassic Fishes and Tetrapods

There can be no argument with the statement that plants are critical to all food webs. Likewise, there is much to be said for the concept that it is the little things (the insects) that make the world go round. Even so, in this particular examination of the Triassic world, I make no excuses for the fact that the vertebrates will be central to the discussions. Who can question the spectacular beauty of an articulated vertebrate skeleton? For many, it is a vertebrate fossil that first lured them into the prehistoric world. Furthermore, the Mesozoic is often referred to as the Age of Reptiles, so perhaps not surprisingly, the reptiles will often be the focus of the discussions. Yet despite my obvious bias, I shall not ignore the fact that if it were not for the plants and terrestrial invertebrates, we would be unable to discuss the lives of those most noble creatures of the land: the tetrapods.

To understand a person fully, it is necessary to understand where he or she came from—background, family history, genealogy, and even what skeletons may be hiding in the closet. In a similar way, it is equally important to have a working understanding of who is related to whom among the main players of the Triassic. Of course, one of the reasons that the Triassic is so captivating is the fact that many of the animals are bizarre, and in many cases, we do not fully understand their phylogenetic relationships with respect to today's groups, such as crocodiles, lizards, birds, and mammals. Nevertheless, I shall attempt to put them into perspective by providing an overview of the different groups, together with some of the more controversial issues concerning phylogeny.

Triassic representatives of modern groups were in all probability unlike their extant relatives. The modern stereotype of sharks as "Jaws"-like creatures infesting the oceans of the world will be shattered by the tiny freshwater sharks that inhabited many inland pools and watercourses. Likewise, forget the picture of sluggish crocodiles and alligators sprawled out along the banks of rivers and swamps. The Triassic crocodilians were fleet-footed animals with a more erect posture that habitually chased their prey across open land. During the Triassic, a different group of reptiles took on the role of modern-day crocodile: the phytosaurs were the couch potatoes of the Triassic poolside.

When reconstructing ancient communities, inevitably, we must speculate. We rely heavily on drawing analogies with present-day environments and ecosystems. Behavior patterns, including uses of color patterns and sound, do not fossilize, yet we can be fairly certain that color played a role in camouflage and mating rituals and that sound was important in communication. Trackways of tetrapods offer some clue about behavior and the composition of the community, but the relative timing of footprints in a large trackway site

Plate 2.1. In an Early Triassic freshwater lake, a small shark swims in front of the nose of the amphibian *Batrachosuchus*. *Zamites* and *Pseudoctenis* foliage line the lake floor.

19

can be difficult to decipher. Do parallel sets of tracks indicate animals moving together, or were they just random sets of tracks following a geographical feature such as a shoreline? Large stridulatory organs on fossil insects strongly imply that they chirped in a manner similar to living crickets and katydids. Whether the song was intermittent, continuous, and so on is impossible to judge and is therefore a matter of speculation. Clearly the artistic restorations of this book contain elements of speculation, but at the same time, we want to emphasize that they are still based on solid discoveries from the Triassic fossil record.

On the other hand, examining phylogenetic relationships is a more exact science. By searching for specialized characters and assessing their distribution among a variety of different animals, we can test phylogenetic hypotheses and thereby corroborate (or contradict) those hypotheses. By way of a simple example, we can look at something like the distribution of the following small suite of characters among vertebrates: multicusped, multirooted teeth, and a lower jaw comprising just a single bone, the dentary. These are all characters that one would expect to find in mammals, but they are not necessarily exclusive to mammals. For instance, multicusped teeth are also occasionally found in such a diverse array of animals as pterosaurs, prolacertiforms, and even crocodiles. Of course, there are many more characters that one could include for mammals, and the more data one has, the more reliable the results ought to be. Although this all seems straightforward, nothing is ever that simple when it comes to science. In reality, fossils are incomplete, at least to some extent, and therefore missing some data. This often makes a direct one-to-one character comparison between one animal and another impossible.

Equally problematic are conflicting data. For example, the occurrence of egg laying in monotremes, such as the platypus, is not consistent with all other mammals. However, in this instance, many other characters found in the platypus also occur in all other mammals, including bearing fur, suckling their young, and having a single bone in the lower jaw. The weight of evidence clearly favors inclusion of the platypus within Mammalia. Moreover, the egg-laying character is not really very informative because it is best regarded as a "primitive" feature and present in a wide variety of vertebrates including birds, reptiles, and amphibians. Conflicting data tend to become more problematic when we are forced to examine more restrictive characters. For instance, the enlarged distal end of the pubis that characterizes theropod dinosaurs also occurs in rauisuchians. A fragmentary rausuchian fossil might conceivably be mistaken for a theropod. Despite these inherent difficulties, phylogeny is still one of the cornerstones of paleontology, and therefore I shall begin by placing some of the key groups into some kind of taxonomic framework.

Although this book is devoted to the terrestrial environment (including the freshwater realm), the division between the land, air, and sea is seamless. It is therefore essential to consider the intermediate environments where the land borders on the sea (see in particular chapters 4 and 6), and aspects of brackish estuarine and nearshore environments will be examined in later chapters.

Fishes

Sharks

Sharks have a long fossil record going back to the Late Devonian, but because their skeletons are cartilaginous, not bony, their fossil remains are often restricted to isolated teeth. In the rarer instances where sharks are known from beautifully preserved specimens, such as the Carboniferous Bear Gulch Limestone of Montana and the Early Carboniferous localities of central Scotland, they reveal some bizarre forms, with a variety of crests and spines behind the head. Unfortunately, such wonderful preservation is the exception rather than the rule, and the limited, albeit sometimes abundant, remains of shark teeth can constitute a major difficulty for assessing shark phylogeny. Because of sometimes marked variation in tooth form along the tooth row, it can prove to be an onerous task to differentiate among isolated teeth of different taxa. This is even true for species of Tertiary sharks that have extant genera, so the problems are inevitably greatly compounded for Paleozoic and Mesozoic taxa. Consequently, many Triassic sharks are "form taxa" that are based on the shape of isolated teeth, and they do not necessarily represent discrete biological taxa. Even so, such form taxa can be useful in comparing different strata to correlate ages and paleoenvironments. Moreover, there are sufficient reasonably complete shark skeletons from the Paleozoic and Mesozoic to make possible some basic distinctions. The Xenacanthida and Ctenacanthiformes are known from the Devonian through the Triassic. The Xenacanths were entirely freshwater forms. The modern sharks, Neoselachii, arose during the Triassic, but their remains are rare. Another Triassic group, the hybodonts, was similar to the neoselachians in having streamlined bodies with fully heterocercal tails, but they were much more diverse.

Actinopterygians

The ray-finned fishes also arose in the Devonian, and they subsequently underwent three major radiations: (1) the chondrostean radiation from the Carboniferous to the Triassic; (2) the holostean radiation in the Triassic and Jurassic; and (3) the teleost (modern-day bony fishes) radiation from the Jurassic to the present. The chondrosteans and holosteans are grades of organization, but they are nevertheless useful descriptors of the types of fishes that were prevalent in the waters of the Paleozoic and Mesozoic. The chondrosteans and holosteans are clearly the most important in considerations of Triassic fish assemblages.

Many of the ray-finned fishes of the Triassic were at least superficially similar to modern-day teleosts. Perhaps most importantly, there was a great diversity of forms that exhibit strikingly similar adaptations to extant fishes. Thus the greatly enlarged pectoral fins of *Thoracopterus* parallel the condition seen in the flying fish, *Exocoetus*. The elongate body of *Saurichthys* is reminiscent of modern-day pipefishes.

Coelacanthinii

Of all the fishes living today, the coelacanths are regarded by many as among the most archaic. This is of course because until the last century, this group of lobe-finned fishes was considered long extinct. In the Triassic, they seem to have been fairly common in lakes and streams. Complete fossils are readily recognizable from their relatively short bodies and prominent array of spines on each fin, including the symmetrical tail fin and its prominent median projection that gives the group their common name of tassel-tails (fig. 2.1).

Dipnoi—Lungfishes

The modern African and South American lungfishes live in areas of seasonal drought and are well known for their ability to estivate. They can burrow into the mud and remain dormant there for more than a year. Permian lungfishes have also been found preserved in estivating burrows, and it is certainly not unreasonable to assume that this was a habit that also occurred in the Triassic. This would be a particularly valuable adaptation in climate zones characterized by pronounced seasonality. The best-known Triassic lungfishes belong in the Ceratodidae. Their characteristic tooth plates are relatively common components of many Triassic vertebrate assemblages.

Amphibians

Living amphibians (frogs, toads, caecilians, and salamanders) are dependent on water in their everyday life. Typically they live in freshwater, or at least in damp and moist ground,

Figure 2.1. Skeletal restoration of the coelacanth *Diplurus*.

or within luxuriant vegetation in humid climates. For reproduction, they are totally reliant on water. But in the past (particularly in the Paleozoic), things were different. Many so-called amphibians seem to have been quite capable of living well away from water, except presumably during the breeding season. Although all the modern-day animals that we consider amphibians may indeed form a monophyletic group (the Lissamphibia), that is not true for the host of extinct taxa that are traditionally included within the Amphibia. They represent only a grade of organization, because from within their midst came the ancestors of the reptiles, which, incidentally, should be also in turn considered a paraphyletic group. Indeed, because of their paraphyletic status, it is now more usual to refer to all amphibians and reptiles, plus their descendant lineages the mammals and birds, simply as Tetrapoda. However, for ease of discussion, the Amphibian and Reptilian grades of organization will be retained here.

Many authorities split Amphibia into two major groupings: the reptiliomorphs and the batrachomorphs. As their name implies, the reptiliomorphs include most of the forms that were obviously adapted to life on land. Animals like *Diadectes* and *Seymouria* were heavily built forms with stout and robust limbs. *Seymouria* even assumed an elevated posture with the body held off the ground. Somewhere within this lineage, the first amniotes (reptiles) are thought to have originated. Although the group includes a diverse array of animals, a unifying theme among the reptiliomorphs is a mobile basal articulation between the braincase and palate, as well as a specialized retractor pit on the front of the braincase for the eye muscles. By the end of the Permian, all these forms had died out, likely giving way to the flourishing groups of reptiles. By Triassic times, only representatives of the batrachomorph lineage were left. Batrachomorphs (including the Lissamphibia) can be characterized by the loss of a finger in each hand so that only four remain, and a fused skull roof with no movement possible between the bones of the cheek region. Batrachomorphs also typically possessed reduced limbs and broad, flat skulls, and had lifestyles that approached those of modern amphibians. Some of the larger batrachomorphs may have been piscivores.

Lissamphibia

Lissamphibia are definitively represented in the Triassic by *Triadobatrachus* (fig. 2.2) from Madagascar (Rage and Roček 1989). It clearly has the hallmarks of the Salientia (frogs and toads), with markedly shortened ribs, elongate ilium, toothless lower jaw, and fused frontoparietal bones. By implication, the sister groups to the Salientia, the Caudata (salamanders) and the Gymnophiona (caecilians), must have also been established by this time, but no records for these groups are currently known. Another probable lissamphibian is *Triassurus* from Kyrgyzstan. This tiny specimen could be a larval form; doubts exist about its exact relationships. Other than *Triadobatrachus* and *Triassurus*, there are only a couple of isolated bone fragments that might belong to Triassic lissamphibians.

Temnospondyls

All remaining Triassic amphibians fall into the so-called temnospondyl grade of organization. Temnospondyls were a diverse lot and not confined to the Triassic. In the Permian, for example, they included forms like *Eryops* that led an alligator-like existence. However,

Plate 2.2. In Middle Triassic waters, the lethargic form of the lungfish *Ptychoceratodus* is framed by cycadophyte leaves and horsetails (*Neocalamites*).

Figure 2.2. *Triadobatrachus*, a possible early frog from the Triassic of Madagascar (after Estes and Reig 1973).

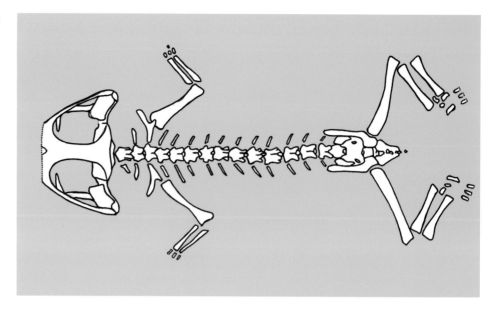

many temnospondyl families disappeared at the close of the Permian, and most of those that survived into the Mesozoic show adaptations to a fully aquatic existence. They continued to decline throughout the Triassic, so that toward the end of the period they are rather rare, with only the capitosaurs, metoposaurs, and plagiosaurs still occurring in decent numbers. By the close of the Triassic, practically all temnospondyl groups had died out, leaving just two lineages extending into the Jurassic, and only one into the Cretaceous.

The trematosaurids formed a particularly diverse family in the early part of the Triassic and had a global distribution, but after the Ladinian, records are scarce. The slender-snouted, gharial-like *Tertrema* (fig. 2.3A) is one of the better-known forms. The plagiosaurs were probably bottom dwellers that used a suction-gulping technique to snare their food. They had exceptionally short, very wide skulls with large, dorsally facing orbits. *Gerrothorax* (fig. 2.3B) and *Plagiosaurus* itself were two typical members. *Plagiosuchus* apparently retained external gills as adults. Another group of suction gulpers were the chigutisaurids. Characterized by short, deep skulls, they probably lived in the open water. Chigutisaurids ranged throughout the Triassic and include forms such as *Pelorocephalus* from the Ischigualasto Formation. The metoposaurids were larger animals, with many exceeding 2 meters in length. Their skulls were flattened like alligator skulls, although their eyes were positioned much further anteriorly. Like alligators, metoposaurs were probably semi-aquatic, spending their days lurking around lake margins and entering the water to catch fish. *Metoposaurus* (fig. 2.3C) and *Buettneria* are just two genera within this family. The capitosaurids and mastodonsaurids were also rather alligator-like and somewhat similar to each other. With their eyes set well back on the skull, they can easily be distinguished from the metoposaurids. The large mastodonsaurids would have been particularly fearsome, with some attaining lengths in excess of 6 meters. Although not quite reaching the monstrous proportions of mastodonsaurids, capitosaurids still included 3-meter-long beasts, and they are probably the most widespread and best known of the Triassic temnospondyls. *Cyclotosaurus* (fig. 2.3D), a common genus in the Upper Triassic of Germany, is known from several different species.

Reptiles

The Mesozoic is sometimes informally described as the Age of Reptiles, and they were unquestionably a diverse lot right from the beginning of the era. It is therefore worth

spending some time examining the many different types and the bewildering array of body plans.

Reptiles, just like amphibians, are really a paraphyletic group, for from within their ranks are derived not only the mammals, but also the birds. Traditional classifications of reptiles relied heavily on the nature of the temporal region and the presence and numbers of openings, or fenestrations, in the side of the skull. It was a generally held opinion that openings in the side of the skull permitted jaw musculature to "bulge" without any physical restrictions as they contracted. The absence of any such openings was typically seen as the primitive condition for reptiles, whereas those groups with openings were considered more "advanced" because they had supposedly developed a more complex and more powerful jaw musculature. This viewpoint was supported by the fact that the oldest reptiles lacked any lateral temporal openings. This configuration of the temporal region is referred to as anapsid (literally, "no arches") (fig. 2.4A). Because chelonians are anapsid, they were widely viewed as the most "primitive" group of living reptiles.

Historically, three fundamentally different arrangements of the fenestrations were recognized (fig. 2.4). Skulls with a single fenestra positioned high up on the side of the skull

Plate 2.3. The somewhat tadpolelike plagiosaur *Gerrothorax* swims up to the surface of a pond to take a gulp of air. In the background a larger *Gerrothorax* rests, and a *Semionotus* swims lazily through the rotting logs and branches cluttering the floor of the pond.

Figure 2.3. Skulls of a variety
of Triassic temnospondyls
show great variability in
form. (A) Trematosaur,
Tertrema. (B) Plagiosaur *Gerrothorax.* (C) Metoposaur
Metoposaurus. (D) Capi-
tosaur *Cyclotosaurus.*

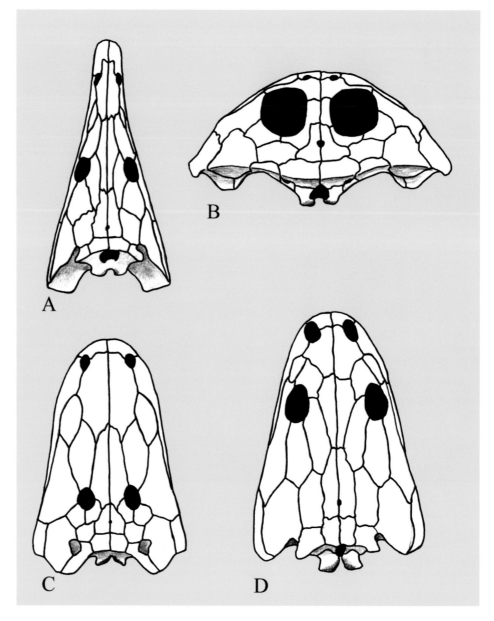

and largely bounded by the parietal, postorbital, postfrontal and squamosal bones were
termed *euryapsid* (fig. 2.4B). These typically would include forms such as the plesiosaurs.
Skulls characterized by a single fenestra in the lower half of the temporal region and
largely bounded by the postorbital, squamosal, and jugal were termed *synapsid* (fig. 2.4C).
This configuration is typical of mammal-like reptiles. Finally, skulls possessing two lateral
temporal fenestrae were referred to as *diapsid* (double arch) (fig. 2.4D), and, among others, these typified the crocodiles and dinosaurs. In addition, it was thought that a host of
different forms, including the lizards and snakes, were derived from this double-arched
condition, and they too were referred to as diapsid even although at least one of the two
arches was missing.

   This basic division of reptiles has changed radically in recent years. First, it has been
recognized that muscles do not really bulge significantly when they contract. Instead, it
seems that the bony margins of the fenstrations provide a more secure attachment point
for muscles than a flat plate of bone (Frazetta 1962). But much more importantly, with the

cladistic revolution, some major phylogenetic studies have been undertaken in which large numbers of characters and character states have been analyzed. It now seems that the position of the temporal openings is just one of a great number of variable osteological characters, and by themselves, the openings are not singularly critical to phylogenetic analyses. Many other features of the skull and postcranium indicate that traditional euryapsids such as the plesiosaurs and ichthyosaurs are part of the traditional diapsid radiation. Thus, in this instance, it would seem that the presence of an opening in the upper part of the temporal region is of more significance. The absence of a lower opening may simply be a secondary closure or representative of an earlier stage in the development of the "true" diapsid condition. More surprising is the fact that several studies indicate that chelonians might also be part of that same diapsid radiation. It is even conceivable that the lack of temporal openings in chelonians might be a derived condition within that radiation. In addition to studies based on traditional osteological data (deBraga and Rieppel 1997; Rieppel and Reisz 1999) pointing to diapsid affinities for chelonians, a study based on mitochondrial DNA (Zardoya and Meyer 1998) concluded that the turtles' closest living relatives are birds and crocodiles, with lizards and snakes being more distantly related. Moreover, other studies based on neontological characters have also claimed support for archosauromorph affinities of turtles (e.g., deBeer 1937; Løvtrup 1977, 1985; Ax 1984; Gardiner 1993), although many of these arguments have been questioned (Rieppel 2000).

Bearing in mind these changing ideas, let's consider the reptiles under three broad categories: parareptiles, diapsids, and synapsids.

Parareptilia

Practically all parareptiles exhibit an anapsid (lacking openings) temporal region. Although some authors argue that they are really a paraphyletic assemblage and really nothing more than an odd assortment of primitive "reptiles," others argue that pareiasaurs and procolophonids really do comprise a monophyletic group.

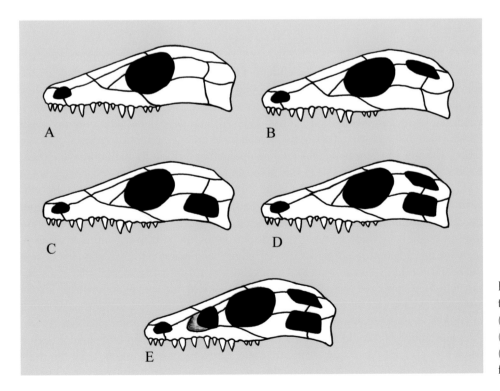

Figure 2.4. Fenestration patterns in the skull of tetrapods. (A) Anapsid. (B) Euryapsid. (C) Synapsid. (D) Diapsid. (E) Diapsid with antorbital fenestration.

Procolophonids

Of the parareptiles, the procolophonids are the most important in any consideration of the Triassic world. Procolophonids are typically relatively small Permian and Triassic herbivores that possess robust, often transversely broadened marginal teeth with a tendency to reduction in number of teeth. I like to think of them as the equivalent of modern-day horny toads (although horny toads are insectivores rather than herbivores)—rather lazy, squat reptiles with broad heads, constantly seeking shelter and shade from the sun. Of course, we have no way of telling if this actually was the case for procolophonids, but there is nothing in their anatomy that hints at daily visits to the Triassic gym!

In most forms, the orbit is extensively developed posteriorly, and in many Upper Triassic forms, it takes on a keyhole shape (fig. 2.5). In addition, many Triassic members display spiky outgrowths of the quadratojugal. Like the chelonians, the lack of fenestrations in the temporal region of procolophonids relegated them to a primitive group. However, it is clear that the extension of the orbital margin into the temporal region achieves the same development of bony margins for the firm attachment of jaw musculature that discrete fenestrations provide in diapsids and synapsids.

The phylogenetic relationships of procolophonids have been the subject of much debate. Some have argued a sister group relationship with the pareiasaurs, a group of large, heavily built Permian herbivores, whereas others strongly favor a sister-group relationship with the turtles. If the latter ultimately proves to be the best-supported hypothesis, then it could be that the procolophonids simply represented one of the early offshoots of the diapsid radiation. My own feeling is that the case for pareiasaur affinities is reminiscent of the one once used to argue for a close relationship between rhynchosaurs and sphenodontians. Here were two groups different in size and overall build, yet were principally united on the basis of one key feature: the beak-shaped premaxillary region. Pareiasaurs and procolophonids are likewise very different in size and shape, and the bony protuberances on the skull and a similar herbivorous dentition may merely reflect homoplasy.

Diapsida

There are two main diapsid lineages, the Lepidosauromorpha and the Archosauromorpha. Setting aside the question of relationships of the Chelonia, of the living reptiles, the lizards, snakes, and the tuatara fall within the Lepidosauromorpha. The crocodiles are the sole surviving nonavian members of the Archosauromorpha. It has been postulated that the principal differences between these two major groups are attributable to basic differences in posture and locomotion (Carroll 1988). By and large, lepidosauromorphs might be considered as sprawlers that move by means of anterior posterior excursions of the limbs, which have a strong mediolateral component as well. At the same time, the backbone is strongly flexed from side to side as the animal moves. One of the characteristics of lepidosauromorphs that can be linked with this type of gait is the presence of a large sternum. Jenkins and Goslaw (1983) showed that the sternum prevents the shoulder from moving backward when the forelimb is brought backward. By contrast, the archosauromorphs might be considered more upright citizens of the past! In the ultimate development of the archosauromorph locomotor pattern (dinosaurs), the backbone was kept rigid, and the hind legs moved backward and forward in a parasagittal plane. Unfortunately, these supposed distinctions become fuzzy when one tries to compare the stance and posture of some of the early archosauromorphs (such as the prolacertiforms) with that of contemporary lepidosauromorphs. Equally, the living archosauromorphs—the crocodiles and alligators—could hardly be exemplified as having an upright stance. Except when they exercise their "high walk," they adopt the classic sprawling pose on the riverbank. Finally, the sternum is not a structure that ossifies very well, and as a consequence,

Plate 2.4. Seen directly from above, a procolophonid displays its broad skull, so characteristic of the group.

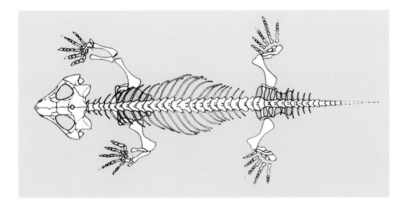

Figure 2.5. Restoration of the Late Triassic procolophonid *Procolophon* in dorsal view.

it is generally poorly preserved in the fossil record. One is therefore forced to rely on more technical details, such as the presence of a specialized opening, the ectepicondylar foramen, in the distal end of the humerus, or perhaps the shape of the quadrate, to determine whether a diapsid is a lepidosauromorph or an archosauromorph.

Before discussing the lepidosauromorphs and archosauromorphs further, there are a couple of other significant diapsid lineages that we shall mention here.

Chelonia (Turtles)

Although some beautifully preserved specimens of Triassic turtles are known, particularly from Germany, they do not form a major component of Triassic terrestrial assemblages. They are included here as diapsids, but it is not considered necessary to go further into their putative affinities within Diapsida. Since their Triassic origins, the basic turtle body plan has remained pretty consistent, with a toothless skull, ribs lying external to the girdle elements, and, of course, the trademark shell. The best-known Triassic turtle is *Proganochelys* from Germany.

The "Euryapsids"

There is nothing conclusive regarding the specific position of the euryapsids within Diapsida. Some authorities (DeBraga and Rieppel 1997; Rieppel and Reisz 1999) consider sauropterygians (plesiosaurs and their allies) to be basal relatives of the lepidosauromorphs, whereas another view supports sauropterygians and ichthyosaurs as basal archosauromorph relatives (Merck 1997).

For many of us, the ichthyosaurs ("fish-reptiles") can be viewed as the Mesozoic ecological equivalents of the whales, or at least the odontocetes (toothed whales). There is certainly no disputing the remarkable similarity in the body outline of classic Jurassic ichthyosaurs like *Ichthyosaurus* and *Ophthalmosaurus* to that of a bottlenose dolphin. This is also true for Triassic forms like *Cymbospondylus*, *Mixosaurus* (see chapter 4), and even the giant deep-bodied *Shonisaurus* from the Upper Triassic of Nevada (fig.2.6). It is widely accepted that ichthyosaurs were completely unable to come out on dry land, and there is substantial evidence to indicate that ichthyosaurs gave birth to live young at sea. Occasionally specimens from the Jurassic have been found with embryos preserved still within the mother's body, and there is at least one individual that apparently died during childbirth, with the baby stuck in the birth canal because it was born tail first. Both front and hindlimbs are modified to form paddle-shaped organs, and they probably had a large dorsal fin and a large lunate tail fin, like that of a modern-day tuna.

The sauropterygians comprise three distinct groups: the placodonts, nothosaurs and plesiosaurs. The plesiosaurs are best known as the "Loch Ness monsters" of the Mesozoic seas, but they only diversified after the Triassic and need not concern us here. The placodonts and nothosaurs, on the other hand, were entirely Triassic.

I like to think of nothosaurs as mini–Loch Ness monsters: they have certainly been regarded in the past as predecessors of the plesiosaurs. They are characterized by long necks and paddle-shaped limbs. They are particularly abundant and diverse in the Middle Triassic of Europe and China.

The third lineage of sauropterygians comprises a group of superficially turtlelike animals, the placodonts. Like the turtles, some placodonts developed a series of bony plates over the body that would have afforded these rather slow-moving marine animals some measure of protection from predator attacks. Placodonts were specialized mollusk feeders with procumbent (forward-jutting) front teeth that were used to pry their food off the seafloor. Large molariform teeth positioned toward the back of the jaws crushed the shells. Some typical members include *Henodus*, *Paraplacodus*, and *Placodus* itself.

Lepidosauromorphs

Living lepidosauromorphs are almost exclusively lizards or snakes (Squamata). However, there are two extant nonsquamate lepidosauromorphs, both referred to the genus *Sphenodon*. Called sphenodontians (or sometimes rhynchocephalians), *Sphenodon punctatus* and *Sphenodon guentherii* are today strictly confined to a handful of isolated islands off New Zealand. However, in the Late Triassic, the sphenodontians were widespread—apparently the most ubiquitous of the lepidosauromorphs. Although no squamate has been positively identified from the Triassic to date, they must have been present because they are almost certainly the sister group to sphenodontians. It may be that they were relatively rare and limited in their distribution. Nor is it unreasonable to presume that the earliest squamates had not acquired all the characteristics that we recognize in squamates today, and consequently, they will not be easily recognized from fragmentary material.

Plate 2.5. The early turtle *Proganochelys* struggles onto the bank of a small pond as a herd of the prosauropod dinosaur *Plateosaurus* ambles by.

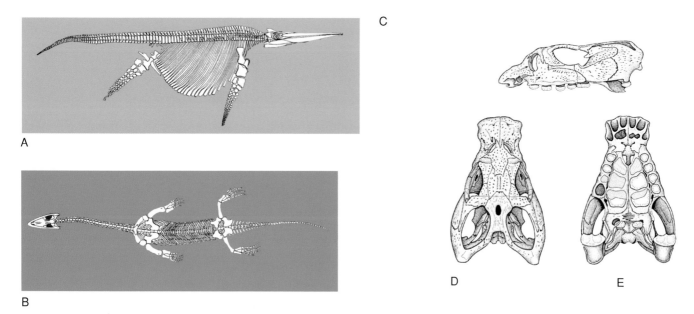

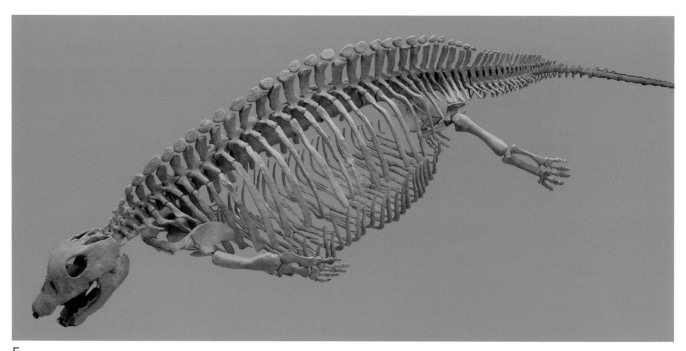

Figure 2.6. Marine reptiles. (A) Skeletal restoration of the Triassic ichthyosaur *Shonisaurus*, showing its dolphinlike form (after Camp 1980). (B) Ventral view of the restored skeleton of the nothosaur *Ceresiosaurus* (after Kuhn-Schynder 1963). The massively built skull of the placodont *Placodus* in (C) lateral, (D) dorsal and (E) ventral views (after Rieppel 1995). The huge, grinding, platelike teeth are clearly displayed in ventral view. (F) Skeleton of a placodont on display in the Paleontological Institute in Zürich.

Plate 2.6. (*opposite page*) Dappled light plays on a pod of *Shonisaurus* swimming near the surface of coastal waters off North America.

The sphenodontians have acrodont teeth—that is, the teeth are fused to the summit of the jaw, rather than set in sockets—and they typically possess a robust lower jaw with a characteristic process on the dentary that extends back well beyond the level of the coronoid process. They are thus easily recognized, even on the basis of quite fragmentary material. The sphenodontians are discussed in more depth in chapter 10.

Plate 2.7. Like modern-day whales, it might be expected that ichthyosaurs like these *Shonisaurus* were occasionally prone to mass deaths through beaching.

Archosauromorphs

With the recognition of the discrete archosauromorph division of the diapsids, it became apparent that many of those Permian and Triassic taxa that had traditionally been viewed as precursors of the squamates (lizards and snakes) were in fact more closely related to the archosaurs. *Prolacerta*, or the prelizard, with its elongate neck, immediately became a misnomer. The position of many other lizardlike taxa was also placed in doubt. For certain forms, such as the wonderful gliding kuhneosaurs (see chapters 10 and 11) or the aquatic thalattosaurs (see chapter 6), the position is still unclear. Other remarkable forms from the Triassic, including *Longisquama* and *Sharovipteryx*, are difficult to assess because of a combination of highly modified features obscuring (overprinting) certain characteristics, together with the poor preservation of critical parts of the skeleton.

What is certain is that recognition of many of the bizarre long-necked reptiles such as

Macrocnemus and *Tanystropheus* (fig. 2.7) as archosauromorphs, together with the ubiquitous rhynchosaurs, has resulted in an even greater claim for this lineage to be regarded as the "ruling party" of the Triassic.

Taxa such as *Prolacerta*, *Macrocnemus*, and *Tanystropheus* are currently placed in a group of their own, the Protorosauria. However, their relationships with each other are equivocal, and it seems likely that with more research, they will be separated into at least two distinct lineages. Most of them were aquatic or semiaquatic animals, and they were especially widespread during Middle Triassic times.

The rhynchosaurs, once aligned with the sphenodontians, were another widespread group in the Middle Triassic and are also a major component of many Late Triassic terrestrial assemblages, although they seem to have died out sometime before the close of the Triassic. They were large quadrupedal herbivores with a barrel-shaped body. They used their beaklike jaws to snip off plant material. *Hyperodapedon* (fig. 2.8) is a typical member of the rhynchosaurs, and it is discussed in more detail in chapter 8.

One particularly strange group of animals that has recently come to light is the Drepanosauria. They are characterized by a very thin, rodlike scapula, well-developed neural spines in the region of the shoulder girdle that typically fuse, a barrel-shaped trunk, pronounced neural and hemal spines in the tail vertebrae (giving the tail a compressed leaflike appearance), and sometimes a clawlike bone at the end of the tail. Some may even have had a prehensile tail. Although certain members, such as *Megalancosaurus* (chapter 6), appear to have been highly adapted to life in the trees, others, like the tiny *Hypuronector* (chapter 11), also informally known as the deep-tailed swimmer, have been regarded as fully aquatic forms. Their position within Diapsida is unclear, but they appear to be more akin to the archosauromorphs than the lepidosauromorphs (Dilkes 1998).

Another archosauromorph of uncertain position is *Trilophosaurus*. It was described by Gregory (1945) on the basis of excellent material from the Dockum of Texas (chapter 7). The premaxillae and anterior portion of the dentaries are edentulous (toothless), and in life, it may have had a horny "beak." Further back were

Figure 2.7. Skeletal restoration of the Triassic protorosaur *Tanystropheus* (after Wild 1973).

Figure 2.8. Skull of the rhynchosaur *Hyperodapedon* showing beaklike jaws (after Benton 1983b).

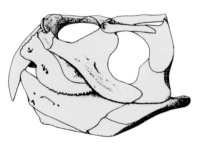

transversely broadened teeth forming sharp shearing surfaces. In the temporal region, only a supratemporal opening is present, and the cheek region is composed of a solid plate of bone. This would seem to be a case of secondary loss of the lower temporal opening—perhaps in a similar manner to the turtles?

Archosauria

A more exclusive grouping of the archosauromorphs is the Archosauria, or "ruling reptiles." As the name implies, these were the dominant tetrapods of the Mesozoic era. They first appeared at the beginning of the Triassic, subsequently radiating rapidly. Along with the dinosaurs and pterosaurs, a number of lineages have been recognized, occupying a great variety of niches and including some bizarre animals. Not surprisingly, archosaurs are central to many of the discussions in this book, and it is therefore important to have a good understanding of their phylogeny as it is currently known. However, there is certainly not a consensus among current workers on all the details of their phylogeny, and indeed, some of the more vigorous debates in vertebrate paleontology center on the division of the archosaurs.

Originally, the key defining feature of the archosaurs was considered to be the presence of an antorbital fenestra (fig. 2.4E). The function of this additional opening in the skull has been the subject of much debate. It has been suggested that it permitted the accommodation of greater muscle mass, but alternative suggestions include the accommodation of a salt gland or an extension of air sinuses. Recently, in a series of elegant studies, Larry Witmer (1997) has argued strongly for the last of these. With one or two exceptions, most authors still accept that the antorbital fenestra is a valid character that probably evolved just once and helps to define a monophyletic clade of organisms. However, most authors do not include the early (basal) members of the clade within the Archosauria. Gauthier (1986) defined Archosauria to include all descendants of the most recent common ancestor of birds and crocodiles (fig. 2.9A), and as such, this excluded the Proterochampsidae, Erythrosuchidae, and Proterosuchidae, even though these three families have an antorbital fenestra. This view was also adopted by Parrish (1993). The more inclusive group he termed Archosauriformes (fig. 2.9B).

Some readers may feel that this juggling of names is a frivolous exercise in semantics. However, there is some merit for the intense discussion over the use of names. As new discoveries are made and new interpretations put forward, phylogenies change. Consequently, different workers move taxa from one group to another, and diagnoses and definitions of taxa are altered and refined. A precise universal language is crucial to taxonomy, and we must strive to avoid constantly changing the definition of any given group. If not, we shall be guilty of continually moving the goalposts so that our scientific language lacks the precision necessary to communicate effectively with each other.

Sereno (1991) provides a good basic starting point and discussion of the division of the basal archosaurs. As he points out, it has long been recognized that early in archosaur history, there was a major split into two major clades. These two have traditionally been based on an understanding that a rotary-style ankle joint evolved independently in two apparently separate lineages. These are the so-called crocodile-normal and the crocodile-reversed types of ankle joint, and the two lineages have been termed the Pseudosuchia (suchians and parasuchians) and the Ornithosuchia (Ornithosuchidae, pterosaurs, and dinosaurs), respectively. However, in an extensive analysis of a wide range of cranial and postcranial characters, Sereno (1991) suggested that in fact the rotary style ankle joint evolved just once, and that the Ornithosuchidae should be grouped together with suchians and parasuchians (fig. 2.9C). For this group, Sereno used the term Crurotarsi. With the exclusion of the Ornithosuchidae, the second lineage essentially comprises just the Pterosauria and Dinosauromorpha, and for this group, Sereno used the term Ornithodira.

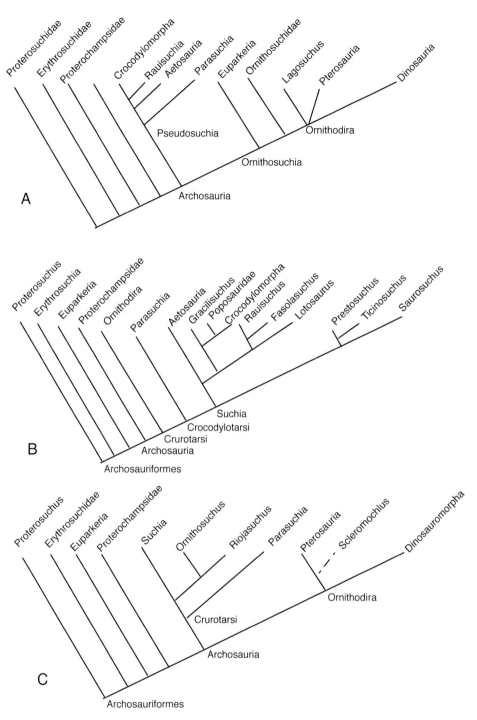

Figure 2.9. Cladogram showing the relationships of different archosaurs according to (A) Gauthier (1986), (B) Parrish (1993), and (C) Sereno (1991). As in any active research area, the relationships of the different groups are not always very stable. New finds can alter our perspective. Nevertheless, these trees all share some things in common, not least of which is the sister-group relationship of pterosaurs and dinosaurs.

Pseudosuchia

In the Triassic, the principal pseudosuchians are the phytosaurs (the parasuchians of some authors), rauisuchians, and stagonolepidids (or aetosaurs). However, the interelationships of these groups is the cause of considerable debate. Both the phytosaurs and aetosaurs seem to have practically appeared out of nowhere, so that there are no intermediate forms that might link them to the other groups. Just as problematic is the constitution of the rauisuchians, and many authors argue that they have become a wastebasket for the reception of a wide variety of, at best, quite distantly related forms.

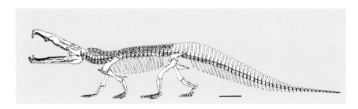

Figure 2.10. Restoration of the skeleton of the phytosaur *Rutiodon* (after Long and Murry 1995).

The superficially crocodile-like phytosaurs are major components of many northern hemisphere assemblages, although there is at least one documented occurrence in South America (Kischlat et al. 2002). Like today's Nile crocodile or American alligator, they were semiaquatic animals that typically possessed elongate rostra, and the jaws were lined with acutely conical serrate teeth (fig. 2.10). They appear to have filled a similar niche to living crocodiles and alligators. For much of their day, they probably lazed on the edges of rivers and lakes, but they took to the water to lie mostly submerged, waiting for unsuspecting prey to come to the water's edge. Like crocodiles, they possessed bony scutes in the skin. However, unlike crocodiles, the external nares were not positioned at the end of the snout but had shifted way back to a point on top of the skull, so that in many instances, the nostrils were close to the eye sockets. As a consequence, when the animal was resting at the surface, all that might be visible protruding above the water would be the eyes and the outline of the nostrils just in front.

Of the pseudosuchians, only the aetosaurs (fig. 2.11) were herbivorous. These armored quadrupeds are restricted to Upper Triassic sequences. They bore leaf-shaped teeth, but both the upper and lower jaws were edentulous anteriorly. They may have rooted around in the dirt like pigs in their search for roots and stems. At approximately 3 meters long, *Longosuchus* (see chapter 7) and *Stagonolepis* (see chapter 8) were typical in size for aetosaurs.

The rauisuchians were mostly large, quadrupedal carnivores that assumed a semierect posture with the limbs drawn under the body in a manner similar to dinosaurs. Some of these large carnivores, such as *Postosuchus* from the American Southwest (chapter 7), may have been bipedal. In dinosaurs, the erect posture is achieved through an offset proximal head on the femur that turns into a laterally facing socket (acetabulum) in the hip girdle. In the rauisuchians, on the other hand, the femur does not have an offset head, but the acetabulum is angled downward. But like theropod dinosaurs, the elongate pubis has a boot-like foot. Indeed, rauisuchians typically share many other characters with dinosaurs, such as an increase in the number of sacral vertebrae and the presence of a brevis shelf on the ilium. They thus provide us with a good example of convergent evolution.

Of course a group that has been referred to as having a "crocodile-normal" ankle joint includes the crocodiles themselves. Although they were not common members of Triassic assemblages, crocodiles first appear in the fossil record at this time, and they include some interesting forms. No matter how fearsome we know a crocodile to be, it is hard to get away from that inanimate object we see lounging motionless poolside at the zoo. But the Triassic crocodiles adopted a very different lifestyle indeed. These early crocodiles were highly active, land-living, cursorial (adapted for running) beasts (fig. 2.12). *Terrestrisuchus* (see chapter 10) was about the size of a beagle and was a fairly typical Triassic form. It is even possible that some of these early crocodiles were habitually bipedal. Despite being fleet-footed creatures, they still possess most of the key features that characterize crocodiles today, including elongate wrist bones (radiale and ulnare), an anteriorly inclined quadrate contacting the prootic, some degree of pneumatization of the cranial bones (having internal air spaces), and an attenuated posterior process on the coracoid bone.

Finally, the ornithosuchids were large, predatory archosaurs that may have been habitually bipedal. It is likely that forms such as *Ornithosuchus* (fig. 2.13) were very like theropod dinosaurs, assuming the role of top predator in the paleocommunities they inhabited.

Ornithodira

Pterosaurs (fig. 2.14) are a group of ancient vertebrates characterized by an enormously elongate fourth finger that supported a membranous wing. They are sometimes broadly

separated into two groups: the rhamphorhynchoids and the pterodactyloids. The former is something of a ragbag assemblage whose members are loosely characterized by an elongate tail, a relatively short neck, and a small head. Only the rhamphorhynchoid grade is known from the Triassic; the pterodactyloids do not appear until the Late Jurassic. Like the crocodiles, pterosaurs are unknown before Late Triassic times, but several new forms have been described in recent years, and it is clear that they diversified quickly.

Pterosaurs and dinosaurs have been grouped together for a long time, and today, the prevalent view among vertebrate paleontologists is still that they do indeed form a natural group of animals. However, there is some skepticism about this grouping on the part of a few workers. These views have been most strongly articulated by Rupert Wild (1978) and most recently Peters (2000). Both authors have argued that the pterosaurs were more closely related to the prolacertiforms. Certainly, as pointed out by Bennett (1996), some of the principal features shared by dinosaurs and pterosaurs are associated with the hindlimb, and as such could simply be a result of convergence toward a similar stance and locomotor pattern. Admittedly, it would still be necessary to invoke the origin of the antorbital fenestra on at least two separate occasions.

That leaves us with the dinosaurs. There is clearly something about dinosaurs that excites the imagination of children—a fascination that for many, I am happy to say, continues well into adulthood. It is fair to say that some of this interest is merely an extension

Plate 2.8. A phytosaur propels itself through the clear waters of Upper Triassic Germany.

Plate 2.9. Aetosaur in Petrified Forest National Park.

of what some would argue is a morbid fascination with catastrophes, as witnessed by the
popularity of the genre of disaster movie such as *Armageddon*. The apparent cataclysmic
demise of the dinosaurs at the end of the Cretaceous, perhaps brought on by some horri-
fying impact of an extraterrestrial body, certainly fulfills all the requirements of such a
movie. But, as Jack Horner says, why are we so consumed with the death of these wonderful animals when their lives must have been so much more interesting? I couldn't
agree more, and where better to start than with the birth of the dinosaurs in the latter part
of the Triassic?

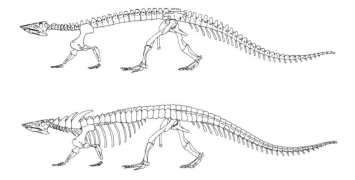

Figure 2.11. Aetosaur *Desmatosuchus* (after Long and Murry 1995).

The Earliest Dinosaurs

The Triassic was a pivotal time in the origin of modern terrestrial ecosystems. Practically all the major groups of modern-day tetrapods trace their beginnings back to the Late Triassic. However, the Triassic was also a critical period for many well-known vertebrates that are not traditionally regarded as modern forms. These include ichthyosaurs, pterosaurs, and, of course, the dinosaurs. But it is widely held among paleontologists that the birds are nested within the Theropoda, and therefore it is not unreasonable to include the dinosaurs among the ranks of "modern" vertebrates. It was perhaps the evolutionary novelties of the Dinosauria—particularly the fully erect posture—that was responsible for ushering in the new age of terrestrial vertebrates.

Dinosauromorphs, Pseudodinosaurs, or Dinosaur Wannabes

Interestingly, all three major divisions of the Dinosauria—the Sauropodomorpha, Theropoda, and Ornithischia—are now known from Triassic strata, suggesting that there was an explosive radiation somewhere within the Carnian, or perhaps the Ladinian. Accepting, as most paleontologists do, that dinosaurs are monophyletic, then somewhere within the world's Middle or early Late Triassic strata there might be remains of some of the most basal dinosaurs—animals that are true dinosaurs but do not possess the special characters that categorize them as theropods, sauropods, or ornithischians. But how would we go about recognizing these fossils? What traits should they possess? In other words, exactly what does it mean when we say an animal is a true dinosaur?

This question is very like one we shall examine later when we look at the first mammals and their possible sister taxa. A whole suite of characters defines groups such as Mammalia or Dinosauria. For the former, it includes, among many other things, a single element in the lower jaw (linked with the quadrate-squamosal jaw articulation and the three bones in the middle ear), the presence of fur or hair, and a single aortic arch on the left side. However, we also know that these characters were not all acquired together overnight: they appeared in some sort of sequence. So just which is the critical character that bestows "mammalness"? Is it the presence of fur, the three bones in the middle ear, or multicusped and multirooted teeth? Likewise for dinosaurs, there is never any doubt that *Diplodocus, Tyrannosaurus, Stegosaurus,* or *Triceratops* are dinosaurs, but precisely what is it about these animals that confers the status of dinosaur upon them? The answer in both cases, of course, is that there isn't one single defining character, and by asking the questions in the first place, we immediately expose the shortcomings of a classification system that uses these higher-order groupings.

Figure 2.12. Restoration of the skeleton of the sphenosuchian crocodile *Terrestrisuchus* (after Sereno and Wild 1992).

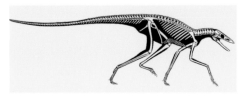

Plate 2.10. Two fully terrestrial crocodiles, *Pseudohesperosuchus* and *Hemiprotosuchus*, warily stalk each other among the tracks of a *Riojasaurus* in Argentina.

Ignoring for the time being the problems raised above, let's return to the question of what constitutes a dinosaur. It is certainly possible to envisage trends in the accumulation of specific characters, and most authorities would agree that *Lagerpeton* and *Marasuchus* (fig. 2.15) from the Chañares reflect stages in the lineage leading toward dinosaurs. They have an ankle joint that is very like that of dinosaurs and at the same time unlike that of any other animal: the astragalus has an ascending process on it, and the articulation of the foot is between the proximal tarsals and the distal tarsals. The proximal ankle bones functioned as part of the lower limb and moved as a unit with it. In other respects, though, *Marasuchus* and its kin fall well outside the parameters (albeit rather artificial) that we use to describe Dinosauria: they lack a perforate acetabulum and have only two sacral vertebrae.

Another animal that clearly falls somewhere within the dinosauromoprh/dinosaur lineage is *Agnostiphys* from the British fissure deposits (see chapter 10). It has an ascending

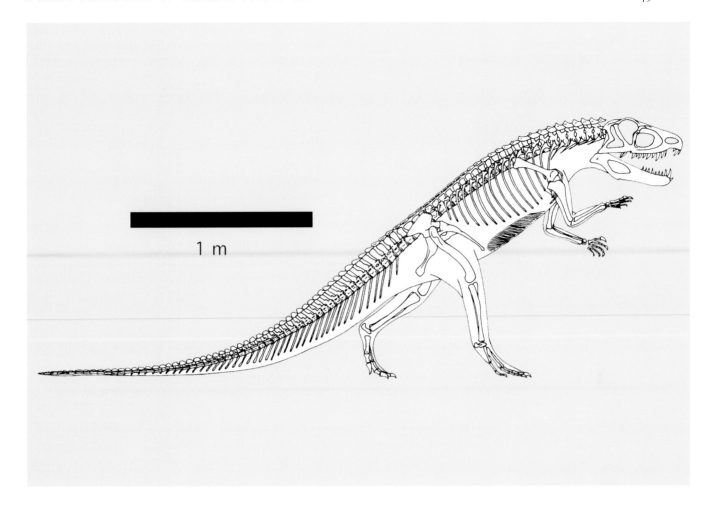

process on the astragalus, a perforate acetabulum, a brevis fossa on the ilium, and a humerus with the deltopectoral crest extending at least one third of the length of the shaft. However, it only has two sacral vertebrae. I believe that the brevis fossa is an important diagnostic feature of the dinosauromorph lineage. It is a variably developed depression or trough just behind the acetabulum, and it is thought to have served as an attachment point for major muscles associated with the tail and hindlimb, and is presumably linked to the upright posture.

The Petrified Forest (chapter 7) also produced its own dinosaurian enigma in the form

Figure 2.13. Ornithosuchid *Ornithosuchus* (after Walker 1964).

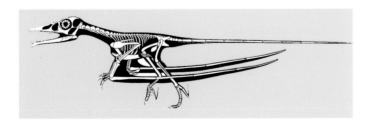

Figure 2.14. Restoration of the skeleton of the pterosaur *Eudimorphodon* showing it in a bipedal mode (after Sereno 1991). Other authorities consider it more likely that in all pterosaurs, the wing membrane attached low down on the hindlimb, thus making such a posture impossible. If correct, then pterosaurs must have adopted more of a waddling motion.

Figure 2.15. Restoration of *Marasuchus* (after Sereno 1991).

of *Chindesaurus.* Long and Murry (1995) described tetrapod remains from Chinde Point in the northern section of Petrified Forest National Park as a herrerasaurid. Although the elements were also associated with a rauisuchian, it is immediately apparent that there are substantial portions of an animal akin to an early dinosaur. Certainly the astragalus is very like that of *Agnostiphys,* but it would also seem that it shares with *Agnostiphys* the presence of just two sacral vertebrae (Long and Murry 1995). Furthermore, it completely lacks a brevis shelf and fossa on the ilium. Of course, as we shall see later (chapter 9), there is still not complete agreement on whether *Herrerasaurus* itself and even *Eoraptor* are true dinosaurs.

To compound our difficulties in dealing with these dinosaur intermediaries, many of the characters used to diagnose Dinosauria, or subgroups within Dinosauria, are also variously distributed in outgroups. Thus a perforate acetabulum is also found in sphenosuchian crocodiles such as *Terrestrisuchus.* A subnarial fenestra, an increased number of sacral vertebrae, a brevis shelf on the ilium, and a bootlike pubic foot are all features that also occur in rauisuchians.

What does it matter whether *Herrerasaurus, Agnostiphys, Chindesaurus,* and *Eoraptor* are true dinosaurs or not? Did the fact that they lacked one or two of the features that we regard as diagnostic for Dinosauria really make that much difference to the animal's lifestyle? The answer to the latter is probably no; they all seem to have been highly active, bipedal carnivores. But the answer to the first question goes right to the root of a number of issues concerning the value of taxonomy and our understanding of the world and our own environment. It does matter because we must have a universal language in taxonomy that does not permit ambiguities. The important thing here is that we recognize the uniqueness of individual species and the great diversity and disparity of past and present ecosystems. Sometimes slight nuances such as color patterns or behavior differentiate between populations of otherwise very similar organisms. This was just as true for these early dinosaurs or dinosauromorphs. After all, what makes our own world so interesting is its tremendous variety. In the same way, in order to appreciate the Triassic world, we need to recognize the great diversity of life at that time. Of course what we don't know is how significant were some of the characters of true dinosaurs that are not preserved in the fossil record, including major physiological features such as warm-bloodedness.

And there's another question. How confident can we be in assigning incomplete and fragmentary remains to particular groups of animals? The answer is that it very much depends on the exact nature of the remains. We are all familiar with the wonderful flights of fancy that are occasionally depicted on the basis of a scruffy fossil represented by some rather insignificant bones such as the digits and ribs. On the other hand, one or two elements can sometimes hold a wealth of information, as is the case with the braincase, wrist bones, and coracoid of a crocodile.

The Real McCoy

Triassic dinosaurs are not limited just to one or two fringe members and taxa based on fragmentary and dubious material. There are some undisputed Triassic dinosaurs known from exceptionally abundant and very well preserved material. Indeed, one of the most spectacular accumulations of dinosaur bones anywhere in the world occurs within the Chinle Formation at the famed Ghost Ranch site in New Mexico. Here the remains of at least 1000 individuals of a small theropod, *Coelophysis,* have been documented (see chapter 7). Moreover, similar accumulations of another closely related form, *Syntarsus*, are known from Zimbabwe. A third renowned death assemblage of Triassic dinosaurs occurs at Trössingen in the region of Baden-Württemberg, Germany, where numerous specimens of the prosauropod *Plateosaurus* were first discovered in 1911 as part of an extensive bone bed.

The Trössingen site was worked on three separate occasions: first in 1911 and 1912 by workers from Stuttgart under the direction of E. Fraas, then during 1921–1923 by crews from Tübingen led by Friedrich von Huene, and finally in 1932 when a team from Stuttgart returned again to excavate the bone bed. On this last occasion, the work came to an abrupt end with the unfortunate death of a worker when a quarry wall collapsed. Careful study of all the bones, including some fairly complete skeletons, indicates that they are representative of a single species of plateosaur, *Plateosaurus engelhardti*, although there does appear to have been some sexual dimorphism.

Plateosaurus was a relatively large dinosaur for the Triassic, reaching adult lengths of 20–25 feet. Von Meyer described the first fragmentary remains in 1837. The jaw joint is situated below the level of the teeth, and the teeth have coarse serrations running down the length of the crown.

Huene restored *Plateosaurus* in an upright bipedal posture (presumably following Marsh and his reconstruction of the Early Jurassic genus, *Anchisaurus*, from the Connecticut Valley). However, it is interesting to note how unstable this posture appears. It is quite possible that their long necks and trunks would have produced a tendency to topple forward. Certainly they do not appear to have perfect balance at the hips in the bipedal position. Galton pointed out that the arms and shoulders are quite long and robust and definitely capable of bearing weight. More importantly, although the fingers could be flexed to grasp objects, they could also be hyperextended (in other words, bent backward). In this way, the fingers of digits II, III, and IV could rest on the ground like the toes of a foot. The enlarged claw on the first finger was probably held clear of the ground.

Von Huene suggested that the bone bed resulted when herds of migrating *Plateosaurus* died of drought. By contrast, Sander (1992) felt that the large numbers of plateosaurs died when they were miring. Sander envisioned these relatively heavy animals coming down to shallow lakes and ponds that occupied the floodplain, but then sinking in the slimy, viscous mud. The more they struggled to free themselves, the deeper they became entrapped, until finally they perished, unable to escape from the ooze. Undoubtedly their putrefying carcasses attracted the attention of a variety of smaller scavengers that were able to cross the mud without the threat of being sucked into the morass.

These occurrences of mass dinosaur deaths strongly suggest that at least certain dinosaurs were highly gregarious, a behavior pattern that is perhaps more frequently associated with herbivores. Thus the occurrences of *Coelophysis* and *Syntarsus* are somewhat unusual. It seems likely that these large congregations of meat-eating dinosaurs only occurred for relatively short periods, perhaps for migration or at breeding times. On the other hand, the mass accumulation of *Plateosaurus* is more in keeping with its presumed herbivorous habit. Later on (chapter 11) we shall see more evidence for communal behavior in dinosaurs on the basis of trackway sites.

Although we may only have a good picture of a handful of Triassic dinosaurs, it is becoming increasingly apparent that the diversity of Triassic dinosaurs and dinosauromorphs was actually much greater. A variety of isolated teeth from the American Southwest provide particularly strong evidence of a richer dinosaur fauna in the Late Triassic. Furthermore, new finds in places such as in Madagascar, Thailand, and Italy continue to be made. Consequently, we now know that the sauropodomorphs extended back at least to Carnian times. Langer et al. (1999) described the genus *Saturnalia* on the basis of a well-preserved, semiarticulated skeleton from the Carnian Santa Maria Formation of Brazil. These beds are equivalent in age to the beds from northwestern Argentina that yielded

Plate 2.11. *Coelophysis.*

Herrerasaurus and *Eoraptor*. At lengths somewhere between 6 and 8 feet, *Saturnalia* was certainly not large by later sauropodomorph standards, and it was also quite a gracile animal. All three major lineages of dinosaur—Theropoda, Sauropodomorpha, and Ornithischia—are known from Carnian deposits. This certainly fuels suspicions that they had an earlier origin, perhaps sometime in the Middle Triassic. Indeed, several descriptions of trackways from Middle Triassic rocks have been attributed to dinosaurs (Demathieu 1989). In fact, Demathieu and Haubold (1978) even suggested that certain tracks from the Lower Triassic could have been made by dinosaurs, or at least "predinosaurs." So currently, the oldest distinct dinosaurian tracks are older than the oldest dinosaur remains. Who knows what other dinosaurs will turn up in Triassic sediments? One thing is certain: there is no question that the basal members of the modern avian clade were also well established during Triassic times.

Plate 2.12. Stricken by drought, a group of *Plateosaurus* carcasses provide much-needed food for the ceratosaur *Liliensternus*.

Synapsids and the Mammalian Line

The synapsid (mammal-like) reptiles are a significant component of many Early–Middle Triassic continental vertebrate faunas. Indeed, many sections in the lower part of the Triassic have been dated on the basis of key synapsid taxa (see appendix 1); this is particularly true of the southern hemisphere (Gondwanan) assemblages. Synapsids are all thought to have derived from ancestors in whom there was a single fenestration positioned low in the lateral wall of the temporal region of the skull.

Dicynodonts

Dicynodonts were rather squat, herbivorous animals with short tails that ranged from the Permian into the Triassic. The anterior end of the skull was shortened considerably, giving them the appearance of having run into a brick wall. Except for a pair of tusks in the upper jaw, they typically lacked teeth. By comparison with Permian diversity levels, the Triassic dicynodonts were quite limited, with fewer than 30 recorded genera (for the Permian *Cistecephalus* zone alone, 35 genera have been reported; King 1990). Dicynodont diversity levels show a general decrease through the Triassic, although there was a temporary reversal of the trend in the Middle Triassic. Even so, dicynodonts still form significant components of certain Triassic terrestrial assemblages right up to Norian times. Occasionally, as at the famed *Placerias* Quarry in the Chinle of the southwest United States (chapter 7), they are the dominant forms, and many authors consider them to have been gregarious.

Cynodonts

Cynodonts are a diverse assemblage of synapsid reptiles that were widespread during Triassic times, and which are widely held to be closely related to the ancestry of "true" mammals, but the exact relationships have been the subject of considerable debate in recent times. The group includes both herbivorous and carnivorous forms, and in the past cynodonts were broadly divided on the basis of their diet (e.g., Hopson and Kitching 1972). This division implied that the "advanced" cynodont condition had been achieved independently in two separate lineages, which is clearly not parsimonious. For that reason, modern classifications typically recognize the Eucynodontia ("advanced"-grade cynodonts) as a natural monophyletic group consisting of both herbivores and carnivores.

The eucynodonts are characterized by several features, many of which are associated with the shift in emphasis away from a jaw articulation directly between the quadrate and articular bones to an articulation between the squamosal and dentary elements. Thus they exhibit a greatly enlarged dentary that approaches—and eventually in mammals, forms—the jaw articulation. There is a reduction of the postdentary bones, including the reflected lamina of the angular (reduced to ossicles of the middle ear and the ectotympanic in mammals), and there developed a secondary articulation between the ventrolateral process of the squamosal and surangular (dentary in mammals).

In fact, in the last 30 years or so, four cynodont families have been proposed as being the sister group to the Mammalia: the Thrinaxodontidae (Hopson 1969; Hopson and Crompton 1969; Barghusen and Hopson 1970), the Probainognathidae (Crompton and Jenkins 1979), the Trithelodontidae (Hopson and Barghusen 1986; Shubin et al. 1991; Crompton and Luo 1993), and the Tritylodontidae (Kemp 1982; Rowe 1988; Wible 1991). These changing views reflect the fact that new material continues to be found, which in turn has the effect of overturning previous hypotheses. Currently the weight of evidence tends to favor the tritheledonts as the sister group to all mammals (Luo 1994), but the tritylodonts are not without their supporters, and Wible (1991) and Wible and Hopson (1993) contend that tritylodonts and mammals are more closely related to each other than either is to the tritheledonts.

Cynodonts known from Latest Permian sediments of Russia and southern Africa are characterized by forms such as *Procynosuchus*. The procynosuchids were succeeded in the early part of the Triassic by a group traditionally called the galesaurids. Galesaurids are typified by the well-known genus *Thrinaxodon*, a small, lightly built carnivore that was about 50 centimeters long (fig. 2.16A). It had an elongate body and relatively short limbs that probably gave it the appearance of certain mustelids (weasels and their kin). Unlike the procynosuchids, *Thrinaxodon* had a solid secondary palate, and the dentition was similar to that of early true mammals, bearing four upper and three lower incisors, canines in both the upper and lower jaws, and seven to nine cheek teeth. In view of these features it is not surprising that Hopson (1969) suggested that the Thrinaxodontidae were closely related to the common ancestry of all mammals.

Thrinaxodon is one of the characteristic carnivores of the *Lystrosaurus* zone, but in the overlying *Cynognathus* zone, a much larger and more heavily built cynodont occurs: *Cynognathus* itself. In *Cynognathus* the dentary is more like that of a mammal than that of *Thrinaxodon*, having a high coronoid process and deep masseteric fossa, and forming a much greater proportion of the mandible: the articular, prearticular, angular, and surangular are much reduced. Nevertheless, the jaw articulation is still between the articular and quadrate. *Cynognathus* possessed large sectorial (adapted for cutting) postcanine teeth but lacked cingulum cusps.

Another group of cynodonts that are prominent components of certain Middle Triassic continental assemblages are the chiniquodonts. These are mostly small carnivores that are typified by the genera *Chiniquodon* and *Probelesodon* (fig. 2.16B). One family, the Probainognathidae, is a monogeneric family for *Probainognathus* from the Chañares. Some earlier authors included *Probainognathus* within the Chiniquodontidae. Crompton and Jenkins (1979) considered the possibility that Probainognathidae were the closest cynodonts to the mammals.

The so-called gomphodonts were a major group of Lower Triassic herbivorous cynodonts, and they are typically divided into two groups: the Diademodontidae and the Traversodontidae. Diademodontids occur in China and eastern and southern Africa. They are broadly similar to *Cynognathus* except for possessing transversely broadened cheek teeth. *Massetognathus* from the Chañares is typical of the traversodonts (fig. 2.16C). It too has transversely expanded cheek teeth, but the canines are greatly reduced and the snout is blunt—not constricted like *Diademodon* or *Cynognathus*—and the broad expansion of the maxilla lateral to the tooth row is suggestive of fleshy cheeks in the living animal (similar to ornithischian dinosaurs; see Weishampel and Norman [1989], but see Witmer [2001]) for a contrasting view). Traversodontids are known from the Lower to Upper Triassic of South America, the Middle Triassic of east Africa, and the Upper Triassic of North America, India, and southern Africa. Until descriptions of new forms from the Newark Supergroup and Europe, the traversodont cynodonts were regarded as exclusively southern hemisphere forms.

But another group of herbivorous cynodonts are still considered by some authorities to be the closest sister group to Mammalia. The Tritylodontidae are first known from Late Triassic sediments, and, persisting well into the Middle Jurassic, they are also the last known of the therapsids. One of the best-known taxa is *Oligokyphus* from the British fissure deposits (earliest Jurassic) (fig. 2.16D). Certainly there is no doubting that in many respects they are indeed mammal-like, even reminiscent of modern rodents: the large temporal opening is confluent with the orbit, the cheek teeth have multiple roots, and the two tooth rows of the upper jaw are typically arranged parallel to each other with no posterior divergence. Furthermore, as Kemp (1983) indicated, the postcranial skeleton bears a remarkable resemblance to that of early mammals, including the disposition of the proximal femoral trochanters, an ulna with an enlarged olecranon, elongate distal caudal vertebrae, and the configuration of the proximal tarsals. Although these similarities led earlier workers to consider tritylodontids as mammals, the dentary is clearly of a nonmammalian type and did not articulate directly with the squamosal. Nevertheless, Kemp (1983) did suggest that they were the sister group of Mammalia, a view that still retains considerable support (e.g., Rowe 1988; Wible 1991).

The Trithelodontidae is the final group of Triassic cynodonts sometimes considered as the sister group of Mammalia. Trithelodonts (sometimes known as ictidosaurs) were small carnivorous forms typified by the genera *Pachygenelus* from North America and *Therio-*

Figure 2.16. Reconstructions of the skeletons of a number of different mammallike reptiles. (A) *Thrinaxodon,* after Jenkins (1984). (B) *Probelesodon,* after Romer and Lewis (1973). (C) *Massetognathus,* after Jenkins (1970). (D) *Oligokyphus,* after Kühne (1956).

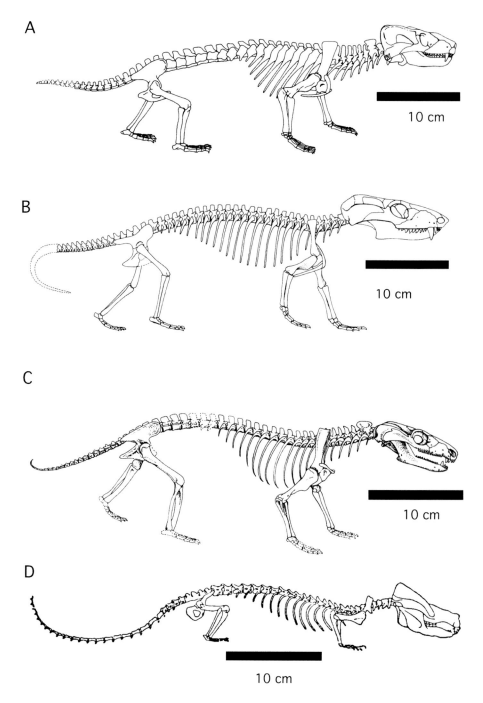

herpeton from South America. Their skulls were small, generally not exceeding 5 or 6 centimeters, and they possessed rather broad cheek teeth. Tritheledonts extended into the Early Jurassic.

Mammals

It is all too easy to regard mammals as a modern group of vertebrates that essentially have only a limited history before the Cenozoic. However, as so eloquently stated in the title of a book edited by Jay Lillegraven, Zofia Kielen-Jaworowska, and William Clemens (1979), the first two thirds of mammalian history occurred in the Mesozoic, with the earliest "true" mammals dating back to the Late Triassic. The word *true* is placed in quotation marks because a contentious debate surrounds the constitution of the Mammalia. The question of just what constitutes a mammal might at first seem silly. After all, today, it is obvious which animals are mammals and which are not. Even the platypus, an animal apparently put together from a variety of spare parts, fits the straightforward criteria of suckling its young and bearing fur. But these characteristics would be impossible to decipher in 99.9 percent of fossils. Moreover, were such characters acquired together, or, as is most likely, were they acquired sequentially? If the latter, just what was the order in which they appeared, and which is the specific feature that designates true "mammalness"? The reader is referred to Gauthier and de Queiroz (2001) for a discussion of such issues.

To some, the set of monophyletic taxa that should be designated as Mammalia seems arbitrary. However, in much the same way as for definitions of Dinosauria, there are different ways at looking at the definition of Mammalia. Traditionally, most authors would include Morganucodontidae and Kuehneotheriidae within Mammalia.

The Morganucodontidae is best represented by *Morganucodon* from the British fissure deposits (chapter 10). Although all the remains of *Morganucodon* have thus far been confined to earliest Jurassic sediments, isolated teeth of *Kuehneotherium* have been recorded in infills considered to be latest Triassic in age (Fraser et al. 1985).

Although the material of *Morganucodon* from the fissure infills comprises completely disassociated remains, it is incredibly abundant. Moreover, some articulated material is known from China, and this has permitted a detailed description of the skull (Kermack et al. 1973, 1981). Although only 3 centimeters long, the skull is large relative to the length of the body, and it was therefore another very small animal. Compared with some of the carnivorous cynodonts, the skull is longer and narrower, and the zygomatic arch (jugal and squamosal) is arched dorsally. Like all mammals, the otic capsule (ear region) is composed of a single ossification, the petrosal (fused opisthotic and prootic). However, the quadrate and articular bones, although much reduced, still enter into the jaw articulation and have not been coopted as auditory ossicles. On the other hand, the teeth are sharply differentiated into incisors, canines, premolars, and molars. Furthermore, the molars were never replaced, and the incisors, canines, and premolars were only replaced once. Thus, in its dentition, *Morganucodon* had attained a uniquely mammalian condition, one that permitted a precise and specific occlusion pattern between the upper and lower jaws.

Kuehneotherium is known from much sparser material. It was another small insectivorous animal. What is unique about *Kuehneotherium* is the arrangement of the cusps on the molariform teeth. Instead of a linear pattern, the three cusps form a shallow triangle. Thus in the lower jaw the central cusp lies buccal (more toward the cheek) to the other two principle cusps, and in the upper jaw the central cusp lies lingual (closer to the tongue) to the other two cusps.

One enigmatic group of Triassic mammals is the Haramiyidae. For a long time the group was only known on the basis of numerous isolated teeth, mostly from Europe. Two principal taxa were recognized, *Haramiya* and *Thomasia*. These teeth were multicusped and not unlike those of multituberculate mammals, but they lacked clearly defined roots, and it even proved difficult to ascertain their orientation within the jaw. Indeed, some authors even questioned their mammalian affinities. However, the discovery of haramiyid dentaries and a maxilla from the Upper Triassic in Greenland (Jenkins et al. 1997) showed unequivocally that haramiyids are mammals, but more significantly, they revealed that haramiyids had a novel pattern of puncture/crushing occlusion that was different from the grinding or shearing mechanisms of other Early Mesozoic mammals. The new form was named *Haramiyavia clemmenseni*. This discovery confirmed suspicions expressed as early as 1946 by Parrington and reiterated by later workers, including Crompton (1974) and Sigogneau-Russell (1989), that *Haramiya* represented upper teeth and *Thomasia* lower teeth of a similar or even the same animal.

The upper and lower teeth were offset and thereby provided an interlocking occlusion

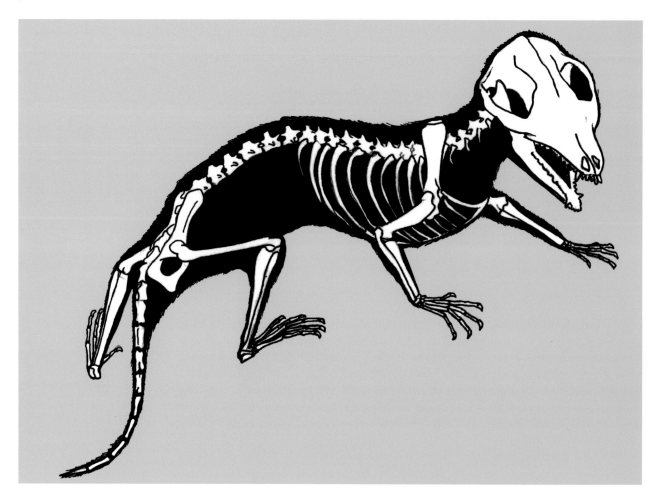

Figure 2.17. Rendition of an
early mammal based on
Megazostrodon.

between the two jaws. However, Jenkins et al. (1997) noted that the nature of the jaw and its occlusion differs not only from that of multituberculates, but also from that of other well-known Triassic mammals, the Morganucodontidae, and the Kuehneotheriidae. These authors also noted features of the *Haramiyavia* jaw that were highly suggestive of a basal position for the group among mammals. Yet at the same time, its exceptionally specialized dentition pointed to the real possibility that mammals arose and diversified much earlier in time—perhaps even as early as the Middle Triassic. The recognition of *Adelobasileus* as a Carnian mammal (Lucas and Luo 1993) tends to underscore this hypothesis.

Terrestrial Invertebrates

In some respects, it is fair to say that arthropods are the pivotal components of biological communities, although of course without the primary producers, the plants, there would be no communities in the first place. Nevertheless, the role of terrestrial arthropods as pollinators, scavengers, detrital feeders, and parasites certainly make them "the little things that run the world." Yet we know relatively little about the diversity of modern-day arthropods. For example, although there are approximately a million named species of insect, estimates of the total number of modern insect species ranges from between 5 and 30 million. If we are ignorant of the majority of insects inhabiting the earth today, what hope is there for understanding insect diversity over 200 million years ago? Indeed, with the exception of a handful of sites worldwide, the record for Triassic insects, and terrestrial arthropods in general, is embarrassingly slim, particularly when compared with the terrestrial arthropod fossil record for earlier and later intervals in geologic time. In the Devonian, several localities document the appearance of mites, centipedes, springtails, and spiderlike arachnids. Then in the Carboniferous, there was another evolutionary explosion that involved the evolution of giant primitive relatives of the dragonflies and mayflies. By Permian times, we have records for a number of living insect orders, including the true dragonflies (Odonata), plant hoppers and sucking bugs (Hemiptera), and the beetles (Coleoptera). For the Jurassic, there are a host of rich insect localities, including the lakebeds at Karatau, Kazakhstan, and the Issyk-Kul in the Tien Shan mountains of Kyrgyzstan. Many of these have a fairly modern assemblage, including flies and hymenopterans. If we go much further up the geologic column, we are treated to the truly spectacular Eocene beds at Florissant (Colorado), complete with lepidopterans with color patterns preserved. But as we shall see in subsequent chapters, we now know that the Triassic spawned the earliest members of many modern orders and families, including the Diptera (true flies), Thysanoptera (thrips), and possibly Tricoptera (caddis flies).

In addition to the body fossils, numerous trace fossils have also been attributed to a variety of invertebrate groups. On the basis of certain trace fossils, there have even been suggestions that social insects, such as termites, existed in the Triassic (Hasiotis and Dubiel 1995). However, the evidence for this is certainly far from conclusive.

The Supporting Cast—The Vegetation

We normally regard the supporting cast of any production as playing a secondary role, but nothing could be further from the truth in the natural world. Here, the plants are the primary producers, and they literally support the animal communities.

Land plants of the Triassic were certainly more diverse than those of the Permian. Although they included various holdovers from the Paleozoic, in addition, there were many new families as well as one new order, the Bennettitales. In fact, just like the tetrapods, the Triassic vegetation can be considered a mixture of ancient and modern. Indeed, some Triassic species are considered to be members of living genera, whereas others bear no resemblance whatsoever to modern-day taxa. There was a clear distinction during the Triassic between the land floras of the northern and southern hemispheres. Certain ferns grew in both hemispheres, but in general, there were relatively few taxa common to both the north and the south.

Building a picture of plant life during the Triassic is difficult because plants are rarely preserved with all their identifying features intact. During a plant's life cycle, certain characteristic features are subject to separation and dispersal from the parent organism. For example, many plants lose their foliage in winter. More importantly, fruiting structures and seeds are even more temporary, so that in the fossil record, they are rarely found attached to the main plant. True, the characteristic appearance of many pollen and seeds, combined with their great abundance, often provides us with a useful stratigraphic framework for dating sediments, and it is important that we provide names for each type of seed, fruiting structure, and foliage. But the downside is the inevitability of assigning different binomial names for different parts of the same plant. The seed fern *Glossopteris* (see below) is a good example of this. Although leaves may be referred to *Glossopteris*, stems might be referred to *Araucarioxylon*, roots to *Vertebraria*, seed-bearing structures to *Scutum*, dispersed seeds to *Pterygospermum* or *Stephanostomata*, and pollen-bearing structures to *Glossotheca* or *Eretmonia*! And spotting errors can be difficult.

Recognizing these inherent difficulties, the ICBN (International Commission on Botanical Nomenclature) code states that all plant fossils are to be treated as morphotaxa (form taxa) for nomenclatural purposes. Of course, the contrast with animal life is not so stark. Plants are designed as modular organisms whose components normally separate as part of the proper functioning and life cycle of the organism. This is not the

Plate 2.13. During the Triassic, parts of Greenland were
well vegetated (Cape Stewart
flora).

case for most anatomical structures in tetrapods. When an animal dies, it does so with
all its characteristic body parts together in one place. Unfortunately, that doesn't mean
that we always avoid assigning more than one name to the same animal. Scavengers and
weathering may widely separate parts of an individual skeleton, so that new taxa are typically erected on less than complete material. Even so, the problems are less prevalent
in paleozoology. Moreover, there is a better chance that we shall eventually recognize
previous errors!

For our portrait of the Triassic, foliage cover provides an immediate impression of the
nature of the environment—whether it was lush, arid, aquatic, or somewhere in between.
Thus in the following overview we shall concentrate on the leaf morphotypes. As mentioned in chapter 1, it is not unusual to see the Triassic portrayed as dominated by a mostly
barren, rocky landscape with a few cycads and distant conifers. At the same time, the scene
would be liberally sprinkled with a variety of tetrapods, including large herbivores. How
these large herbivores could have survived in such an environment is not made clear. Although angiosperms, especially the grasses, did not exist during the Early Mesozoic, there
were many other Triassic and Jurassic foliage taxa that would likely have formed dense
ground cover. Herbaceous seed ferns and extensive stands of horsetails and ferns were
probably common. There is certainly every reason to think that open, verdant hillsides,

wooded and forested slopes, and hot and steamy forests and swamps were very much part of the Triassic landscape.

Club Mosses

The club mosses (lycopods) are widely distributed today in subtropical and tropical forests. They are typically herbaceous plants with trailing shoots. An appropriate common name for living lycopods is "ground pine." Many of them form short, upright branches that closely resemble dense stands of pine seedlings (fig. 2.18). But probably the best-known lycopods are the gigantic treelike forms of the Carboniferous such as *Lepidodendron*. During the Triassic, herbaceous lycopods were widely distributed, but they do not constitute major parts of any floras. Typically these herbaceous forms seem to have been closely related to the main living genera: *Lycopodium, Isoetes,* and *Selaginella*. Ash (1972) even referred Triassic specimens from Arizona to *Selaginella*. As for the treelike lycopods, although they are known from the Triassic, they are rare.

Figure 2.18. Restoration of the lycopod *Pleuromeia* (after Retallack 1975).

Horsetails

Today, horsetails (Equisetales) are often considered in the category "living fossil." Certainly they were widespread in Paleozoic and Mesozoic communities, and Triassic horsetail remains are common. Many Triassic representatives were herbaceous plants that were similar to the living genus *Equisetum*. The sedimentary sequences that contain such remains frequently contain spores that also closely resemble the spores produced by modern-day *Equisetum*. Consequently, many of the Triassic remains are referred to *Equisetum*. Like the lycopods, the horsetails are also well known for their enormous treelike forms from the Carboniferous, in particular *Calamites* (fig. 2.19). During the Triassic, there were also examples of giant horsetails. These are generally referred to the genus *Neocalamites*. Although *Neocalamites* did not quite attain the truly spectacular dimensions of *Calamites* (up to 20 meters tall), they did reach impressive proportions. For example, Holt (1947) described specimens from Colorado in growth position that were almost 6 meters tall and 30 centimeters in diameter. The long, straight stems of horsetails are characterized by distinct nodes, with a marked longitudinal ridged pattern extending between nodes. Leaves are typically unveined and occur in distinct whorls. Horsetails are usually associated with swampy or marginal aquatic habitats, and this seems to hold true for the Triassic. No major evolutionary changes occurred within the Equisetales during the Triassic, although there was an apparent reduction in the diversity of treelike forms.

True Ferns

After a major decline at the end of the Carboniferous that continued through Permian times, the true ferns (Filicales) were comparatively rare at the beginning of the Triassic. But the Triassic witnessed a resurgence in fern diversity and the appearance of several new families. By and large, Late Triassic ferns can be readily assigned to living families (there is a trend toward fern assemblages of modern aspect).

One of the largest Early Mesozoic families was the Osmundaceae, with *Cladophlebis* one of the more widespread genera.

Figure 2.19. Reconstruction of a *Calamites* plant (after Boureau 1964).

Figure 2.20. Foliage of the fern *Dictyophyllum*.

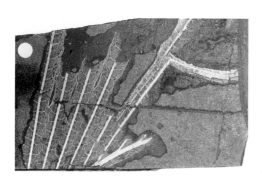

The families Matoniaceae and Dipteridaceae both comprise generally large and showy ferns. *Phlebopteris* is the oldest member of the Matoniaceae, a family that today is restricted to Borneo, Indonesia, and New Guinea. *Dipteris* is also found today in the same region of the world and is the only surviving genus of the Dipteridaceae. Two Triassic genera of dipteridaceous ferns are *Clathropteris* and *Dictyophyllum*, both of which are particularly widespread in Late Triassic rocks (fig. 2.20). By Early Jurassic times, the Dipteridaceae were major elements of northern floras (Corsin and Waterlot 1979). The fronds of dipteridaceous ferns branch in a dichotomous fashion, and the veins form a characteristic reticulate pattern.

Seed Ferns

There are no living members of seed ferns (Pteridospermophyta), but they were widespread during the Paleozoic and extended into the Mesozoic. In fact, Mesozoic seed ferns have attracted a good deal of attention because many authorities believe that we might be able to trace the ancestry of the angiosperms (flowering plants) to somewhere within this group. It is clear from preserved fossils that in life, pteridosperms were fernlike; indeed, for a long time, many were actually considered to be true ferns. But with the discovery of seeds attached to leaves, it was necessary to revise their classification, and they are now considered to be part of the gymnosperm radiation.

One of the most important families of pteridosperms is the Glossopteridaceae. Members of this family occur only in the southern hemisphere, and they rose to prominence during the Permian, with the most common fossil being that of the spatulate leaf, *Glossopteris*. Gould and Delevoryas (1977) reconstructed *Glossopteris* as a large tree, but several different species are recognized, and it may well be that not all were arborescent (treelike). *Glossopteris* leaves, such as *Glossopteris sidhiensis* from India, are known from Triassic sediments, but there is a definite decrease in their abundance toward the end of the Triassic, so that the Glossopteridaceae were extinct before the beginning of the Jurassic.

The family Corystospermaceae was almost exclusively Triassic. Although it comprises a relatively small group of plants, it includes the foliage form genus, *Dicroidium* (fig. 2.21). Some measure of their diversity can be gained from Retallack (1977), who recognized 15 species of *Dicroidium* leaves and 16 varieties. Along with other members of the family, *Dicroidium* is restricted to Triassic sediments in the southern hemisphere and in India, where it occurs widely. Unlike the *Glossopteris* plants, the Corystosperms were small and woody.

Two other families, the Peltaspermaceae and the Caytoniaceae, have Triassic members. Leaves assigned to the Peltaspermaceae include *Lepidopteris* and *Scytophyllum*, and both are common in the Triassic of the southern hemisphere. By contrast with the other Triassic seed ferns, the Caytoniaceae seems to be most common in western Europe, although they do occur worldwide. This family extends into the Jurassic and even becomes widespread in the northern hemisphere.

Cycads

The cycads (order Cycadeles) range from the Upper Carboniferous to the present but really attained their acme during the Early Mesozoic, when they achieved a global distribution. Today they have a much more restricted distribution, occurring in Central and South America, South Africa, Australia, and parts of Asia and Mexico.

Some are arborescent, resembling palms; others have short, squat trunks (fig. 2.22). One of the widespread Triassic foliage types that may be referred to the cycads is *Taenopteris*. A variety of seed bodies in Triassic sediments testify to cycad diversity throughout the period.

Bennettitaleans

The order Bennettitales (sometimes called the Cycadeoidales) first appeared in the Triassic and became an important component of many Mesozoic floras before the explosion of angiosperm diversity in the Cretaceous. These plants are found in both the northern and southern hemispheres. The leaves and stems of bennetitaleans are similar to those of cycads, but detailed study of the leaves has shown that there were basic differences in the epidermis and the form of the stomata (pores in the epidermis) (Thomas and Bancroft 1913). There are also differences in the reproductive structures, with small bisporangiate cones being typical. The nature of these cones has led to the proposal that progenitors of flowering plants (angiosperms) can be found in this group (Doyle and Donoghue 1986; Crane 1988). Fossil leaves common in the Triassic that are referred to the Bennettitales include *Zamites*, *Otozamites*, and *Pterophyllum*.

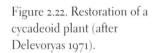

Figure 2.21. Restoration of the foliage of the seed fern *Dicroidium* (after Taylor and Taylor 1993).

Conifers

The conifers (Coniferales) include large woody trees as well as small shrubs. The taxonomic position of some of the Triassic families is confusing to say the least. The genus *Voltzia* (see chapter 4), and other similar forms such as *Aethophyllum* (Miller 1982), are some of the most widespread foliage types in the Middle Triassic, yet their taxonomic status remains unclear. Some authors, such as Miller (1977), recognized a distinct family, the Voltziaceae, whereas others (Taylor and Taylor 1993) prefer to merely treat them as form genera and do not assign them to a specific family. On occasions, conifers have even been placed in a separate order, the Voltziales, which is sometimes referred to as the "transient conifers" (e.g., Ash 1986a).

Figure 2.22. Restoration of a cycadeoid plant (after Delevoryas 1971).

One of the largest families of Mesozoic conifers is the Cheirolepidiaceae. As Taylor and Taylor (1993) note, this family includes a diverse array of taxa, both arborescent and herbaceous, which undoubtedly inhabited a variety of ecological niches. The single unifying character would seem to be a unique type of pollen referred to as *Classopolis*. In the past, the foliage of the Cheirolepidiaceae had been assigned to a variety of different families. Only when specimens are recovered with attached reproductive organs can definitive referrals be made. Thus we have some examples of the foliage form genus *Brachyphyllum* (leaves helically arranged on the vegetative shoot) and *Frenelopsis* (leaves in whorls) referred to the Cheiroleipidiaceae. But in the absence of associated reproductive structures, sterile foliage (as opposed to foliage bearing fruiting bodies) is difficult to classify, and *Pagiophyllum*, *Podozamites*, and *Elatocladus*, as well as some examples of *Brachyphyllum*, may have been de-

Figure 2.23. Today, large logs
of the conifer *Araucarioxylon*
lie where they fell over 200
million years ago in the Triassic woods.

rived from a whole spectrum of different conifer families, including the Cheirolepidiaceae and Araucariaceae.

The two largest families of extant conifers are the Cupressaceae (cypresses) and the
Pinaceae (pines, firs, spruces, hemlocks). There is evidence that both were present in Triassic times, although they were certainly not common. Detached cones and some leafy
twigs from the Upper Triassic of France are indicative of Cupressaceous conifers, and an
ovulate cone, *Compsostrobus*, which has been described from the Newark Supergroup
(Delevoryas and Hope 1987) is widely accepted as one of the earliest representatives of the
Pinaceae.

A large number of Upper Triassic fossils have been referred to the family Araucari-
aceae. Principal among these fossils are pieces of secondary wood placed in the genus
Araucarioxylon. However, this type of wood is not just limited to the Araucariaceae, and
Ash (1986a) noted that it may be a basic wood type for Coniferales in general. Moreover,
Araucarioxylon wood has also been associated with the Glossopteridaceae (Gould and
Delevoryas 1977), so referral of fossil wood to *Araucarioxylon* does not necessarily mean

that it is wood from an araucariaceous conifer. Some modern-day araucarians are extremely tall trees, and on the basis of the logs from Petrified Forest National Park, the same also held true for the Triassic (fig. 2.23).

Ginkgoes

The sole surviving member of the ginkgophytes is the maidenhair tree, *Ginkgo biloba*. This beautiful tree is frequently planted in temperate parts of the world, and its resilience to pollution and insect and fungal attack make it popular in urban areas. Because the group is thought to extend back to the Late Paleozoic and there is only a single surviving species, *Ginkgo biloba* is another of those organisms encumbered with the label "living fossil." Some Triassic leaf fossils so closely resemble the modern taxon that they have been

Plate 2.14. Fallen ginkgo branches provide shelter for a variety of Triassic fish such as this *Saurichthys* and *Myriolepsis* (chondrosteans) in the Lower Triassic of Australia.

placed in the same genus or *Ginkgoites*. Other common foliage types referred to the ginkgophytes are *Sphenobaiera* and *Eretmophyllum*. The former take the form of wedge-shaped leaves repeatedly divided into narrow ribbonlike sections.

Flowering Plants

Although flowering plants (angiosperms) are always thought of as having a Late Mesozoic origin, there are occasional hints that they may have first appeared in the Jurassic, or even in the Triassic. One of the most intriguing fossils in this respect is *Sanmiguelia*, first described on the basis of palmlike foliage from the Late Triassic of Colorado (Brown 1956) (fig. 2.24). Tidwell et al. (1977) believed that it had affinities with the monocots, but Read and Hickey (1972) considered it more likely to be the foliage of a cycadophyte. However, additional leaf material was later found from a late Carnian sequence in Texas (Cornet 1989), but more importantly, these beds are also known to contain reproductive structures, which are thought to be representative of *Sanmiguelia*. The pollen organs have been called *Synangispadixis*, and the ovulate organs *Axelrodia*. The former comprise a large number of helically arranged microsporophylls, each bearing a pair of pollen sacs. The latter bear small structures interpreted by Cornet (1989) as carpals, and he believed that pollination was carried out by insects. Cornet regarded *Sanmiguelia* as an early angiosperm that combined features of both dicots and monocots, but this is a controversial view. Other fossils that have been debated as part of the same issue are leaves known as *Furcula* from the Latest Triassic (Rhaetian) of Greenland, and *Pannaulika* from the Carnian of eastern North America (Cornet 1993). *Furcula* exhibits a reticulate venation pattern that is reminiscent of some modern dicots, but it has also been argued that the pattern is consistent with certain seed ferns (Scott et al. 1960). The venation pattern of the single described specimen of *Pannaulika* has also been considered to closely match modern dicots,

Figure 2.24. Restoration of the Chinle plant *Sanmiguelia* (after Tidwell et al. 1977). This plant holds a controversial position in paleobotanical circles because it has been argued by Cornet (1989) that it is has close affinities with the angiosperms. The conventional view holds that angiosperms did not arise until 100 million years later.

but a recently discovered dipteridaceous fern leaf from the same locality suggests that *Pannaulika* is really a rather atypical fern leaf.

In summary, the Triassic floras do not show quite the marked shift in emphasis from ancient to modern that is exhibited by the terrestrial vertebrates. Nevertheless, it can be said that by the end of the Triassic, many of the plants exhibit some similarity to modern forms, and there is a definite hint of a modern flavor to the floras.

Plate 2.15. The early morning light picks out the beguiling "flower" of *Sanmiguelia*, while in the background a *Coelophysis* skulks in the shadows.

chapter three

Starting Out in the Triassic

Given the undoubted size and severity of the Permian mass extinction, one would be forgiven for thinking that the Early Triassic world was one bereft of life. But that was certainly not the case—diversity levels declined noticeably, but the fossil record indicates that there were many places where life flourished.

The Karoo

Despite suffering widespread losses at the end of the Permian, the synapsids survived the end-Permian mass extinction and were the most common tetrapods at the beginning of the Triassic. The dicynodont *Lystrosaurus* was particularly widespread and so common that it forms a useful index fossil. Its name is practically synonymous with the lowest unit of the Triassic, the Lootsbergian (see appendix 1).

Another name that conjures up images of the Triassic (as well as the Permian) is the Karoo. This large sedimentary basin comprises some massively thick sequences stretching across South Africa and Lesotho. The sediments of the basin actually extend from Upper Carboniferous glacial deposits through a variety of Permian sequences, including marine sediments, and into the classic dicynodont-bearing terrestrial deposits of the Beaufort Group. These are Latest Permian and Triassic in age and contain hundreds of vertebrate fossils. Above the Beaufort Group is the continental Stormberg Group comprising the Molteno, Elliot, and Clarens formations. Although the upper part of the Elliot and the Clarens formations are Jurassic in age, as we shall see later (chapter 9), the Molteno also has some important Triassic (albeit Late Triassic) fossils. Capping the whole sequence are extensive Jurassic basalts. But here we shall concentrate on the Triassic sediments of the Beaufort Group.

Without question the most significant animal in the Early Triassic of the Karoo Basin was *Lystrosaurus*. This was an animal that was not just confined to one small part of the world. *Lystrosaurus* was a true cosmopolitan, with extensive records from the Lower Triassic of India, Australia, China, Antarctica and Russia, as well as south Africa. This dog-sized herbivore was typical of all dicynodonts in having two well-developed tusks in an otherwise edentulous mouth. However, it had a particularly reduced, and consequently exceptionally short, facial region (fig 3.1). As such, it is often considered to be one of the most specialized of all dicynodonts. Cope (1870) first proposed the generic name, and at one time over 30 species were included in the literature. Now the count has dropped to

Plate 3.1. *Tetracynodon* (Permian mammal-like reptile) eyes a tasty morsel in the form of young *Lydekkerina* (temnospondyls).

Plate 3.2. *Moschorhinus* (therocephalian) in the Katberg Formation.

Plate 3.3. Some authorities think that *Lystrosaurus* had a lifestyle like that of modern-day hippos, and that they may have spent long periods wallowing together in ponds and small lakes.

around 10, and the number may still be falling because even some of these may be synonymous.

Many authors considered *Lystrosaurus* to have been quite at home in the water. This notion dates back to the work of Watson (1912), who stated that the postcranial skeleton was adapted to an aquatic existence. Watson believed that the large ischium supported powerful hindlimb musculature that generated a strong swimming stroke. He felt that the worn tusks, a flexible neck, and strong head-depressing muscles all indicated that *Lystrosaurus* obtained its food by digging using its tusks, with the wide-open mouth acting as a shovel. Furthermore, Watson suggested that the massive jaws pointed to an ability to crush resistant food material such as mollusks. In contrast, recent studies on jaw function (Crompton and Hotton 1967; Cluver 1971) indicate that the jaws were incapable of a grinding action. First of all, it was not possible for the flat dentary plates to contact the palatines. Moreover, the edges of both the dentary and the maxilla are sharp and form something more akin to a beak. Instead of grinding, during retraction, the flat outer surfaces of the lower jaw would have contacted the deep margins of the maxilla. The sharp edges of both jaws would have acted in concert to form an effective guillotine, a tool that

A

Figure 3.1. Characteristic blunt-faced skull of the ubiquitous dicynodont genus *Lystrosaurus*. (A) Skull from the Triassic of Asia on display in the Paleontological Institute in Moscow. (B) Restoration of the skull (after King 1988).

B

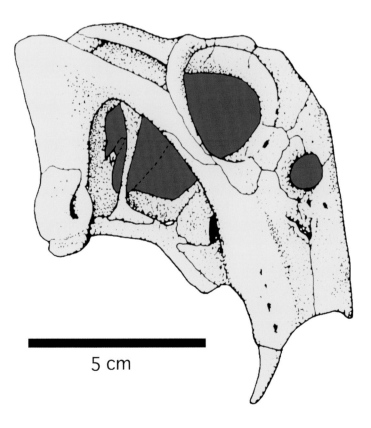

5 cm

would have been more adept at slicing leaves. Even so, Cluver (1971) believed that the foreshortened face was an adaptation to feeding underwater. The mouth could be submerged below the water surface while maintaining the nostrils above water level.

Although many have considered the habits of *Lystrosaurus* to have been something akin to those of modern-day hippos, others have shunned the idea that it spent much time in the water. Barry (1968) documented small rods of bone in contact with the stapes of several individuals of *Lystrosaurus*. These he considered to represent the hyoid apparatus (bones in the throat). This arrangement of bones is strongly suggestive of the capability of picking up ground vibrations through the lower jaw and passing them on through the hyoid and thence via the stapes to the inner ear. Such a system is unlikely to be present in an animal that spent a great deal of its time in water. King and Cluver (1991) and Retallack and Hammer (1998) have also argued against an aquatic habit. In documenting remains of a *Lystrosaurus* skeleton from the lower part of the Fremouw Formation of Antarctica, Retallack and Hammer (1998) noted that the bones were scattered on a paleosol, which is inconsistent with inundation by a catastrophic flood. Retallack believed that horsetails found in similar paleosols were part of an understory beneath trees, and he thought that *Lystrosaurus* was fully terrestrial and capable of burrowing. However, *Lystrosaurus* occurs in the same horizons as lycopods such as *Isoetes*. These herbaceous plants probably had shallow corms and are considered to have been aquatic plants. They are locally abundant in Early Triassic rocks, but they decline significantly in the Late Triassic along with the dicynodonts. Perhaps *Lystrosaurus* browsed on these plants in the shallows and dug up the underground corms.

On occasion, occurring together with *Lystrosaurus* is another dicynodont, *Myosaurus*, but this taxon is nowhere near as common. Its jaws lack any crushing surfaces, and the animal is considered to have fed on rather soft material. Whether it was herbivorous or not is unclear. Both these dicynodonts occur in South Africa and Antarctica, together with the procolophonid *Procolophon* (a small herbivore), and various amphibians and cynodonts.

One of the best-known carnivorous tetrapods found in the Lower Triassic (Lootsbergian) of South Africa is *Proterosuchus*. This long-snouted reptile was a little like a modern-day crocodile. Unlike the semierect to erect posture of many later archosaurs the proterosuchids had a sprawling posture, which is indicative of their basal position within Archosauria. *Proterosuchus* had a distinctly downturned tip to the snout, and its long jaws were lined with curved and serrated teeth. It may well have fed on some of the smaller reptiles, such as the procolophonoids. In South Africa there are at least two species of *Proterosuchus*, *P. vanhoepeni* and *P. fergusi*. Another carnivore was the synapsid *Thrinaxodon*.

In the overlying Burgersdorp Formation (equivalent to the Nonesian), *Lystrosaurus* is replaced by the larger dicynodont, *Kannemeyeria* (fig. 3.2), and *Proterosuchus* also disappears, but the heavily built *Erythrosuchus* is an even more formidable-looking predator. Typically it reached over 25 feet in length and had a disproportionately large skull set on a short neck. With its stout limbs and bulky body, it was not an agile animal, and it has been suggested that it lived in marshes, perhaps using an ambush technique for capturing prey.

But the most characteristic animal of the Nonesian in South Africa is *Cynognathus*. This robustly built cynodont replaced the Lootsbergian synapsid *Thrinaxodon*. It would thus seem that there was a trend for both prey and predators to increase in size in the Nonesian. Nevertheless, there were a variety of smaller contemporaneous herbivores. A number of procolophonids have been described from the Burgerdorp Formation. All were originally referred to the genus *Thelegnathus* (Gow 1977). However, recent work (Modesto and Damiani 2003) has shown that the original *Thelegnathus* material has no diagnostic features and the name is therefore considered a nomen dubium—that is, the name is taxonomically meaningless because new material can never be unequivocally referred to this genus. Additional study of the Burgersdorp procolophonids indicates that there are at least four different genera, and these have been named *Thelerpeton*, *Theledectes*, *Thelephon*, and *Teratophon* (Modesto and Damiani 2003).

Plate 3.4. Mother *Thrinaxodon* stands guard as her brood scuffle in the dirt and bare their sharp teeth. The fossil record doesn't tell us whether these mammal-like reptiles were actually mammal-like in their behavior, but at some point in their evolutionary lineage, such behavior patterns developed.

Plate 3.5. Sapped by a shortage of food, *Cynognathus* slumps on its hindquarters in the sand. The shrubs in the background are the seed ferns *Dicroidium*. The foliage in the foreground represents *Taeniopteris*.

It is in the same beds as *Erythrosuchus* that the much-discussed *Euparkeria* is also found. This small archosaur has been variously regarded as a basal form or an early member of the ornithodiran clade. Currently, most authors (e.g., Sereno 1991) place it within the basal part of the archosaurian radiation. This small carnivore, which grew to about 2 feet long, may have been at least facultatively bipedal.

Beyond the Karoo

Another interesting picture of Lower Triassic dicynodont life comes from footprints. Trackways from Bellambi in southeast Australia are considered to represent dicynodont tracks, possibly even *Lystrosaurus* (Retallack 1996). The tracks show claw impressions at the ends of the toes and scales on the soles of the feet. Moreover, surrounding the individ-

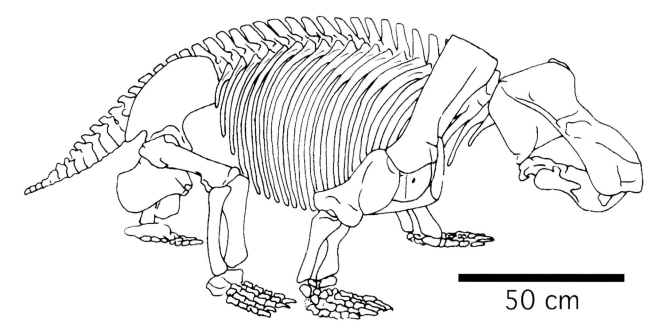

50 cm

ual tracks are fine striation patterns that radiate outward. These bear a definite resemblance to the impressions that hair might make in wet mud. However, there are alternative explanations to hair, such as the presence of a wrinkled cyanobacterial scum. Furthermore, on other parts of the block, scratch marks testify to the fact that the excited original collector got a little carried away when trying to clean up the tracks! It is therefore not inconceivable that the hairlike impressions are nothing more than artifacts of poor preparation. If the impressions are indeed the result of hairs trailing into the mud, this would represent significant support for the idea that therapsids were warm-blooded (Bakker 1986). On the same slab are invertebrate tracks that, by a process of elimination, Retallack considered to be the traces of insects. Rare insect remains do occur in the same horizon.

Although the synapsids share common species between Antarctica and Africa, the amphibians are only similar at a higher taxonomic level (family). The inference is that although there were no barriers to the movement across land between Africa and Antarctica, the intervening terrain was not conducive to the free dispersal of amphibians.

China

Despite the great distance separating China and Africa, it is remarkable how similar the Lower Triassic faunas of the two continents are. China has a particularly good Lower and Middle Triassic terrestrial vertebrate record, with the fossiliferous units being concentrated in two principal basins: the Junggur Basin in the northwest part of the country, and the Ordos Basin situated toward the center of China. Because these two regions are separated by some considerable distance, the rock units naturally have completely separate names, yet they do represent similar time frames.

In the Junggur Basin, the lowermost Guodikeng Formation yields the characteristic *Lystrosaurus*, although only from the upper part of the formation, and the overlying Jiucaiyuan Formation produces a prolific vertebrate fauna that is dominated by *Lystrosaurus* fossils. *Proterosuchus* (*Chasmatosaurus*) is another interesting component of the Jiucaiyuan vertebrate assemblage, although it is smaller than the South African *Proterosuchus* species.

In the Ordos Basin, the lowermost formation is the Liujiagou, but to date it has not

Figure 3.2. Heavy-set skeleton of the dicynodont *Kannemeyeria* (after King 1990).

Plate 3.6. Startled in the mud, and with a sudden burst of energy, the procolophonid *Teratophon* momentarily becomes airborne as it attempts to evade the jaws of the therocephalian, *Bauria* (*Watsoniella*).

produced any vertebrates. Nevertheless, palynomorphs and plant megafossils are present, and these indicate an earliest Triassic assignment. The overlying Heshanggou Formation does, however, have a rich vertebrate fossil assemblage that includes abundant procolophonids, erythrosuchid material (*Fugusuchus*), a possible proterosuchian, and a therocephalian. The procolophonids are particularly abundant and have all been referred to the genus *Eumetabolodon* (Li 1983). With its large orbits and approximately triangular skull outline in dorsal view, *Eumetabolodon* is similar to *Procolophon*. Like *Procolophon*, *Eumetabolodon* has transversely broadened teeth in mature individuals, but in juveniles, the cheek teeth are conical. This may be indicative of different feeding habits, with the young having a more cosmopolitan diet than the adults. It is even possible that the young *Eumetabolodon* were totally carnivorous, as seen in many reptiles today (Pough 1973).

The absence of dicynodonts in the Heshanggou Formation is particularly striking, a feature made even more remarkable by their reappearance in the overlying Ermaying Formation. Either dicynodonts are present but have yet to be discovered, or alternatively, the Heshanggou Formation represents deposition in a slightly different paleoenvironment—one unsuitable for dicynodont populations.

The Ermaying Formation is particularly notable for its vertebrate fossils. Like the equivalent upper part of the Burgersdorp Formation of South Africa (Nonesian), the lower part of the Ermaying Formation yields large dicynodonts in the form of *Kannemeyeria* and *Parakannemeyeria*. It is also known for procolophonids (*Pateodon*), euparkeriids (*Turfanosuchus* and *Halazhaisuchus*), and therocephalians (*Ordosiodon*), as well as an erythrosuchid (*Vjushkovia*) and a proterosuchian (*Guchengosuchus*). The upper part of the Ermaying Formation yields a different faunal assemblage dominated by kannemeyeriid dicynodonts, but these are almost certainly Middle Triassic (Perovkan approximately = Anisian), and they are considered in chapter 5.

Elsewhere in Asia, *Lystrosaurus* is found with other Lootsbergian tetrapods in the Panchet series of India, including the early archosaur *Proterosuchus*. Likewise, *Kannemeyeria* has a wide distribution, occurring in Russia and India as well as South Africa.

Cruickshank (1972) noted that fish were absent in the localities yielding *Proterosuchus*

Plate 3.7. Skittishly, two *Euparkeria* sniff at the carcass of a *Trematosuchus* that has fallen victim to the powerful *Erythrosuchus*. The abundant vegetation in the background is *Dicroidium*. The single frond of foliage that lies in the foreground is *Pterophyllum*.

Plate 3.8. In the Lower Triassic of China, a herd of *Lystrosaurus* clamber through
the lycopods (*Pleuromeia*)
surrounding the water's edge.

and *Lystrosaurus*. This is remarkable, given that so much attention has been given to the
idea that *Lystrosaurus* was semiaquatic in habit. Moreover, Cruickshank also considered
amphibians to be rather rare, and again, this is somewhat surprising, given that conditions
during earliest Triassic times in this region were supposedly more humid than at the end
of the Permian. Broili and Schröder (1934) regarded *Proterosuchus* as another aquatic animal, something akin to modern-day crocodiles. However, the lateral position of the external nares and the well-ossified limbs are actually more suggestive of a fully terrestrial
habit. Thus the suggestion that many *Lystrosaurus* assemblages are marginal aquatic is in
some doubt. Consequently, Anderson and Cruickshank (1978) prefer to talk about
"plains" communities in referring to the *Lystrosaurus* assemblages.

Higher Latitudes

On their own, the African and Chinese Lower Triassic sequences portray a pretty uniform
picture of the landscape in the Early Triassic (Lootsbergian through Nonesian). But we
do get hints of some departures from the norm when we venture into higher latitudes. Although *Lystrosaurus* and *Kannemeyeria* were widespread, they were unable to conquer
the whole world. Some regions of the world had different land vertebrate faunas—faunas
dominated by temnospondyls such as *Luzocephalus*, *Tupilakosaurus*, and *Wetlugasaurus*,
and almost completely devoid of *Lystrosaurus*. That the *Lystrosaurus*-dominated assem-

blages are indeed equivalent in age to these temnospondyl assemblages is suggested by the overlap of *Lystrosaurus* and *Tupilakosaurus* in the Urals.

Lower Triassic of Russia

The majority of earliest Triassic tetrapod remains from European Russia occur in coarse fluvial sediments, with temnospondyl amphibians constituting by far and away the dominant faunal constituents. The rather eel-like *Tupilakosaurus* is one of the dominant temnospondyls, but there is some geographic variation in the abundance of taxa, and no single genus is as dominant as *Lystrosaurus* is in Africa and China. Elsewhere in Russia, the capitosaurid *Wetlugasaurus* is more abundant than *Tupilakosaurus*. Although temnospondyls are common, a number of reptiles are also known, including procolophonids (typically forms such as *Phaanthosaurus* and *Contritosaurus* that possess simple peglike teeth as opposed to the transversely broadened, sometimes molariform-like teeth of later forms), archosaurs (such as the proterosuchian *Chasmatosuchus*), and prolacertiforms, but therapsids are rare. Almost inevitably the therapsids include *Lystrosaurus*, but even it is uncommon. Along with these tetrapod assemblages are abundant fish remains, with elements of the dipnoan *Gnathorhiza* being particularly abundant. Although the assemblages seem to represent mostly aquatic environments, paradoxically, the overall picture for this region during earliest Triassic times is one of aridity. Many authors (e.g., Shiskin et al. 2000) have commented on the small size of the tetrapods as an indicator of aridity, but more importantly, there are also thick sequences of eolian sands. When it comes to correlating these faunas with assemblages elsewhere in the world, we find that *Tupilakosaurus* also occurs along with *Luzocephalus* in eastern Greenland (Wordy Creek Formation). Interestingly, the Wordy Creek Formation is regarded as a nearshore marine sequence and contains characteristic ammonites that allow the terrestrial sequences to be tied into the marine stratigraphy.

Gosford, New South Wales

The Early Triassic exposures near Gosford in New South Wales have revealed a plethora of fish fossils. For instance, at the Somersby fish site close to Gosford, nearly complete xenacanthid sharks have been recovered. Xenacanthids were primarily freshwater predators. Another well-known fish from the area is the lungfish *Gosfordia truncata*. It was originally described from the Hawkesbury Sandstone by the well-known British paleontologist Arthur Smith Woodward. *Gosfordia* is thought to have close affinities to the lineage of *Neoceratodus*, the modern-day Australian lungfish. *Gosfordia* possessed a deep body and grew up to 50 centimeters long. It is postulated that it was a relatively good swimmer, and in all likelihood not adapted to estivation, like the South American and African lungfishes.

But the Early Triassic of Australia is not just noted for its fishes. Numerous amphibian remains have been found in parts of Queensland and New South Wales, and include mastodonsaurids, capitosaurids, and chigutisaurids. The Australian Lower Triassic sequences show more signs of humid conditions than their African and Chinese counterparts.

As discussed in chapter 2, the flora of the Gondwanan continents during Early Triassic times was in transition. The *Glossopteris*-type flora that had been dominant was in decline, giving way to a *Dicroidium*-dominated flora. It would seem that *Glossopteris* was better adapted to the cooler conditions of the Permian and was less tolerant of the warmer climate that is thought to have prevailed through much of Triassic time. Although in the southern continents lycopods and horsetails are quite common, there would seem to be some reduction in their abundance. Swampy, damp environments were apparently less widespread than they were in the Permian. Nevertheless, in mid- to high latitudes, the cli-

Plate 3.9. Watching from a grove of *Pleuromeia* stalks, the perleidiform (chondrostean) fish *Tripelta* darts forward to snap at an insect flying just above the surface of an Early Triassic lake in Australia (Gosford).

mate was still generally rather humid, and the land was probably dominated by rivers, broad floodplains, and lakes, together with riparian forests and woods. The Cis-Ural region represented a departure from this generalization.

In an examination of a number of terrestrial Early Triassic faunas, Ochev (1993) supported the concept (e.g., Robinson 1971) that there was some kind of twofold regional faunal division, with on the one hand therapsid-dominated assemblages and on the other nontherapsid ones comprising primarily amphibians and sauropsid reptiles. His study took into account primary data from the Ural region that had not been previously considered. Ochev considered the nontherapsid faunas to be centered in the Ural and Australian regions, whereas the so-called therapsid-dominated assemblages were widespread in South Africa, Antarctica, India, and northern China. How this distribution translates to maps of the Triassic world depends on which paleogeographic reconstruction models one follows. Certainly plotting this distribution on the Smith et al. (1994) Anisian map lends some credence to the idea that nontherapsid-dominated assemblages are associated with higher latitudes.

Part Two

A COMFORTABLE MIDDLE AGE
The Middle Triassic

chapter four

A Hint of the Sea

What makes a period instantly recognizable for geologists is that it has a very definite beginning and end. Something extraordinary must inevitably have occurred at the transition from one period into another that makes it relatively easy to recognize a period boundary worldwide. Its not surprising that a corollary of this tends to be that during "between" times, nothing particularly remarkable appears to have been happening. So it is with the Triassic. From what we know from the fossil record, the Middle Triassic (Anisian and Ladinian approximately = Perovkan and Berdyankian) was a time of comfortable, even mundane, middle-age spread with a general trend of increasing diversity levels. That is not to say that the fossil assemblages documenting this time period are mundane—in fact, as we shall see, quite the reverse is true. Middle Triassic terrestrial faunas are perhaps less widespread and somewhat less well-known than those of the Lower and Upper Triassic, and so any totally terrestrial fauna from the Middle Triassic immediately has the potential to be important. Probably the most prolific Middle Triassic assemblages are in China and Russia, and we shall continue moving up section in the Junggur and Ordos Basins in the next chapter. In the southern hemisphere, there is the extensive Molteno Formation of southern Africa and the famous Chañares Formation in South America. There are also one or two less celebrated assemblages, and we shall also look at a couple of these in chapter 5. But let's begin our look at Middle Triassic life in central Europe. Here, the Middle Triassic is largely characterized by marine sediments, and there are some particularly exciting marginal environments in France.

The *Voltzia* Forests

The Grès à Voltzia

Today, the gently rolling, conifer-clad slopes of the Vosges Mountains in Alsace-Lorraine form a quiet haven for tourists seeking relaxation. The well-ordered vineyards arranged on the lower slopes provide sustenance of a different sort—one that paleontologists attracted to the area are also sure to appreciate! However, for the paleontologist, the Vosges contain an even greater attraction. The Voltzia Sandstone of eastern France corresponds with the upper Bundsandstein of Germany. Gall (1971) considered it to be an important link between the terrestrial deposition of the Lower Triassic and the marine sediments of the Muschelkalk. Certainly the combination of marine and terrestrial forms is rather unusual.

Plate 4.1. A forest of *Voltzia* trees clothe volcanic plugs overlooking Lake Narrabeen, Gosford, Australia.

Plate 4.2. Conifers growing in
the salty sea air of coastal
Tethys.

Add that to the fact that the Voltzia Sandstone documents an interval in the Triassic that is not particularly well represented elsewhere, and top it off with the awesome preservation of many of the specimens, and you have a truly magnificent fossil assemblage that is critical to the Triassic story. Although a number of workers have added to our knowledge of the Grès à Voltzia, we can attribute much of what we know about the Voltzia Sandstones to the painstaking work of Louis Grauvogel, Léa Grauvogel-Stamm, and Jean-Claude Gall.

What's in a Name?

Of the Lower to Middle Triassic plants, one of the most widely distributed genera is *Voltzia*. Unfortunately, for a variety of reasons, the taxonomic status of *Voltzia* is confused. Brongniart (1828) originally gave the name *Voltzia brevifolia* to conifer foliage, and as such, the generic name *Voltzia* is based on sterile material, even though there is some fertile material associated with the original material in the form of lobed cone scales. Additional conifer foliage has been assigned to the same genus. Another common species is *Voltzia heterophylla*, and this too is based purely on sterile foliage.

As we discussed in chapter 2, the problem with plant fossils is that rarely do we have a good understanding of the whole plant and all the different aspects of its life cycle. There are many disparate structures—fruits, foliage, and wood—and typically they are not found all together as one complete fossil plant, so it is impossible to assign them to the same taxon with any degree of confidence. Even in a living plant, how often do we get all the parts growing on the plant at any one time? In deciduous plants, the leaves are not present all year round, fruits typically occur after the flowering structures, and seeds may be only briefly attached to the parent plant.

Under ICBN practice, in which all plant fossils are treated as form taxa, the name *Voltzia* only has validity as a genus of sterile conifer foliage. Unfortunately, ICBN practice has not been consistently followed. For example, Magdefrau (1953) described sterile leaves from the Upper Triassic of Germany as *Voltzia coburgensis*. He also described specimens of the foliage with attached female cones bearing lobed cone scales (all the lobes are about the same size; and two are sterile and three are fertile). By itself, this would not have been a problem, had he erected a new genus for the cones as a completely separate morphotaxon. Instead, Magdefrau included them as part of the type *Voltzia coburgensis* material.

Delevoryas and Hope (1975) named a female cone from the Pekin Formation (see chapter 10) *Voltzia andrewsii* even though this cone was not associated with foliage of any kind. Nevertheless, they justified using the name *Voltzia* purely on the cone's similarity to the cones described by Magdefrau as *Voltzia coburgensis*, a sterile conifer branch. This certainly might be considered a real overextension of the concept of *Voltzia*, and in fact, Delevoryas and Hope (1987) later reconsidered the referral to *Voltzia*. They noted that among members of the Voltziaceae, subfamily Voltziodeae (sensu Miller 1977), the poorest known ovulate cones are those of the genus *Voltzia*. Yet this is inevitable, given that strictly, *Voltzia* should be applied only to sterile foliage. Delevoryas and Hope, recognizing the problems, wisely chose to transfer the Pekin fossil into a new genus, *Florinostrobus*. Nevertheless, isolated cones are still frequently referred to *Voltzia*. For instance, as we shall see later (chapter 5), specimens from the English Midlands are still called *Voltzia heterophylla*.

The problems don't end there. More recently, Schweitzer (1996) described wonderful material of attached female cones on a foliage type originally described as *Voltzia hexagona*. The curious thing is that *Voltzia hexagona* was first named on the basis of leaves that were first considered to be those of lycopods. Nevertheless, Schweitzer kept the name *Voltzia* for the whole plant. Furthermore, the cone scale lobes on the new material are not the same size as in most voltzias, but instead, they have large lateral lobes and a large central lobe. Moreover, the cone is indeterminate—that is, vegetative growth continues distal to the cone apex. The point here is that in several characters, the cone is much like those of the family Majonicaceae rather than the other so-called voltzias. There thus appears to be no real consensus among paleobotanists on the *Voltzia* issue, and some current workers still recognize the genus *Voltzia* as a natural taxon and are applying the name to cones as well.

What does this all mean for our picture of Early and Middle Triassic conifer assemblages? One clearly has to be circumspect when using the name *Voltzia*. Perhaps the name should be used only informally to refer to whole plants, including cones, and for cones alone only when they can be shown to be associated with the foliage morphotype. When the leaves that dominate a floral assemblage conform to the morphospecies description, terms such as *Voltzia* forests and Voltzia Sandstones provide us with a clear picture of the original floral assemblage. On the other hand, it is difficult to accept "Voltziaceae" as a real monophyletic group, or any other evolutionary/ecological entity beyond the status of a grab bag of mostly poorly understood conifers.

Voltzia Forests in the Vosges and the Muschelkalk Transgression

As noted earlier, the so-called Grès à Voltzia of northeastern France is equivalent to the upper Buntsandstein (approximately equivalent to the Perovkan land vertebrate faunachron [LVF]). In this case, there is no question that many of the numerous conifer shoots conform to the *Voltzia heterophylla* morphospecies, and that we can justifiably talk about *Voltzia* forests. However, the fossils from the Voltzia Sandstones provide us with a picture of more than just conifer forests. They also give us some insight into changes that were taking place as the Muschelkalk Sea transgressed north and west across what is today France.

There are two basic members of the Voltzia Sandstone: the lower Grès à Meules and the upper Grès Argileux (Gall 1971). The Grès à Meules is considered to represent a deltaic deposit with actively meandering channels in a floodplain dotted by lagoons and abandoned channels. This member contains four quite distinct facies. First, there is the sandstone that was commercially quarried. It is fine grained and well sorted, ripple marks are common, and it can be interpreted as the result of sand deposition at the mouth of a delta. Next there are gray sandstones with amphibian and plant remains occurring as channel fills. These are thought to represent channel deposits of rivers while in spate. Green and red shale layers occur between these sandstone beds. The shales are frequently fossiliferous with three main assemblages: (1) crustacean-rich assemblages, (2) assemblages with brachiopods and bivalves, and (3) a terrestrial assemblage. It would appear that the shale lenses show a gradual change from marine conditions to environments that were more terrestrially influenced. Moving upward through the section, first the crustaceans and then the bivalves tend to become fewer, and eventually plant fossils increase dramatically. We can envisage over time an environment of lagoons and brackish pools giving way first to tidal flats and then to salt marshes. Finally, carbonates are found as channel fills intercalated between the sand and shale layers. These carbonates would have been precipitated by evaporation in saltwater pools.

Taken together, these four facies paint a picture of a deltaic region with meandering channels cutting through the floodplain, with a number of pools and lagoons. Judging from the abundant vegetation in some units, the hinterland was probably a well-wooded area dominated by conifers of the *Voltzia* morphotype.

There is a distinct boundary between the Grès à Meules and the upper Grès Argileux members, with the upper portion of the Grès à Meules marked by a great density of plant fossils. Within the Grès Argileux, there are three distinct facies. First, the Grès à Meules consists of thin sandstone beds with extensive bioturbation. Lingulids (brachiopods) and bivalves are common. Second, red and green shales with abundant trace fossils were deposited in a tidal-flat type of environment. And third, dolomitic sandstones containing numerous marine fossils point to the complete transgression of the region by the Muschelkalk Sea.

When looked at together, the two members paint a picture of an oscillating shoreline. Beginning at the base of the Grès à Meules, marine conditions change first to a coastal and tidal environment at the top of the Grès à Meules, then revert back to marine at the base of the Grès Argileux before a tidal phase returned in the middle Grès Argileux, only to be finally replaced by fully marine conditions once more with the complete inundation of the floodplain by the shallow Muschelkalk Sea. Consequently, the fossils of the Voltzia Sandstone are a combination of terrestrial and marginal marine forms.

The array of different animals is staggering and for the marine forms ranges from jellyfish to worms, mollusks, echinoderms, crustaceans, and fishes, and for the terrestrial animals includes reptiles and amphibians, as well as a wonderful array of insects and arachnids. The plants include foliage, seeds, fructifications, and spores representing such diverse groups as ferns, horsetails, and gymnosperms. In addition, trace fossils provide some indication of animal activity and include egg clusters of certain insects, and various examples of plant-insect interactions.

Beyond the Naked Eye

At the base of many food chains are vast numbers of microscopic organisms. Today's oceans teem with all kinds of amazing microorganisms such as the single-celled dinoflagellates, foraminiferans, and radiolarians. The ancient world was no different. Careful preparation of many sediments will reveal wonderful secrets under the microscope. The Voltzia Sandstone has yielded several different foraminiferans—the single-celled organisms that secrete a hard calcium or, rarely, silica test. The Voltzia Sandstones have also yielded large numbers of palynomorphs—spores and pollen of different plant forms. Apart from offering clues regarding the types of vegetation that occurred in the region, palynomorphs are also of great use as biostratigraphic index fossils.

Rarely Preserved Soft-Bodied Animals

It is widely known that such delicate organisms as jellyfish rarely appear in the fossil record. One has only to recall amorphous blobs washed up on a beach to realize that jellyfish quickly lose all of their beautiful form. Nevertheless, such is the preservation in the Voltzia Sandstone that occasionally medusae are preserved. One such taxon is *Progonionemus*, a saucer-shaped medusa that has been assigned to the Limnomedusae. One living genus of Limnomedusae, *Gonionemus*, bears tiny suckers on the numerous tentacles that enable it to "walk" over seaweed. There is, however, no evidence for such suckers in *Progonionemus.*

Inarticulate brachiopods, or lamp shells, are abundant, and, perhaps rather surprisingly, these have been placed in the genus *Lingula.* This is one of the better known of today's brachiopod genera, but clearly the forms in the Grès à Voltzia are not the same as the living members. Nevertheless, they share the same basic shell morphology, and it seems likely that they may have also shared a similar lifestyle. Although many brachiopods are anchored permanently to rocks by their stalks, the modern-day lingulids often live in mudflats, where they burrow into the substrate with their muscular stalks. The animal lies in its shell near the top of the burrow, filtering out food particles from the water column. If disturbed, the stalk contracts, thereby drawing the animal back down to the safety of its burrow. Such an existence for the Grès à Votzia lingulids is consistent with our interpretations of the paleoenvironment.

Because they lack any hard-part anatomy, annelid worms are generally rare in the fossil record, but they do occur in parts of the Voltzia Sandstones. Two errant polychaetes

have been described. *Eunicites* grew up to 35 millimeters long; *Homaphrodite* was a smaller form attaining a length of approximately 10 millimeters. They are characterized by their prominent parapodia bearing bundles of chaetae. It is likely that both of the Grès à Voltzia forms actively moved about, burrowing into soft sediment in search of food. They might have superficially resembled present-day "ragworms." Although many of the living polychaetes are fully marine, there are also some forms, such as *Nereis diversicolor*, that live in estuaries. This may also have been true for *Eunicites* and *Homaphrodite*. As well as these active polychaetes, there are also examples of sedentary annelids in the form of the characteristic limey tubes of serpulid worms. These fossils have been assigned to the genus *Spirorbis*, but in the absence of impressions of the actual animal, some question should remain about this referral.

Living members of the genus are often found attached to pebbles and rocks, but some of the smaller species, such as *S. borealis*, commonly occur on seaweed. The animal lives

Plate 4.3. In a shallow, brackish lagoon, the crayfish *Clytiopsis* crawls along a sunken log while in the background a horseshoe crab creeps across the muddy bottom. A *Dipteronotus* swims among vegetation washed into the lagoon by a recent storm.

Plate 4.4. Seen from above, a small fish (*Pericentrophorus*) casts its shadow over a submerged fern frond (*Pteridophylla*). The arthropods *Triops* and *Euthycarcinus* move about on the lagoon floor, searching for food.

in the protected "shell," but it can project a series of tentacles from the head end and uses these to catch food particles wafting through the water. These tentacles are frequently highly colored. There is no reason to believe that the Triassic serpulids were any different in their lifestyle, but just how closely related they were to the modern-day *Spirorbis* is less clear. Examples of serpulid tubes have been found on vegetation, shells of mollusks, and even fish scales in the Grès à Voltzia.

The variety of mollusks immediately alerts us to the fact that we are dealing with a nearshore coastal environment. They range from sedentary bivalves such as specimens referred to the genus *Modiolus*, to boring types like *Homomya* and probable free-swimming forms assigned to the Pectinidae. Gastropods such as *Naticopsis* are also present.

Today we often associate the beach and back bays with a variety of different crustaceans like crabs and amphipods. Judging from the Grès à Voltzia, the Triassic was no different. There is a wonderful array of crustaceans, including branchiopods, conchostracans, ostracods, isopods, and decapods. Once again, they were rather different than the living taxa, but we suspect that they still filled similar niches. Many of the smaller, almost microscopic forms presumably occurred in large numbers and contributed to a rich planktonic fauna. There were also shrimplike forms, *Antrimpos* and *Clytiopsis*. Other definitive indications of the marine environment include parts of echinoderms and numerous specimens of the merostomatan (xiphosurid) *Limulitella*. The modern representative of this ancient group is the horseshoe crab, *Limulus*. In fact, *Limulitella* closely resembles a dwarf modern horseshoe crab. Merostomatans possess chelicerae rather than mandibles and thus are often united with the other chelicerates, sea spiders (Pycnogonida) and the arachnids.

Perhaps the most intriguing fossils in the Grès à Voltzia are the terrestrial arthropods. These include myriapods, insects, and arachnids. The myriapods are represented by a handful of specimens. They are unequivocally diplopods (millipedes), as two pairs of jointed legs can be seen on each segment, but the specimens lack detail, so more precise referrals remain impossible at this stage.

Plate 4.5. Among the vegetation in the coastal region around the Tethys Sea lurk a scorpion and a millipede. The vegetation includes shoots of various conifers including *Aethophyllum*, a cone of *Albertia*, and the fern *Pecopteris*.

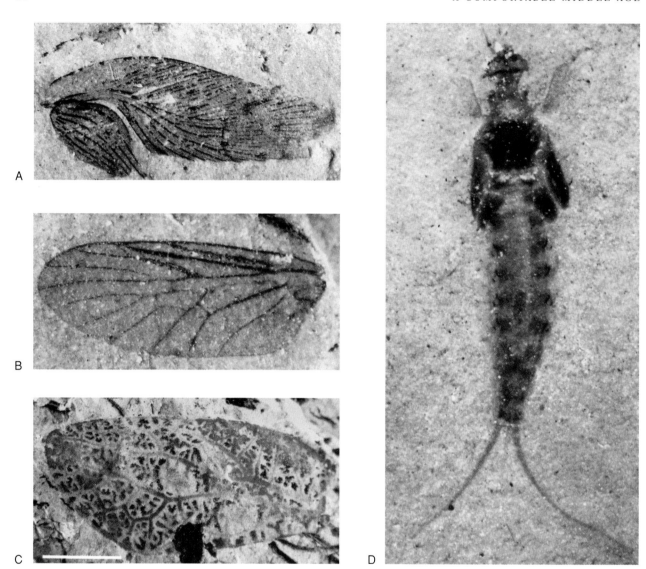

Figure 4.1. (A) Isolated wing of a cockroach. (B) Beautiful preservation of the wing of a fly, which could almost be from a modern-day insect. (C) Remarkable wing clearly exhibiting the preservation of the original color pattern. (D) Magnificently well-preserved mayfly nymph.

The insects (fig. 4.1) were first documented by Louis Grauvogel in 1947. Since then, a variety of different forms have been recorded, so that now more than 5300 insect specimens have been documented (Grauvogel and Gall collection) (Marchal-Papier 1998). The Blattodea (roaches) (Papier and Grauvogel-Stamm 1995), Odonata, Mecoptera, and Orthoptera (Papier et al. 1997; Marchal-Papier et al. 2000) have been most widely studied, but there are also reports of many other orders, including Coleoptera, Diptera (Krzemiński et al. 1994; Krzemiński and Krzemińska 2003), numerous Homoptera (e.g., Lefebvre et al. 1998) and Hemiptera, Titanoptera, and Plecoptera. They all originate in the Grès à Meules sediments. The majority of the insect remains are isolated wings, but the venation pattern is usually clearly expressed, and this has permitted identification of several different families. In addition, there is the occasional preservation of color patterns on the wings, adding even more to the truly remarkable picture of the insect fauna.

The cockroaches are the predominant insect remains, with over 2000 specimens representing nine different species (Papier et al. 1994; Papier and Nel 2001). Next come the beetles with some 500 specimens, but here there is apparently a much larger diversity, with 40 putative species (Gall 1996). Although this might seem like an extraordinarily large number of different forms, it must be remembered that in a small area (one square mile or so) of woodland today, there may be easily 2000 different species.

Arachnids

The order Araneae (spiders) can be divided into two suborders: the Mygalamorpha and the Araneomorpha. Today the mygalomorphs include the trapdoor spiders and the large—and some would say grotesque—"bird" or "monkey" spiders. They are characterized by cheliceral fangs that lie parallel to each other so that they move in the same plane as the long axis of the body. They also have relatively thick legs and always possess two pairs of book lungs. One complete specimen of a mygalomorph spider is known from the Grès à Voltzia. It was described under the name *Rosamygale grauvogeli* by Selden and Gall (1992) and it is the earliest known record of a mygalomorph (fig. 4.2). With a body about 2 millimeters long, it can be considered small for a mygalomorph. The sister group of the mygalomorphs, the araneomorphs, is also known on the basis of a couple of Triassic specimens (see chapters 9 and 11). Araneomorphs are the largest group of living spiders. Typically, they have one pair of book lungs, as well as fangs on the chelicerae that lie at a right angle to the long axis of the body. Their legs are also generally much more slender than those of mygalomorphs.

One of the most exciting terrestrial arthropods in the Grès à Voltzia is a gorgeous specimen of a tiny scorpion (fig. 4.3). Although it is only 5 millimeters long, it is almost complete, and the large pincerlike pedipalps are prominently displayed, together with the

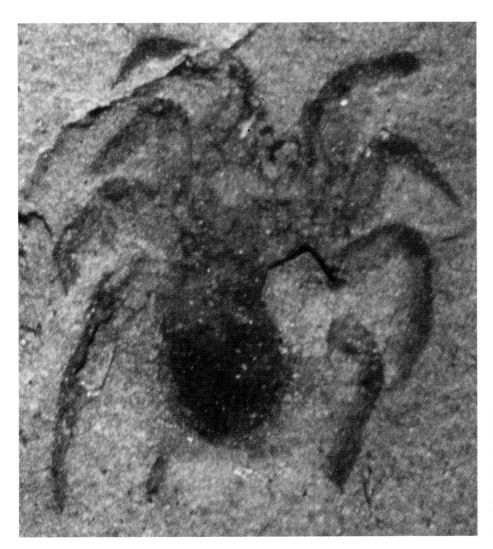

Figure 4.2. Complete spider from the Grès à Voltzia, just one of three spiders known from the Triassic worldwide. The other two are from the Molteno (chapter 9) and the Newark Supergroup (chapter 11).

Figure 4.3. Exquisitely preserved scorpion demonstrating the proximity of a fully terrestrial community to the marine environment.

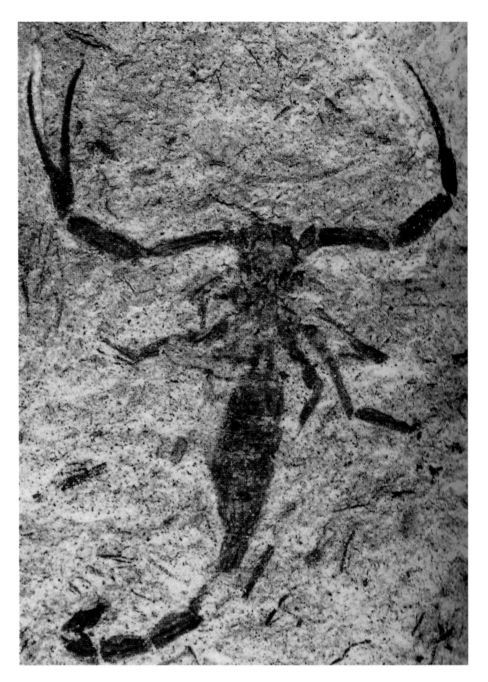

characteristic long and curved abdomen with its terminal stinging apparatus. Although today we often associate scorpions with deserts and dry environments, there are many species living in the tropics that require moist habitats. They typically tend to be secretive and nocturnal, and although generally they range over the tropical and subtropical zones, they have been found in places such as British Columbia and Virginia. There is even one population of scorpions thriving today at a station on the London Underground train system! Thus the presence of a scorpion in the Grès à Voltzia does not necessarily imply anything about the paleoenvironment.

The remarkable preservation of much of the Grès à Voltzia material provides us with a fairly straightforward picture of life in the estuaries, mudflats, salt marshes, and lagoons of the Triassic in Europe. But what is perhaps even more remarkable about these sequences is the tangible evidence for interactions between some of the organisms.

Plant-Animal Interactions

Alongside the numerous insect specimens is an abundant flora, and it might therefore be expected that there is a good chance that some basic record of any interaction between the two might be preserved. The simplest interaction would be insects eating plant material, thereby leaving characteristic bite marks. Plant eating (phytophagy) in insects goes back at least to the Pennsylvanian (Southwood 1973), and today there are several taxa that are obligate phytophages, including one entire order, the Phasmatodea. The majority of lepidopterans are also plant feeders. Nevertheless, only a third of the major insect orders include phytophagous species, so that it might be expected that plant-eating insects would be limited in the Triassic. This is exactly what Grauvogel-Stamm and Kelber (1996) found in a thorough examination of a large collection of foliage remains from the Grès à Voltzia. Despite the co-occurrence of groups such as the Orthoptera (grasshoppers and allies), which are commonly phytophagous today, there is little evidence of Triassic orthopterans chomping their way around the margins of leaves. As Grauvogel-Stamm and Kelber (1996) noted, one possible reason for this could be that some of the plants had a particularly thick cuticle and were thus immune to insect attack. Thick cuticle, which resists water loss, is found in plants living in estuaries and salt marshes. In any, case none of the large *Yuccites* leaves exhibit any sign of insect damage.

However, two specimens of the fern *Neuropteridium* do show some evidence of having been eaten by insects. In both specimens, there is significant reduction of the margins of the pinnae, with scalloping of the edges that is consistent with insect biting. Although *Schizoneura* (a sphenopsid) foliage occurs in the Grès à Voltzia, none of the specimens apparently shows evidence of insect feeding traces. However, elsewhere, *Schizoneura* leaves do have damaged margins that are indicative of insect activity. One specimen from the lower Keuper of Franconia consists of fused leaves in a single leaf sheath. There is a prominent oval-shaped gap partway down the junction between the two leaves. Kelber and Geyer (1989) interpret this as the direct result of a cockroach, or possibly a feeding gallery through which insects entered to access the parenchyma of the leaf. A second specimen, also from the lower Keuper of Franconia, is a single leaf that dramatically shows the activity of an insect in a series of bite marks down one side. The edges of these bite traces are darker, which Grauvogel-Stamm and Kelber (1996) believe corresponds to a wound reaction that in turn indicates the leaf was alive at the time of the insect attack. From such specimens, it is easy to picture a Triassic scene where an insect, perhaps an orthopteran, systematically moves down the leaf, nibbling along the margins as it goes, before jumping onto the next leaf and repeating the action. Only the marginal portions of the leaves have been attacked in this way. Presumably the occurrence of lignified vascular tissue and tannins protected the central part of these leaves from insect attack, thereby acting as a defense mechanism against complete destruction.

Taeniopteris, a foliage type affiliated with the cycads or bennettitaleans, is another of the plants in the lower Keuper that displays clear signs that insects chomped along the leaf margins, but this particular form genus is not known in the Grès à Voltzia. Although in these instances ichnotaxa have not been erected for the feeding traces, this has been done: Van Amerom (1966) erected the generic name *Phagophytichinus* for feeding traces on *Neuropteris* leaves. Such a practice has to be questionable, given that in most cases there are no definitive patterns of bite marks.

A far more common type of insect-plant interaction is indicated in the Grès à Voltzia sediments by the occurrence of numerous egg clusters in intimate association with plant material. In fact, thousands of egg clusters attributed to insects have been documented from the Grès à Voltzia (Grauvogel 1947; Gall and Grauvogel 1966; Gall 1971). But these

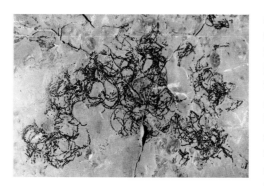

Figure 4.4. *Monilipartus tenuis* is the name given to presumed insect eggs that are arranged in characteristic long strings.

egg clusters are not always associated with plant material. Even where they are entangled with plant debris, it is sometimes not clear whether they are merely accidental associations. It is quite possible that the adult insects simply chose a quiet body of water in which to lay their eggs, and that leaves floating near the surface simply wafted in among the egg clusters. The clusters often have characteristic shapes, and Gall and Grauvogel (1966) felt that they warranted erecting form genera and species. Accordingly, they recognized two distinct types. *Monilipartus tenuis* is characterized by eggs arranged in a long string that was apparently maintained by some kind of jellylike coating (fig. 4.4). The eggs are oval, each measuring about 0.25 millimeters long, and each cluster contains about 1500 individual eggs. A second type, *Clavapartus latus*, comprises cigar-shaped egg clusters containing between 500 and 2000 eggs. The clusters are typically 20 to 30 millimeters long and about 5 millimeters across. Where they do occur, the egg clusters commonly occur in large numbers, which suggests group oviposition—swarms of insects laying their eggs on the water surface all together.

Gall and Grauvogel (1966) compared these egg clusters directly with those of Trichopterans (caddis flies) and the Chironomidae (a family of small flies). However, with the exception of an occasional isolated wing, adult dipterans are not common in the Grès à Voltzia, and there are no documented records of caddis flies. But Gall and Grauvogel (1966) also noted a resemblance to eggs of some living odonate (dragonflies and damselflies) taxa, in particular anisopterans (dragonflies) such as *Tetragoneuria* (so-called climber dragonflies). This genus lays strings of eggs that may be free-floating or fixed to plants. Given that odonates are relatively common in the Grès à Voltzia, it seems more likely that the egg clusters also belonged to odonates, although there is no hard evidence to support this opinion.

No egg clusters securely attached to leaf surfaces have been found in Grès à Voltzia sediments, but they are known from the lower Keuper of Franconia. Geyer and Kelber (1987) and Kelber and Geyer (1989) described compact egg masses attached to the leaf surfaces of *Equisetites*. These particular egg clusters are also associated with numerous tubes belonging to the sedentary polychaete, *Spirorbis* (see above). This in turn suggests that the eggs were laid underwater. Some caddis flies lay their eggs underwater and cement them to a substrate in this fashion. However, modern caddis flies are strictly freshwater forms, and it seems improbable that they would be associated with serpulid worms.

There are some Triassic examples of so-called endophytic eggs—that is, eggs that have been laid within the plant tissue rather than being attached to the surface. But once again, these are in the lower Keuper of Franconia and elsewhere, and there are no definitive examples from the Grès à Voltzia. Visible on several leaf sheaths of *Equisetites* are elongate oval bumps with their long axes parallel to the stem axis. These scars are frequently in a rather ordered pattern and in high densities. In fact, some authors have considered such scars to be part of the original ornamentation of the plant. Roselt (1954) erected a new species of horsetail, *Equisetites foveolatus*, on this basis, and Weber (1968) did the same with a specimen of *Neocalamites*. Later it was shown that the surface really was scarred and that for the *Equisetites*, the leaf sheaths could be assigned to a well-known species, *Equisetites muensteri* (Kelber 1988). In the case of the *Neocalamites*, Kelber considered the scars to either be the result of some disease in the plant or some unspecified kind of insect interaction. Some living odonates deposit eggs in plants that grow in water, and the resultant egg scars are similar in shape. However, the scars on the Triassic examples typically range from 2.5 to 6.0 millimeters long and 1.5 to 2.5 millimeters wide, with some reports of even larger scars. The eggs of present-day Odonata are noticeably smaller, typically measuring less than 2.0 millimeters in length. This might suggest that the insects emerging from the Triassic endophytic eggs were much larger than modern dragonflies. Grauvogel-Stamm and Kelber (1996) considered it more likely that protodonates were responsible for these egg scars. Protodonates are probably best known for the giant meganeurids of the Upper Carboniferous, with some species having wingspans of approximately 70 centimeters. Some entomologists actually consider the protodonates to represent a suborder of the Odonata rather than a separate order. They are certainly closely related. Wings of a protodonate, *Triadotypus guillaumei*, do occur in the Grès à Voltzia (Grauvogel and Laurentiaux 1952), and the same species has also been documented from the Muschelkalk of Franconia (Reis 1909; Grauvogel-Stamm and Kelber 1996) and the middle Keuper of the French Alps (Larentiaux-Viera et al. 1952). *Triadotypus* was a large insect with a wingspan exceeding 30 centimeters, and it—or something similar—remains a good candidate for the endophytic egg scars on many of the Triassic horsetails.

Galls (abnormal tissue growth in plants) are perhaps the most widely recognized result of plant-insect interactions. But the growth of galls is induced by a variety of agents, not just insects, so the occurrence of a gall on a fossil plant cannot be taken as definitive evidence of insect activity. Nevertheless, today, many galls have a characteristic morphology that can be directly associated with specific types of insect interaction. Modern beetles such as the Buprestidae and Cerambycidae induce stem galls with swellings, and sometimes they are quite species specific. For example, one species of Cerambycidae, *Saperda populnea*, induces stem galls on willows and poplars (Jacobs and Renner 1988). Stem galls are also induced by flies, hymenopterans, bacteria, and fungi. A characteristic modern gall, called a pineapple gall, is found on *Picea excelsa* and is known to result from the attacks of two species of aphid. Examples of conifers bearing stem galls have been recorded from the Grès à Voltzia. One of the *Voltzia* twigs exhibits a fairly prominent swelling that is surrounded by modified needles and that is superficially like a pineapple gall. Spindle-shaped galls have also been recorded on stems of *Aethophyllum stipulare* (Grauvogel-Stamm 1978; Larew 1992).

One of the most intriguing examples of plant-insect interaction known from the Grès à Voltzia sediments involves mimicry. Exquisitely preserved wings of the orthopteran, *Triassophyllum leopardi*, have been described complete with an obvious color pattern consisting of dark irregular spots. The venation pattern of *T. leopardi* is similar to modern-day Tettigoniidae, which in turn have a fore-wing venation resembling that of angiosperm (flowering plants) leaves. Grauvogel-Stamm and Kelber (1996) question the value of such apparent mimicry in the wing venation pattern of *T. leopardi*, when accepted wisdom states that angiosperms originated much later, with the first examples appearing in the Cretaceous (e.g., Crane 1993; Doyle and Donoghue 1993; Sun et al. 2002). Admittedly, some paleobotanists have suggested that, on the basis of a sister-group relationship with Bennitatales, angiosperms did extend back to the Triassic (see chapter 2), but there is no definitive evidence of the group before the Cretaceous. However, Cornet (1993) documented a remarkable venation pattern in foliage from the Cow Branch Formation (Carnian) of Virginia. This foliage, which Cornet named *Pannaulika triassica*, displays a looping venation that is consistent with a dicot leaf (see chapter 12). This, together with some fruiting structures, convinced Cornet that here for the first time was unequivocal evidence of angiosperms in the Triassic. As it turns out, additional, more complete specimens of this foliage type have since been recovered from the same site in Virginia, and we now know that they represent a rather atypical dipteridaceous fern. Nevertheless, the fact that the venation pattern could be mistaken for an angiosperm suggest that the venation pattern of *Triassophyllum leopardi*, rather than mimicking an angiosperm, may well have been mimicking the leaf of a fern.

What we are essentially seeing from all these insect-plant associations are examples of interactions in a freshwater environment. We need only a little imagination to begin to piece together what life must have been like in this region 230 million years ago.

The Alpine Realm — The Seas of Middle Triassic Europe

Although there was only a single continental landmass during Triassic times, it was centered around the Tethys Sea, a large bay encroaching over what is today central Europe. The Tethys Sea teemed with life during the Middle Triassic. A huge variety of invertebrates including ammonoids, bivalves, and crustaceans inhabited this sea. Shoals of fish swam in the clear waters that covered the carbonate platforms. These included predatory *Saurichthys* and a variety of pycnodonts. In the shallows and the lagoons were a number of nothosaurs that ranged in size from about 20 centimeters to 4 meters long. Further out to sea were some of the earliest ichthyosaurs, such as *Cymbospondylus* and *Mixosaurus*. These fast-swimming, highly predatory animals occupied a similar niche to modern-day dolphins. But the most bizarre animal was surely *Tanystropheus*, which grew up to 4 meters long, over half of this length contributed by a ridiculously long neck. And nearby, in the hinterlands bordering this productive sea, were coniferous woods and almost certainly an equally diverse fauna of terrestrial tetrapods. How can we be so certain of this wonderful mosaic of Middle Triassic life? A variety of fossiliferous units are known in central Europe today that document this time period, but the so-called Grenzbitumenzone and a series of carbonate horizons immediately above it are surely the most influential in contributing to this vivid picture.

The Grenzbitumenzone and associated sediments were deposited in a 10-kilometer-wide marine basin. Today, this series of sediments outcrop in northern Italy and the Ticino region of southern Switzerland, a charming region of old villages and vineyards clinging to steep hillsides above the clear waters of Lake Lugano. One of the most prolific localities is Monte San Giorgio, which consists of a series of localities dotted across a hillside straddling the Swiss-Italian border. Over 70 years of collecting in the region has yielded an incredible array of vertebrate fossils, but one group, the Nothosauria, is present in extraordinarily large numbers. The Nothosauria can be split into two main groups: the generally rather small pachypleurosaurs, and the somewhat larger nothosaurs sensu stricto such as *Ceresiosaurus*. At Monte San Giorgio, the pachypleurosaurs are the most abundant. For example, there are typically more than 200 specimens of each of the four different pachypleurosaur taxa. These specimens include all ontogenetic growth stages, and they thus permit us to make meaningful statements concerning life cycles and the nothosaurs' way of life. Other tetrapods include four genera of ichthyosaur, two genera of placodonts, three thalattosaurs, and the prolacertiforms *Tanystropheus* and *Macrocnemus*, as well as also the rare remains of the pseudosuchian *Ticinosuchus*.

Although the sediments in the Monte San Giorgio region are sometimes collectively referred to as the Grenzbitumenzone, it is only the lowermost fossiliferous units comprising bituminous shales and interbedded dolomites that form the Grenzbitumenzone proper. Above that are a series of massive dolomites and bedded limestones (the San Giorgio Dolomite and lower Meride Limestone) that are followed in turn by the so-called Cava Inferiore, the Cava Superiore, and finally the Alla Cascina. Above these horizons are thick limestone sequences of the upper Meride Limestone, culminating in the Kalkschieferzone. A detailed interpretation of these sediments suggests that the northern part of Tethys during the Triassic comprised an extensive reef system (Zorn 1971), but its southern limit had no obvious boundary, although it is thought to have been partially closed off from the open sea (Rieppel 1992). Thus the basin can be regarded as a complex of reefs, tidal flats, and lagoons reminiscent of the present-day Bahamas.

In all probability there were times when the bottom of the basin was anoxic, so that only epipelagic animals could survive in these waters. Certainly this appears to be true for the period of Grenzbitumenzone deposition. These epipelagic animals included the large ichthyosaur *Cymbospondylus* and the nothosaurids *Ceresiosaurus* and *Paranothosaurus*. The Grenzbitumenzone corresponds to the base of the Ladinian, although on the basis of palynological studies (Scheuring 1978) the top of the Kalkschieferzone corre-

Plate 4.6. *Cymbospondylus* is a longipinnate ichthyosaur from the Middle Triassic of Nevada that grew to over 30 feet long. On the basis of the presence of soft-part preservation in a few specimens from the Jurassic, it is known that some ichthyosaurs possessed a large dorsal fin similar to modern dolphins. However, whether this was a feature of all ichthyosaurs is not known. *Cymbospondylus* is shown here without such a dorsal fin. Compare this with the restoration of *Shonisaurus* in chapter 3. Unlike the Jurassic and Cretaceous ichthyosaurs, Triassic ichthyosaurs like *Cymbospondylus* do not display the marked downward terminal bend, and they probably used a different method of swimming.

Plate 4.7. A school of *Mixosaurus* leaps through the surf chasing shoals of fishes. *Mixosaurus* was widespread in Middle Triassic seas, and has been found from Alpine Europe, the United States, eastern Asia and Spitsbergen.

sponds with the end of the Ladinian (Furrer 1995). Thus, combined, they are approximately equivalent to the Berdyankian LVF (see appendix 1).

Nothosaurs sensu stricto were well adapted to the marine environment, having paddle-shaped fore- and hindlimbs and long, mobile necks. Their jaws bristled with acutely conical teeth that would have efficiently dealt with slippery fish, as well as the smaller pachypleurosaurs such as *Neusticosaurus*. However, on land, they would have been ungainly; perhaps they only ventured onto the shore to lay their eggs. Nothosaurids typically have narrow neural arches and relatively tall neural spines, features pointing to a fairly rigid backbone and reduced capabilities for generating a propulsive force by lateral undulations of the body. The forelimbs are more robust than the hindlimbs, and this suggests that the forelimbs generated the main propulsive force. The skull of *Ceresiosaurus* is wide posteriorly so that the temporal openings are larger than the orbits. Another of the nothosaurids known from the Monte San Giorgio region is *Lariosaurus*, with

but a single juvenile specimen of *Lariosaurus balsami* (Tschanz 1989). However, this specimen comes from beds much higher up in the section in the Kalkschieferzone (upper Ladinian).

Like the nothosaurids, the smaller pachypleurosaurs were highly adapted to life in the water, but they may have been more common in the shallower coastal waters. They may even have ventured into estuaries. In support of this interpretation *Neusticosaurus* remains have been recovered in Germany in association with the amphibian *Plagiosuchus* and the lungfish *Ceratodus*, typically viewed as freshwater, or at least brackish water, indicators. Rieppel (1989a) suggested that the terrestrial middle ear of *Neusticosaurus* precluded it from deep dives and restricted it to the upper portion of the water column. In common with many aquatic animals, certain bones, such as the ribs, were pachyostotic. Pachyostosis (increased bone density) is typically associated with rather slow-moving, aquatic vertebrates, where it functions to achieve neutral buoyancy. As a consequence, little energy is required to move up and down in the water column. Schmidt (1984, 1987) suggested that pachyostosis in nothosaurs would have facilitated walking along the seafloor, and that this was an important means of locomotion in *Neusticosaurus*. How-

Plate 4.8. Two large nothosaurs, *Nothosaurus* and *Simosaurus*, squabble over a *Birgeria*.

Plate 4.9. Gliding through a tidal pool along the margins of the Tethys Sea, a *Dactylosaurus* (nothosaur) passes over the skull of the larger nothosaur *Cymatosaurus*. Littering the seafloor are the empty shells of *Fedaiella*, *Aspidura*, and a *Pecten*.

ever, given that the bottom waters of the basin were anoxic, it seems that *Neusticosaurus* would have been walking along the seafloor only along the reefs. Moreover, the middle ear restrictions would also limit these animals to shallow waters. Thus the function of pachyostotic elements in these pachypleurosaurs remains something of a mystery.

Sander (1989) portrays *Neusticosaurus* and its allies feeding on small fish (fry) and small cephalopods, picking them off with their small, needle-sharp teeth. Prey species would include forms such as juveniles of the chondrostean, *Luganoia*. Another closely related chondrostean, *Habroichthys*, only attained a length of 25 millimeters, even as an adult. Occurring in the Cava Inferiore, this fish was definitely of manageable proportions for *N. pusillus*. On the other hand, the suggestion (Zangerl 1935) that *Neusticosaurus* juveniles contributed to the diet of the adults seems highly unlikely, because at approximately 80 millimeters long, even the hatchlings of *Neusticosaurus* would probably have been too large to ingest whole (Sander 1989). In fact, the single specimen of an embryonic *Neusticosaurus* is close to 40 millimeters long and would have been quite a mouthful! In contrast with the larger nothosaurids, pachypleurosaurs exhibit relatively low neural spines and broad and flat neural arches that would have facilitated undulatory movements of the trunk. Thus, in these smaller animals, side-to-side motion of the body would have provided the main locomotive thrust. It is possible to envisage *Neusticosaurus* and its allies swimming lazily through the reef systems, forelimbs closely pressed to their sides, using the hindlimbs to alter direction, occasionally diving around a corner or into a crevice to snatch a tiny fish.

Sexual dimorphism was apparently pronounced in the pachypleurosaurs. Sander (1989) observed a sharp variation in the form of the humerus in all the Monte San Giorgio pachypleurosaur assemblages. In each case, roughly half of the specimens had a humerus that was markedly differentiated, with a slender shaft set off from a slightly expanded proximal head and a much more pronounced expansion to the distal end. The remaining specimens are characterized by a poorly differentiated humerus with little constriction in the shaft region and weakly expanded proximal and distal ends. In fact, a much greater difference exists between the humeri of different sexes of the same species than between the same sex of the three species of *Neusticosaurus*. Additional, less widespread differences between the two sexes have also been recorded. For example, in *Neusticosaurus peyeri* there is a clear difference in tail length, so that those individuals with the differentiated humeri also possess a longer tail. The question, of course, is which form belongs to which sex? In modern reptiles, it has been observed that the tails of males are often longer than those of females. Why males should have a longer tail is not really apparent. But whatever the reason, on the basis of the *Neusticosaurus peyeri* population alone, we might surmise that in all Monte San Giorgio pachypleurosaurs those individuals with a more strongly differentiated humerus are the males.

The next question is, why should males possess better-developed humeri with well-defined muscle attachment and articular surfaces? It might be to allow the males to grasp the females during copulation in a fashion similar to modern-day anurans (frogs and toads). On the other hand, what if the females possessed the more powerful forelimbs? As it turns out, an equally good case can be made for this hypothesis. If the pachypleurosaurs were oviparous (and there is no evidence to suggest otherwise), then the females would be the ones to go ashore to lay the eggs. It is then easy to argue that the females would then be the sex with the better-developed forelimbs, which they would use to haul themselves up the shore. The males, having no reason to come ashore, would therefore remain in the water throughout life. If female pachypleurosaurs did have the better-developed humeri, then the possession of the longer tail in *Neusticosaurus peyeri* females would be somewhat atypical by comparison with modern reptiles.

The preservation of tetrapods in the Monte San Giorgio deposits is quite spectacular, and on rare occasions, there is even some soft-part preservation. In one specimen of *Neusticosaurus peyeri*, small rhomboidal scales are preserved as bluish apatite on the underside of the skull. The scale morphology is not dissimilar to the scales found in many modern lizards and snakes. A few specimens exhibit a black band of organic matter running parallel to the ventral side of the tail. Sander considered this to represent remains of integument, ligaments, and subvertebral arteries. Similar organic masses, sometimes with a silvery sheen, also occur in the abdominal region and likely represent parts of the gut or even remains of the animal's last meal. This type of preservation is reminiscent of the Solite *Tanytrachelos* specimens (see chapter 11). Studies of this organic matter show that it consists of fossilized bacterial mats. This is typical of other cases of soft-part preservation (e.g., Briggs 1999), and it would seem that, counter to intuition, spectacular preservation of soft tissue typically requires at least some minimal degradation of the original tissue by bacteria. In turn, the bacteria are fixed by replacement of authigenic (formed in place) clay minerals to provide ghost images of the original soft parts.

It is interesting that no two of the four separate pachypleurosaur taxa occur at the same stratigraphic level. Thus *Serpianosaurus* only occurs in the Grenzbitumenzone, *Neusti-*

cosaurus pusillus is restricted to the Cava Inferiore beds, *Neusticosaurus peyeri* comes from the Cava Superiore beds, and a third species of *Neusticosaurus*, *N. edwardsii*, is known only from the Alla Cascina beds. In the absence of any overlap of their temporal range in this region, questions concerning partitioning of food resources among similar animals are moot. The stratigraphic separation does, however, suggest that there is a pattern of evolutionary change, with *Serpianosaurus* being the most plesiomorphic (primitive) and *Neusticosaurus edwardsii* being the most derived members of the clade. The *Neusticosaurus* species are most readily distinguished on the bases of presacral vertebral counts and the ornamentation pattern of the bone surface. Ornamentation is not just limited to membrane bones, as is typical of many tetrapods, but unusually, it is also prevalent in endochondral bone, particularly pachyostotic elements. In *Neusticosaurus pusillus*, for example, the bone exhibits well-defined grooves extending over long distances across the bone surface. By contrast, *Neusticosaurus edwardsii* bones have a surface pattern that Sander (1989) describes as akin to the "skin of an orange." In the Monte San Giorgio region, pachypleurosaurs have been found no higher than the Alla Cascina horizon.

Tanystropheus (fig. 4.5) has long been regarded by paleontologists as one of the world's most improbable creatures. The long neck of most species in this genus is typically composed of 12 extraordinarily elongate cervical vertebrae (although one referred species, *T. antiquus*, has only 8 cervicals, and yet another possible tanystropheid from China has 24 cervicals!). Like the nothosaurs, it seems unlikely that *Tanystropheus* habitually came ashore, except possibly to lay eggs and complete the early part of its life cycle. However, Nosotti (1999) raised the point that the structure of the foot in *Tanystropheus* was rather more consistent with that of terrestrial tetrapods. Certainly there is a high degree of ossification of the tarsal elements, with two distal tarsals in addition to the astragalus and calcaneum. In fact, the original argument for an aquatic habit seems to have been based largely on the difficulties of maintaining neck posture on land; water offers a better solution for its support. The strong marine component of the Grenzbitumenzone and the abundance of *Tanystropheus* specimens also convincingly argue for an aquatic lifestyle. Moreover, most of the neck vertebrae bear a pair of elongate and slender ribs that extend posteriorly well past the subsequent vertebrae. The result is a bundle of overlapping neck ribs that would have acted to keep the neck in an almost permanent horizontal position, so that movement of the neck would have been severely restricted. Indeed, the comparison of the neck ribs to the ossified tendons of hadrosaur tails seems appropriate—both would have provided a great deal of rigidity. It is therefore difficult to imagine how *Tanystropheus* could have existed on land. The purpose of the long neck remains something of a mystery. It has been postulated that perhaps in the darker depths of the seas, fish would only have been aware of a relatively small and apparently innocuous head at the end of the long neck. The recesses of the water would hide the bigger body. All of a sudden it would have been too late, and another fish or cephalopod would become nothing more than a tasty morsel!

There is admittedly nothing in full-sized *Tanystropheus*—such as relatively large orbits—to suggest that they were adapted to detecting prey in low-light conditions. On the other hand, it has been suggested that the young *Tanystropheus* had a different lifestyle, at least in the best-known species, *T. longobardicus*. As noted by Wild (1973), juveniles of *T. longobardicus* have a strongly heterodont (mixed tooth types) dentition, with multicusped cheek teeth in the posterior part of the jaw. This might be seen as an adaptation toward an insectivorous diet, and Wild postulated that early in life, *Tanystropheus* was more terrestrial. Mature individuals, by contrast, exhibit a more homodont (uniform tooth type) dentition composed of acutely conical teeth. Yet even the smallest individuals have extraordinarily long necks, and the overlapping bundles of cervical ribs would again limit much movement away from the horizontal. Consequently, juveniles would likely have faced the same difficulties on land as the adults.

The presence of "postcloacal" or "heterotopic" bones on the tail in some specimens

Figure 4.5. *Tanystropheus longobardicus.* (A) Small skeleton. (B) Crushed skull seen from above.

A

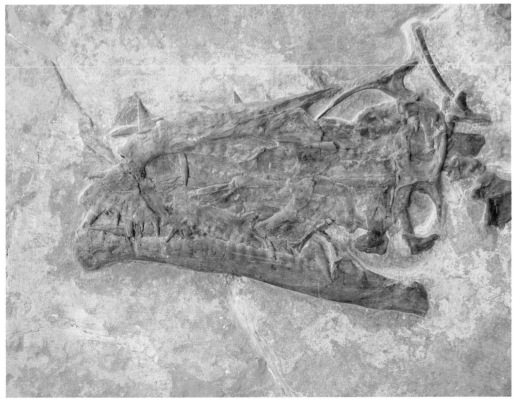

B

clearly shows that there was sexual dimorphism. Perhaps these bones were associated with a copulatory organ, and thus individuals bearing them were probably males. However, the bones are both much more complex structures than those known to occur in lizards, and also much larger—almost too large to be associated with such an organ, being approximately the same dimensions as the pelvic girdle. An alternative theory is that they supported some kind of brood pouch. If the adults were unable to come out onto land, it might be speculated that the females retained the eggs or embryos in some kind of internal pouch and gave birth to live young at sea. However, there is currently no really plausible explanation for the precise function of these heterotopic elements.

Several different *Tanystropheus* species have been recognized, but as with the pachypleurosaurs, they largely seem to have lived at different times. Individual species are recognized on the basis of a variety of relatively minor differences in bone proportions. These include the length and number of the cervical vertebrae, shape of the skull, and slight variations in the dentition. *Tanystropheus longobardicus* comes from the Grenzbitumenzone beds. Another possible species from the Monte San Giorgio region, *T. meridensis*, is restricted to the slightly younger Meride Limestone (Alla Cascina beds). However, it is actually difficult to distinguish between this form and juvenile *Tanystropheus longobardicus* specimens, so that in the future *T. meridensis* will likely become redundant. Elsewhere, *Tanystropheus* extends down into the upper Muschelkalk (lower Ladinian) of Germany, where "*T.*" *antiquus* is recognized. However, this species has distinctly shorter neck vertebrae from all later species and only 8 vertebrae, compared with 12 in the other species. Consequently, it is almost certain that this will eventually be recognized as a completely different genus. At the other end of the *Tanystropheus* temporal range is *T. fossae* from the Norian of the Lombardian Alps (Argillite di Riva di Solto; see chapter 6).

Although not from the Grenzbitumenzone, a relative of *Tanystropheus* has recently turned up in the Middle Triassic of China. *Dinocephalosaurus orientalis* possesses the same elongate neck, but in this case, neck elongation can be attributed to increased vertebral count as well as elongate vertebrae, with at least 24 (Li et al. 2004) and probably 27 cervicals. *Dinocephalosaurus* would seem to have been a fully aquatic animal: the limbs elements, including the carpals and tarsals, are poorly ossified (pedomorphic) and are more akin to those of a plesiosaur. Nevertheless, it has the same characteristic cervical ribs extending across several intervertebral articulations, and Li et al. (2004) postulated that by slight lateral and ventral extension of the ribs, an increase in esophageal volume could be achieved that would permit a rather unusual suction feeding mechanism.

The chondrostean *Birgeria* can also be considered one of the open-water inhabitants. By contrast, many of the animals could only have thrived in the reef environment to the north, and it is highly likely that this was true for many of the fishes. One of the common fishes is *Saurichthys*, a ubiquitous Triassic actinopterygian. The worldwide distribution of *Saurichthys* led to the conclusion that it was a pelagic fish. The fact that, at least the Early Triassic, species apparently lacked an opercular process on the hyomandibula is indicative of a weakly developed opercular pump. Thus it could be argued that continuous movement was necessary in order to maintain water flow over the gills sufficient for respiration. This is also true for modern-day high-speed pelagic fish. By contrast, in the Monte San Giorgio *Saurichthys* specimens, an opercular process is present on the hyomandibula. This, then, implies that the Monte San Giorgio *Saurichthys* were not pelagic fish, but rather ambush predators living in the vicinity of the reef. There are at least four species of *Saurichthys* in the Monte San Giorgio sediments, and at least two, if not three, of these species coexisted (Rieppel 1985, 1992), raising the question of competition and niche separation among the different species. Although there may be size differences for the fully mature individuals, growth stages for the different species undoubtedly overlapped. Spatial separation of the different ontogenetic stages of the different species is best explained by considering them as reef inhabitants; living in such highly structured habitats would have facilitated species segregation.

Although the predominant tetrapods in the assemblages are highly aquatic, one animal is known from the Grenzbitumenzone that was fully terrestrial. *Ticinosuchus* was a moderately large quadrupedal carnivore attaining a length of 3 meters. Although it has been referred to as a rauisuchian, it is not as derived as later so-called rauisuchians such as *Chatterjea* and *Postosuchus* (see chapter 7), and future work will almost certainly result in *Ticinosuchus* being placed in a more basal pseudosuchian group. It exhibits no adaptations whatsoever for aquatic life, and its remains were fortuitously washed into the basin from the nearest landmass before sinking into the anoxic bottom waters. Unfortunately, other terrestrial components are rare, so we can only guess at the lifestyle of *Ticinosuchus*.

One animal that is quite abundantly represented in these Middle Triassic marine sediments, but that has typically been regarded as a terrestrial animal, is *Macrocnemus* (Rieppel 1989b). Several well-preserved specimens of this prolacertiform have been found in the Besano Formation of Monte San Giorgio (fig. 4.6). This rather small tetrapod had a slender, pointed skull atop a long, slender neck. Its gracile build would appear to have been ideal for actively hunting insects and invertebrates, perhaps along the ancient shoreline. However, it is also interesting to note that many of the specimens are articulated and

Plate 4.10. A group of *Helveticosaurus* (primitive placodonts) wait on the edge of the Tethyan surf as the rauisuchian *Ticinosuchus* lurks in the background.

Figure 4.6. (A) Juvenile individual of *Macrocnemus*. (B) Well-preserved, although somewhat crushed, skull of *Macrocnemus*.

exceptionally complete, indicating that they were not transported far after death. Thus, given the undoubted aquatic lifestyle of its closest known relatives, *Tanystropheus* and *Tanytrachelos*, there certainly has to be a question mark regarding the lifestyle of *Macrocnemus*.

It is interesting to note that in at least two *Macrocnemus* specimens, including a juvenile, distinct skin impressions are preserved (Peyer 1937; Renesto and Avanzini 2002). In the region of the pelvic girdle and tail, small overlapping scales are readily visible. They show some variation in form, possessing rounded to rhomboidal outlines. Like *Tanystropheus*, *Macrocnemus* is not just restricted to the Grenzbitumenzone but extends up into the higher beds of the Monte San Giorgio sequence. One well-preserved specimen of *Macrocnemus* was recovered from the Cassina beds. Yet unlike *Tanystropheus*, workers have so far not deemed it necessary to recognize this as a different species, so currently all specimens are referred to a single species, *Macrocnemus bassani*.

The thalattosaur *Askeptosaurus*, at about 2 meters long, is another of the moderate-sized marine tetrapods from the Grenzbitumenzone. Like so many other aquatic reptiles,

Plate 4.11. Splashing through the shallow streams that trickled through the caverns, a full-grown *Macrocnemus* tries to fend off the pursuit of an equally hungry member of its own kind. In all likelihood, they would share the spoils after squabbling in the relative security of the underground grotto.

the body and tail are elongated, but the limbs are short and not specialized as paddles. *Askeptosaurus* possessed a low skull with the external nares placed well back on top of the long snout, and in common with other thalattosaurs, the upper temporal fenestrae are so reduced that they are nothing more than long, narrow slits. Two other thalattosaurs occur in the Monte San Giorgio sediments. *Clarazia* and *Hescheleria* have prominent down-turned snouts, a feature that also characterizes thalattosaurs from western North America (Merriam 1905).

Along with the vertebrates, there is also a wonderful record of invertebrate life in the waters of Tethys. In addition to ammonoids and bivalves, some of the most abundant remains belong to thylacocephalan crustaceans. The taxonomic position of thylacocephalans is equivocal, but superficially, they resembled small amphipods. One of the most interesting discoveries of thylacocephalans was described by Affer and Teruzzi (1999). They described several carapaces from a quarry known as Sasso Caldo, situated within the Besano Formation of Varese Province. This quarry was first opened in 1985 and is still actively being excavated by Fabio Dalla Vecchia (Udine), Georgio Teruzzi (Milan), and their colleagues. The Besano Formation is equivalent to the Grenzbitumenzone of Swiss authors, and in addition to the Sasso Caldo site, crops out in a variety of localities in Varese Province and the neighboring Monte San Giorgio area of Switzerland (Canton Tessin). The Besano Formation was deposited in an intraplatform basin, and the center of this basin was almost certainly anoxic (see Jadoul et al. 1992; Bernasconi 1994). The Sasso Caldo crustacean finds are therefore particularly intriguing. Clearly benthic thylacocephalans could not have been living in the area of deposition. Affer and Teruzzi (1999) noticed that the thylacocephan remains consisted mostly of crushed carapaces. More significantly, they occasionally occur within coprolites. Affer and Teruzzi postulated that all the remains of these crustaceans were transported into the area within the guts of predators such as hybodont sharks or placodonts that had originally ingested them on the reefs and carbonate platforms. They were then incorporated in the fecal material of these predators, which was evacuated while swimming over the anoxic part of the basin.

As with the vertebrates, the invertebrate record is predominantly one of aquatic organisms, but three insect specimens, representing an ephemeropteran and a coleopteran, have been described from the Kalkschieferzone (upper part of the Meride Limestone, and late Ladinian in age). Two specimens have been referred to a new genus and species of mayfly, *Tintorina meridensis*. One of these is a fairly nicely preserved and relatively complete individual missing the head; the other comprises paired wings and a small part of the body. The wing venation is comparable to that of *Litophlebia oplata*, an ephemeropteran from the Upper Triassic of South Africa (Riek 1976; Hubbard and Riek 1977), and an unnamed form from the Voltzia limestones. The coleopteran is a single elytron and was referred to the Cupedidae by Krzemiński and Lombardo (2001). In addition, isolated insect fragments are known from the lower Meride Limestone (Krzemiński and Lombardo 2001). The presence of these insects, particularly the ephemeropterans, confirms that land was nearby and that bodies of freshwater were present.

Cave fills constitute an unusual type of terrestrial deposit. Such deposits are particularly well known for the Pleistocene, and they are invaluable because they often contain enormous numbers of individual elements. However, much older cave and fissure fills are known for high concentrations of bone. The Permian Fort Sill material is particularly notable. For the Triassic, the cavern and fissure deposits of southwest Britain are unquestionably the most famous (see chapter 10), but the Middle Triassic infills of Gliny, Poland, also provide us with another look at life living around the margins of the Tethys Sea (Lis and Wojcik 1958, 1960; Tarlo 1959). A large cave developed in Devonian limestones during Middle Triassic times. Waters washing through the cave deposited sediments and bones of contemporaneous animals. Fishes, nothosaurs, and prolacertiforms (including *Macrocnemus*) are the most abundant elements. Some of the bones may have been washed in from dead and decaying animals lying at the surface; other bones may have come from animals that lived and died in the caves.

From what the French and central European Middle Triassic fossil assemblages tell us, the coastal margins of Tethys apparently enjoyed a fairly balmy and damp climate. There doesn't appear to be any major indication of arid conditions. But that is what we would expect in maritime regions. Although we have some hints of Middle Triassic terrestrial life from parts of central Europe, in order to get a feel for fully terrestrial environments, we need to pick up the story in other parts of the world. What was the Middle Triassic world like further into the continental interior?

chapter five

No Midlife Crisis

England

Although the vagaries of the British climate bring many a complaint and provide a constant source of conversation for the islands' current residents, the same may not have been true for their Triassic predecessors. As we shall see, being only a little removed from the influences of the ameliorating effects of the Tethys Sea may have resulted in a fairly balmy climate. However, there are also some data hinting at aridity more in keeping with modern-day Arizona than the typically wet British bank holiday weekend.

The English Midlands

The English Midlands were home to some intriguing Middle Triassic terrestrial vertebrates. One of the most striking features of the Midland assemblages is the record of a reasonably diverse flora and some fully terrestrial invertebrates, scorpions.

The deposits from the Midlands have a history of research dating back to the early 1820s. William Buckland reported on finds near Warwick that were presented to the Oxford Museum. Soon thereafter, a variety of trackways were discovered in both Cheshire and Warwick. No lesser a name than Richard Owen stamped his authority on some of the findings, describing a variety of reptile and amphibian remains mostly as various species of *Labyrinthodon*. Indeed, in the same way that the ubiquitous Cope managed to collect from an amazing variety of strata in the United States, Owen had his fingers in every paleontological pie in England. Actually, Owen committed a faux pas somewhat akin to Cope's famous error with the placement of the elasmosaur skull. Owen apparently mixed remains of temnospondyl skulls with rauisuchian postcranial material, resulting in images with a resemblance to giant Triassic frogs (Benton and Gower 1997)! Some of the more notable finds came from Grinshill and ultimately turned out to be from a rhynchosaur, *Rhynchosaurus articeps*. The amphibian remains include *Mastodonsaurus* species as well as capitosaurid remains.

At the turn of the twentieth century, L. J. Wills (1907, 1908) reported on a variety of specimens from Bromsgrove. Again mastodonsaurid and capitosaurid (*Cyclotosaurus* species) amphibian remains, together with *Rhynchosaurus*, feature in the assemblage. In addition, fragmentary remains of a prolacertiform, a nothosaur, and a rauisuchian have been recorded (Benton et al. 1994). Along with the tetrapods, a variety of fish, bi-

Plate 5.1. The ctenosauriscid (a derived rauisuchian) *Lotosaurus* with its distinctive "sail back" plods through a grove of horsetails.

valves, arthropods (including fragments attributable to scorpions such as *Willsiscorpia bromsgrovienensis*), plants, and even annelids (*Spirorbis*) were recovered from Bromsgrove. Plant remains include cones referred to *Voltzia* (but see chapter 4 for a detailed discussion of this taxon), as well as a variety of equisetaleans and other gymnosperms. Overall, the fossils and sediments indicate deposition in a freshwater or brackish environment.

Three different species of *Rhynchosaurus* are actually known. These are *R. articeps*, *R. brodiei*, and *R. spenceri*. They occur in slightly different depositional environments. For example, *R. articeps* occurs in strata that also contain raindrop impressions, ripple marks, mud cracks, and halite pseudomorphs. The same beds yield *Rhynchosauroides* tracks as well, although these trackways need not necessarily have been produced by *Rhynchosaurus*. (For more discussion on ichnotaxa nomenclature, see chapter 9.) Taken together, it has been suggested that the depositional environment was a tropical, arid belt, maybe a marginally marine, hypersaline lagoon, but with rivers feeding into this lagoon. *R. brodiei* occurs at Coton End and Bromsgrove, and the two localities may represent slightly different environments. Coton End is suggestive of mature meandering river channels and floodplain complexes. Other elements of the fauna include lungfish like *Ceratodus* and the amphibian *Cyclotosaurus*. Wills (1950) suggested that Bromsgrove represented a freshwater or slightly saline lake, whereas Warrington (1970) favored a region of sinuous streams passing into a marine-influenced plain, possibly with brackish conditions. Finally *R. spenceri* comes from the Otter Sandstone (see below). It occurs in horizons that show numerous calcrete horizons indicative of subaerial soil formation in semiarid conditions. However, there are no signs of salt pseudomorphs or desiccation cracks, and conditions were perhaps not as dry as those of the *R. articeps* environment.

All the species of *Rhynchosaurus* have a large nasal capsule that suggests a good sense of smell. They also have an extensive hyoid apparatus, which points to a powerful tongue. This would have been used to manipulate food. The beaklike premaxillae might have served to dig or at least rake the surface for food, and the narrow, high claws on the foot would have been well suited to scratch digging. Horsetails have been found in association with *R. spenceri*. Occurring together with *R. brodiei* are sphenopsids and pteropsids, plus the spores of lycopsids, cycadopsids, and "coniferopsids." We can therefore envisage a well-vegetated plain inhabited by these rhynchosaurs.

Southern England

The south coast of England is world famous for its Chalk Cliffs and the Jurassic ichthyosaurs collected by Mary Anning along the coast around Lyme Regis. Sidmouth and Budleigh Salterton, by contrast, could hardly be described as world renowned. The two towns and the River Otter lie along the coast road west of Lyme Regis. In this region, a careful examination of the Otter Sandstone might yield a few scraps of teeth and bone. Although they do not compare to the spectacular Lyme Regis finds, the vertebrate fossils of the Otter Sandstone do offer another valuable glimpse of Middle Triassic terrestrial life. The tetrapods found here are very like those from the Midlands and include identifiable remains of mastodonsaurids, capitosaurids, *Rhynchosaurus*, and rauisuchians. One of the most readily recognizable vertebrates is the characteristic capitosauroid amphibian *Eocyclotosaurus* (Milner et al. 1990), which is also found in Germany (upper Buntsandstein) (Ortlam 1970). On the basis of the presence of *Eocyclotosaurus*, Lucas (1998) argued for an Perovkan age for the Otter Sandstone.

Other elements of the Otter Sandstone include a variety of procolophonid remains and several specimens of the paleonisciform (perleidiform), *Dipteronotus*. This deep-bodied fish has a characteristic humped back. Benton et al. (1994) reconstructed the Otter Sandstone as a somewhat arid landscape, with at least some seasonal pools. Milner et al. (1990) also noted the conflicting environmental signals from the assemblage. On the one hand, there are the abundant *Dipteronotus* and three superficially crocodile-like temnospondyl amphibians that were fundamentally aquatic, indicating at least some permanent water bodies. Further supporting this picture are the pteridophytes, which are commonly associated with damp environments. On the other hand, the most significant component of the limited terrestrial forms (rhynchosaurs and procolophonids) were supposedly animals feeding on durable plant material. This in turn might be indicative of semiarid conditions. The plant material also includes rhizoliths interpreted as conifers (Mader 1990) and phreatophytic (deep-rooted) plants that also point to rather dry conditions. This anomaly seems to be a recurrent theme in analyses of Triassic ecosystems. We shall return to this problem in chapter 8. For the time being, it is worth merely stating that we should always be cautious about oversimplifying and overgeneralizing, and that reality often calls for complex resolutions.

Asia

Traveling east around the world to the vast expanses of Asia takes us to present-day Russia, which would have occupied the northern margins of Pangaea, and to modern-day China, which would have comprised the eastern part of the great supercontinent.

The Russian Middle Triassic

Russia is particularly notable for its Middle Triassic terrestrial vertebrates. Ivan Efremov, one of the most celebrated Russian vertebrate paleontologists, conducted extensive research on the temnospondyls and reptiles from European Russia (Cis-Urals). He began a systematic prospecting program and as a consequence was able to identify a plethora of productive continental vertebrate localities ranging from the Lowermost Triassic to the Ladinian. The Lower Triassic assemblages were mentioned in chapter 3, and here I shall discuss only the Anisian and Ladinian vertebrates. As already mentioned, the Middle Triassic saw the worldwide radiation of the large herbivorous kannemeyeroid dicynodonts, and the Cis-Ural Middle Triassic is no exception, with a diversity of described taxa. The skulls of kannemeyeroids are typically about 15 inches long and have long, pointed snouts. The top of the skull is drawn up into a high, narrow crest for the attachment of powerful jaw adductor musculature.

What is confusing is the tendency of different authors to use different names for the same material. Thus genera such as *Rhadiodmus*, *Rabidosaurus*, *Rhinocerocephalus*, *Uralokannemeyeria*, and *Rhinodicynodon* may or may not be valid taxa. Lucas (1998), for instance, considers the first three to be junior synonyms of *Kannemeyeria* itself, and *Rhinodicynodon* to be equivalent to *Shansiodon*. As we shall see, *Shansiodon* is known from several localities in northern China and is much smaller than the kannemeyeriids. It is further characterized by a temporal region that extends into nothing more than a slight crest, and its blunt snout is relatively short and wide. According to Lucas (1998), *Shansiodon* is an important biostratigraphic marker. Co-occurring with these large herbivorous dicynodonts are a variety of carnivorous archosaurs, including the large and massively built erythrosuchids such as *Erythrosuchus magnus*, *Shansisuchus shansisuchus*, and *Garjainia*, and rauisuchids such as *Vjushkovisaurus berdjanensis* and *Dongusuchus efremovi*. Euparkeriids also extend up into Middle Triassic strata, but they are lesser-known taxa such as *Dorosuchus neoetus* and possibly *Wangisuchus tzeyii*.

The Donguz Formation (Perovkan) has also yielded an array of temnospondyls, including *Eryosuchus* and *Plagiosternum*. In the overlying Bukobay Formation, temnospondyls are particularly abundant and include *Mastodonsaurus* and possibly a species of *Cyclotosaurus*. The plethora of amphibians is suggestive of relatively damp conditions. Because the presence of some kind of water body is so frequently a prerequisite for fossilization, it is conceivable that we are missing a picture of life in the driest parts of the Triassic world. Still, it is becoming increasingly difficult to support the classic image of a hot and dry wasteland extending across vast tracts of the Triassic landmass.

China

As we have already seen for the Lower Triassic (chapter 3), China presents a rather different picture from Russia, and this continues to hold true throughout Middle Triassic times. Again, we pick up the story in the beds of the Junggur and Ordos Basins. The upper part of the Ermaying Formation in the Ordos Basin and the lower part of the Kelamayi Formation in the Junggur Basin have produced similar assemblages, but in contrast to the Donguz Formation, they are dominated by dicynodonts. In the Kelamayi Formation, the kannemeyeriid *Parakannemeyeria* is found together with the euparkeriid *Turfanosuchus* and the erythrosuchid *Vjushkovia*. The Ermaying Formation also produces kannemeyeriids, euparkeriids, and erythrosuchids, but they are somewhat different from those from the Junggur Basin. *Parakannemeyeria*, together with another form, *Sinokannemeyeria*, are common to both regions, but both have limited distributions. By contrast, other components of the Chinese Middle Triassic faunas have a much broader distribution. For instance, the kannemeyeriid *Shansiodon* is almost as cosmopolitan as *Lystrosaurus*. It is now widely recognized that *Shansiodon* occurs in China, Russia, Tanzania, Zambia, and South Africa and can be regarded as the standard-bearer of the Anisian.

The abundant dicynodonts were almost certainly preyed upon by large, heavy-set archosaurs like *Shansisuchus*. Other more gracile and somewhat smaller carnivores included euparkeriids. These were probably more adept at tackling smaller prey than dicynodonts. Procolophonid remains, although relatively rare, are occasionally found in the Ermaying Formation, as is another small tetrapod, the rare therapsid *Traversodontoides*.

Although amphibian remains are sometimes encountered in the upper Ermaying, they are uncommon and indeterminate, a marked difference between the contemporaneous terrestrial faunas of Russia and China that most likely reflects broad climatic and geographic differences between the two areas. The Chinese assemblages represent terrestrial faunas that lived in a relatively dry inland and/or upland region, whereas the Russian assemblages primarily represent wet lowland and coastal regions. In broad terms, it is fair to say that the Early to Middle Triassic terrestrial vertebrate faunas of China and Africa are much more alike—this despite the fact that they were widely separated during Triassic times.

The Middle Triassic of China also boasts one of those oddities that characterize the Triassic in general. In this instance, it is the peculiar "sail-backed" archosaur, *Lotosaurus*. The Moenkopi (also Perovkan equivalent) has yielded a related sail-backed animal called *Arizonasaurus*. These animals, known as ctenosauriscids, would seem to be derived rauisuchians (Nesbitt 2003). The greatly exaggerated elongation of the neural spines in the dorsal vertebral series of both these animals is reminiscent of the condition that characterizes the well-known Permian tetrapod, *Dimetrodon*. But a similar feature is also known in a variety of other animals, including the Cretaceous theropod dinosaur *Spinosaurus*. The function of the exaggerated sail on the back of all these animals is unclear. Various ideas have been suggested. Probably the most likely are that it acted as a thermoregulatory structure or some kind of display and recognition feature. Certainly these animals occupied somewhat different ecological niches. Although *Dimetrodon* and *Spinosaurus* were carnivores, other sail backs, such as *Edaphosaurus*, were herbivores, and *Lotosaurus*, with its edentulous jaws, was probably also herbivorous.

Finally, we cannot leave the Triassic of China without remarking on the Nine-Dragon Wall. A Chinese myth relates the importance of the so-called Nine Realms, in which eight fundamental domains radiate from the central one of the sun. The dragon is also prevalent in Chinese mythology. It is typically regarded as a symbol of rain and water and is therefore critical for flourishing crops. As a consequence, the image of nine dragons was

immensely significant to the ancient Chinese. Panels depicting such an image can still be seen today in the Imperial Palace in Beijing. It was therefore a remarkable discovery when scientists from the Institute of Vertebrate Paleontology and Paleoanthropology discovered a natural nine-dragon wall. This spectacular block was collected from the Kelamayi Formation and bears the remains of skeletons from nine juvenile *Parakannemeyria*.

Plate 5.2. In the Triassic of Germany, a close relative of *Lotosaurus*, *Ctenosauriscus*, roams through the undergrowth of a Voltzia forest.

Africa

Lower Triassic sediments of the Karoo Basin in southern Africa are particularly well documented in the literature (see chapter 1), and as we shall see later, the Karoo also has some intriguing Upper Triassic sequences. On the other hand, the Middle Triassic sequences have not been subjected to such close scrutiny. Even so, Africa does have some productive Middle Triassic beds.

The lower Manda Formation of Tanzania is known for the dicynodonts *Shansiodon njalilus* and *Angonisaurus cruickshanki* as well as the rhynchosaur *Stenaulorhynchus*. Not unexpectedly, *Stenaulorhynchus* rather closely resembles *Rhynchosaurus*. In addition there is the pseudosuchian "*Mandasuchus.*" This possible "rauisuchid" related to *Ticinosuchus* (possibly even congeneric) has never been properly named, and "*Mandasuchus*" is therefore a nomen nudum.

Tying these faunas in with the marine succession has proved difficult, but most authors (e.g., Anderson and Cruickshank 1978; Lucas 1998) refer them to the Anisian (=Perovkan). More detailed publications of the Manda tetrapods is required, but in general terms, this fauna seems to be more akin to the typical Chinese Middle Triassic fauna than that of Russia.

South America

Chañares

There can be no question that the most widely studied Middle Triassic terrestrial assemblage to date is that from the region of La Rioja Province in Argentina—the so-called Chañares fauna (late Anisian or early Ladinian). It is famous for its reptile fauna, in particular a number of small archosaurs and mammal-like reptiles. To some, the archosaur assemblage assumes prime importance because it includes forms that are widely thought to form the closest sister group to the Dinosauria.

Most of the early work on this assemblage was undertaken by Alfred Romer together with James Jensen. Some of the small archosaurs, such as *Gracilisuchus* and *Gualosuchus*, have been described on the basis of relatively complete and at least partially articulated material. By contrast, others are much more fragmentary, and their affinities are consequently somewhat equivocal. Romer (1972c), in his description of *Lewisuchus*, reported that until that time six "thecodonts" had been described from the Chañares collection. He noted it was only the careful and extended preparation of a block known to contain parts of a gomphodont cynodont and miscellaneous parts of small thecodonts that brought to light the new form. The high density of remains on certain slabs and the occurrence of different taxa in close proximity can make it difficult to distinguish one taxon from another. This has recently been borne out by the work of Arcucci (1990), who added the new archosaur, *Tropidosuchus*, to the faunal list on the basis of a more detailed analysis of bones initially attributed to *Gracilisuchus*. The block of matrix catalogued as MCZ 4137 in the Museum of Comparative Zoology at Harvard is yet another example of two different taxa in intimate association. The partially articulated right hindlimb of *Marasuchus* lies alongside a left femur of *Gracilisuchus*.

The sediments of the Chañares Formation comprise evenly bedded white to bluish white volcanic ashes that are conformably overlain by the Los Rastros, a series of carbonaceous sediments. Above the Los Rastros is the Late Triassic Ischigualasto Formation, which consists of a series of variegated shales and which is famous for the early dinosaurs *Herrerasaurus* and *Eoraptor* (see chapter 9).

At the classic Los Chañares localities (Romer 1966), there are over 100 individual vertebrates entombed in concretions with matrices of relic glass shards diagenetically replaced by calcite. The vertebrates comprise a variety of different forms and include cynodonts, dicynodonts, and archosaurs. This mass mortality assemblage is notable for the mixture of predators and prey. The arrangement of the carcasses is consistent with accumulation along a strand line, and Rogers et al. (2001) suggested that volcanism led to catastrophic flooding of the area through damming and/or the diversion of local drainages. Uncompacted skeletal elements and relict outlines of glass shards indicate that carbonate concretions formed shortly after the skeletons were buried in reworked volcanic ash. Microbial action brought about the decay of soft tissues, a process catalyzing concretion diagenesis. The excellent preservation of the bones may owe much to the fact that entombment within early diagenetic concretions protected them from subsequent pedogenic and

diagenetic processes that would have been somewhat destructive. Here, then, is a site where the "mandatory" Triassic volcanoes actually existed, and their activity was instrumental in the preservation of a snapshot of life 230 million years ago in what is now South America.

The Chañares is biased toward the preservation of vertebrates, lacking both terrestrial invertebrates and plants, but it is also striking in that it apparently lacks amphibians and, more surprisingly, lepidosauromorphs. Although certain Triassic amphibians were undoubtedly less reliant on water than is typical of modern forms, a complete lack of amphibians (including metoposaurs, mastodonsaurs, capitosaurs, trematosaurs, brachyopids, latiscopids, chigutisaurids, and plagiosaurids) is probably best taken as evidence of a fully terrestrial community living in a relatively dry environment. As we shall see in chapter 7,

Plate 5.3. A lone *Chanaresuchus* stands out against the stark backdrop of ash that has suffocated the surrounding hills and floodplain. From time to time, such catastrophes inevitably wiped out local plant and animal communities. But in the same way that eruptions like Mount St. Helens or the forest fires of Yellowstone cause only temporary blips today, such events in the Triassic typically had no lasting effects.

Plate 5.4. The cynodont *Probelesodon* tenses
behind a tree trunk as it scents a *Marasuchus*
hurrying through the undergrowth.

the Late Triassic Chinle Formation, consisting of waterborne muds and sands mixed to-gether with volcanic ashes, represents an ancient terrestrial environment occasionally in-fluenced by major volcanic activity, but here, amphibians are found, as well as reasonable numbers of lepidosauromorph remains. We might therefore surmise that, as with the pres-ence of amphibians, the occurrence of lepidosauromorphs also points to some degree of humidity in the ancient climate, whereas the absence of both would be highly suggestive of arid conditions.

Unfortunately, it doesn't seem to be as simple as that. For example, the vertebrate as-semblage from the Lossiemouth Sandstones (chapter 8) is typically thought to represent a community that lived under xeric (dry) conditions, yet lepidosauromorphs are not un-common in these deposits. Perhaps the Chañares has only captured a small portion of the vertebrate community living at the time, or it represents an area that, because of some combination of age, geography, climate, or environment, possessed its own unique terres-trial fauna of archosaurs and mammal-like reptiles. But again, this seems to be too sim-plistic an approach because lepidosauromorphs seem to be relatively rare components of Middle Triassic assemblages worldwide. (It is worth noting in passing that basal ar-chosauromorphs are not particularly widespread in the southern hemisphere during this interval either.) This begs a question regarding lepidosauromorph diversity in the Early and Middle Triassic.

Plate 5.5. Then, underneath a canopy of foliage compris-ing *Hausmannia* (small, broad fern leaves), *Johnsto-nia*, and a single leaf of *Chi-ropteris*, the *Probelesodon* pounces on the unfortunate *Marasuchus*.

Plate 5.6. The lower part of the Santa Maria Formation in Brazil was deposited about the same time as the Ischichuca (Chañares) Formation. Here the large rauisuchian *Prestosuchus* challenges a small herd of the dicynodont *Stahleckeria* against a backdrop of *Dicroidium* shrubs.

Although we must assume that the earliest lepidosauromorphs date to the earliest appearance of their sister taxon, the archosauromorphs, is it possible that they failed to really flourish until after the Middle Triassic? I actually suspect that this is unlikely, for as we shall see when we come to examine the Late Triassic assemblages, some of the Triassic sphenodontians, including the oldest known member of that group, are actually nested within the Crown Group Sphenodontia—that is, they are highly derived. This also implies that they (and their sister group, the Squamata) actually originated and diversified at least during the Middle Triassic, and probably even earlier. Indeed, the early history of the Squamata presents us with precisely the same dilemma. Although the oldest members of their presumed sister group, the Sphenodontia, are Carnian, currently the oldest squamate is Middle Jurassic. Where, then, are the Triassic squamates? Have we just failed to recognize them because any specimens that have been collected are too poorly preserved and lack any distinguishing features? Perhaps we are sampling the wrong depositional environments? Maybe we have not looked

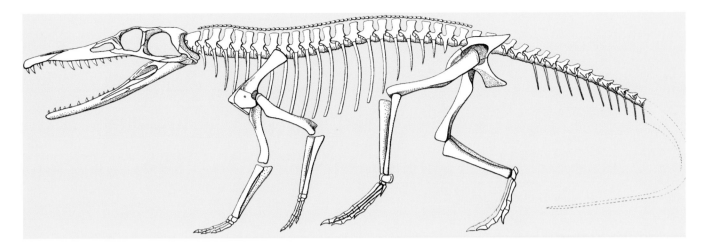

hard enough. Likewise, the reason for the paucity of Middle Triassic lepidosauromorphs must remain for the time being just another unanswered question about Triassic life.

Although the lack of amphibians (and perhaps also the absence of lepidosauromorphs) may be suggestive of dry conditions in the Chañares region, the evidence, if you will excuse the pun, is not altogether watertight. For instance, it has been suggested that at least one of the archosaurs, the dog-sized *Chanaresuchus* (fig. 5.1), was amphibious and occupied a niche similar to that of phytosaurs or modern-day crocodiles (Romer 1971b, 1972a). Romer restored this long-snouted animal in a quadrupedal pose and envisaged it spending much of its time in the water. At the same time, he felt that it might have been capable of adopting a bipedal posture when on dry land, so there are clearly some mixed messages here. Other small- to moderate-sized carnivores, including *Gracilisuchus* (fig. 5.2), *Marasuchus* (*Lagosuchus*), and *Lagerpeton*, were considered by Romer (1971a, 1972b) to be cursorial, probably bipedal animals. Such a view shifts the balance back in favor of a fully terrestrial assemblage of animals with either little or no representation of semiaquatic forms.

The Chañares cynodonts have played an important part in discussions on the evolution of mammals. They include carnivorous forms such as *Probelesodon* and *Chiniquodon* (Romer 1969, 1970) as well as herbivores like *Massetognathus* (Romer 1967). *Massetognathus* possessed a battery of grinding molars (fig. 5.3c) that would have efficiently broken down tough plant matter before swallowing.

Diversity Levels in the Fossil Record

Discussion of the Chañares fauna is an opportune moment to bring up an issue concerning the depiction of the ancient world. Restorations of past life, particularly those gracing museum dioramas, frequently try to cram as many different organisms as possible into the limited space available. This inevitably produces a scene that is totally unrealistic ("the lion lies down with the lamb"). However, in large part, this may mirror the scientific descriptions, or at least the impression these descriptions often give to the reader. Thus descriptors such as the "Chañares reptiles" or the classic "Solnhofen fauna" imply that the animals that constitute these assemblages were all found together at one locality. However, in most cases, nothing could be further from the truth. For instance, the Solnhofen fauna originates from a whole range of small quarries separated by over 40 miles. Likewise, for the Chañares and the Ischigualasto, there are numerous vertebrate-producing sites—none of which has produced every single taxon described from the Chañares or Ischigualasto. Indeed, most localities have produced fewer than half of the total number of

Figure 5.1. Long-snouted archosaur *Chanaresuchus* (after Romer 1972a).

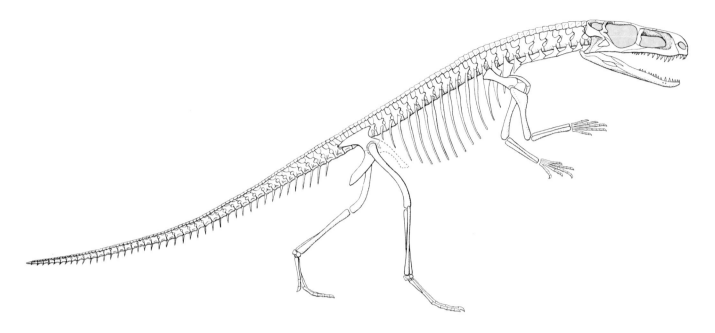

Figure 5.2. *Gracilisuchus* may be closely related to the earliest crocodiles. Here it is restored with a bipedal posture (after Romer 1972b). Some of the sphenosuchians, such as *Terrestrisuchus* (chapter 11), may also have been bipedal.

taxa collectively known from the Chañares, although possibly with one or two taxa common to many of these localities. It is almost certain that within a given area, such as the Chañares Basin, there were several discrete habitats and paleocommunities, each with its own set of characteristics.

When it comes to the Dockum and Chinle Groups, the situation becomes worse, with many sites separated by vast distances and many different stratigraphic levels involved. The vertebrate assemblages from the British fissure deposits (chapter 10) are another example where lumping together has created a false impression. There has been a broad age distinction made between fissure systems that yield only sauropsid reptiles and fissures that, in addition to sauropsids, also contain early mammals and palynomorphs, with the former being regarded as Late Triassic and the latter as Early Jurassic. Unfortunately, it is all too easy to leave it at that and thus ignore the fact that many of the sauropsid-producing fissures are separated by considerable distance and have their own unique assemblages. Even at one quarry, such as Cromhall, there are several separate fissure fills, some of which have different faunal components (sometimes with no more than three or four different taxa). There is evidence to suggest that each of these fissures filled at different times, and that they could represent changing faunas and conditions over time. Grouping all the different animals from each of the different fills together would be misleading.

Our perceptions of ancient diversity levels are influenced by a variety of factors. Primary among these, of course, is the patchiness of the fossil record, which will tend to lower the level of the apparent diversity. Absence of soft-part preservation automatically removes numerous characters that in the living world would clearly distinguish between closely related species. Behavioral traits that might also permit differentiation between species are also impossible to evaluate in the fossil record. Then there are the taxa still out there waiting to be discovered by future generations of paleontologists. And many species lived in environments that were totally unsuited to fossilization processes, and as a consequence, there are probably whole ecosystems missing from our radar screen. Severely underestimating diversity levels of the past is difficult to avoid. Nevertheless, there have been many attempts to redress the balance. For example, there has been a propensity on the part of certain paleontologists, including some of the most eminent names in the field, to erect new taxa on small differences in body form—differences that readily fall within variation within a species. This would give the false impression of higher levels of diversity than actually existed. Thus as recently as 1980 we had as many as 15 species of *Triceratops*, but now it is widely accepted that there is either one or possibly two species.

The incompleteness of so much fossil material unfortunately leads to another pitfall: the occasional inability to recognize that different elements belonged to the same taxon. Thus there is the real danger of assigning more than one name to the same organism. As we have seen, this is particularly problematic with plants, with the possibility of attaching different identities to fruiting bodies, seeds, foliage, and palynomorphs produced by the same species. But potentially the most misleading practice is the grouping of strata and/or different localities together to form a type assemblage or fauna. Thus we talk about the "Elgin fauna," a collection of fossils from several different quarries in Morayshire, Scotland. Although the entire faunal list has not been recorded at any one quarry, they all occur in a similar facies, and it is not beyond the realm of possibility to imagine that they all lived at the same time in one paleoenvironment. On the other hand, the Chinle fauna and flora come from a much larger geographical area, occur in a variety of sedimentary facies, and represent a more protracted period of geologic time. If we were to lump all the Chinle deposits together in the same manner as the Elgin reptiles, there is no question that we would be artificially increasing diversity levels.

It is easy to say that different factors might tend to cancel each other out, and that our picture of life at any given time in a broad area averages out quite well. But such is not the case. In any given situation, one or more of the above factors had a greater effect than the others. It would seem that there was no such thing as an "average" environment and paleocommunity in the Triassic. For this reason, the images in this book do not attempt to integrate every single fossil known from a broad area into an intricate, yet implausible, picture of life in the past. Instead, each image portrays one small aspect of life in a variety of settings during the Triassic.

An Interregnum?—A Malagasy Fauna

One of the most recently discovered Triassic assemblages comes from Madagascar and provides us with an intriguing perspective on the transition between Middle and Late Triassic continental vertebrate faunas. Madagascar has a fascinating modern fauna and flora, but it seems that the island has a long history of unique and exotic faunas that extends back at least to the Early Mesozoic. The Cretaceous sediments from the north are producing some of the most exquisitely preserved and unique vertebrate fossils known. These include a bizarre crocodile with a blunt box-shaped snout and a heterodont dentition, the bird *Rahonavis*, a titanosaur, and the theropod dinosaur *Majungasaurus*. But the Morondova Basin to the south has yielded an equally important Triassic vertebrate assemblage. Known as the Isalo Group, the basal sediments (Isalo II) have yielded an unusual combination of taxa, including abundant remains of two prosauropod dinosaurs, two rhynchosaurs, a sphenodontian, a kannemeyeriid dicynodont, and at least four eucynodonts (Flynn et al. 1999). This fauna appears to be particularly important in that it represents a bridge between the Middle and Late Triassic. Although rhynchosaurs and dicynodonts are not known to occur elsewhere after the Carnian, no sphenodontians and prosauropods are known to be older than the Carnian. (As I have previously emphasized, that is not to say that they, or at least more basal members of their lineages, did not occur before the Carnian.) There are at least three traversodontid eucynodonts in the Isalo assemblage, which is consistent with pre-Carnian diversity levels for this group, but which is well in excess of middle Carnian and later assemblages. One of these forms, *Dadadon*, is similar to *Massetognathus* from the Ladinian (Chañares) of South America, although the resemblance is probably nothing more than that they are both plesiomorphic ("primitive") (fig.

120

Figure 5.3. Skull of *Massetognathus* in lateral (A), dorsal (B), and ventral (C) views (after Romer 1967).

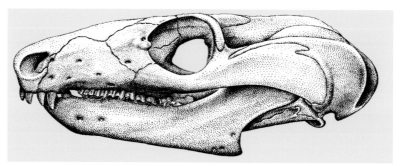

A

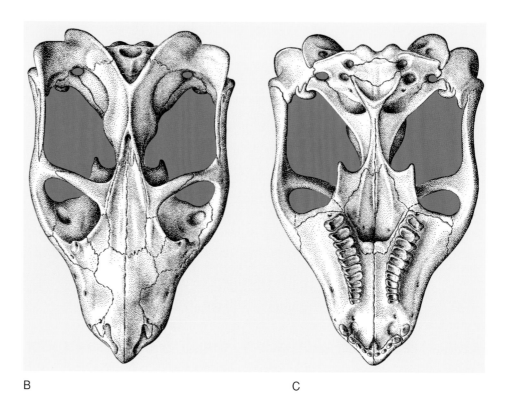

B   C

5.3). Another of the traversodonts resembles forms from the Middle Triassic of Tanzania, and a third, *Menadon* (Flynn et al. 1999), represented by one fairly complete skull, represents a new taxon closely allied with *Scalenodontoides*, *Gomphodontosuchus*, and *Exaeretodon* (whose earliest occurrence is the lower beds of the Santa Maria Formation of Brazil [Ladinian age]; see chapter 9).

From what little we know of Malagasy Triassic tetrapods, they appear to be rather different from those penecontemporaneous tetrapod assemblages elsewhere in the world. As Flynn et al. (1999) note, this may simply be a foretaste of the island's faunal provincialism that gradually came about, starting first with its separation from Africa in the Early Jurassic, from Antarctica in the Early Cretaceous, and finally from India during the Late Cretaceous. It may equally be the case that the Malagasy fauna documents a temporal interval that is not well represented elsewhere (Flynn et al. 1999). Although this type of fauna might have been much more widely distributed, for whatever reason, we have little record of it in Triassic sediments. The Malagasy fauna might be seen as transitional between the eucynodont-rhynchosaur-pseudosuchian (or at least noncrocodilian pseudo-suchian) dominated faunas of the Middle Triassic and the dinosaur-lepidosaur-crocodile-mammal (modern tetrapod taxa) dominated faunas of the Late Triassic.

As we shall see in chapter 11, although it is conceivable that traversodonts might have some use as biostratigraphic indicators, they may also have been intimately associated with humid belts on either side of the equator. Given that Pangaea was drifting northward during Triassic times, the already unclear picture of their distribution both in time and space is made even more complicated.

Part Three

A TIME OF CHANGE
The Early Late Triassic

chapter six

Marginal Creatures in a Marginal World

From time to time, paleontological discoveries are made that defy our comprehension because they do not neatly fit into our ordered classification schemes. Some Burgess Shale organisms are examples that immediately spring to mind, but the Triassic seems to have more than its fair share of enigmas. Two examples from Kyrgyzstan top the list, but not far behind are two tetrapods from northern Italy. In this chapter, we shall discuss these oddities and the environments that they inhabited.

The Madygen Beasts

It is a pity that one of the most important paleoentomological Triassic sites in the world has been overshadowed by the presence of two vertebrate specimens. The Triassic sediments of the Fergana region of Kyrgyzstan are well known to paleoentomologists and paleobotanists for their wealth of insect and plant remains. The deposits consist of thick fluvial and lacustrine sequences. The rapid burial of organisms under very fine-grained sediments is responsible for the fantastic preservation here of both plant and insect remains. The insects in particular led scientists from the Palaeontological Institute in Moscow to make repeated visits to the region. During one such excavation for additional insects, two of the most bizarre vertebrates known from the Triassic were recovered: *Longisquama* and *Sharovipteryx* (Sharov 1966) (formerly *Podopteryx*, or "foot wing": unfortunately the name *Podopteryx* was found to be preoccupied—by an insect—because it was an apt name for this curious creature). I shall discuss these unusual finds below.

At the time of their discovery, the deposits were considered to be Early Triassic, but later detailed studies of the flora (Dobruskina 1980, 1995) indicated that they were more likely to be Ladinian to Carnian in age. Although the unique vertebrates don't permit correlation with the land vertebrate faunachron (LVF) system of Lucas (1998), the paleobotanical data would indicate that the sediments just predate the Otischalkian, sometime around the Ladinian-Carnian boundary.

The rich floral assemblage is dominated by ferns, horsetails, cycads, lycophytes, ginkgoes, and conifers. The insect assemblage would appear to be one of the most diverse known from the Triassic (e.g., Sharov 1968; Vishnyakova 1998; Papier and Nel 2001). For instance, all 11 specimens of true flies (Diptera) described by Shcherbakov et al. (1995) from the Dzhailoucho locality were considered to be quite distinct species. The most spectacular insects are the titanopterans. This uniquely Triassic group is represented by a

Plate 6.1. With limbs outstretched to extend its gliding membranes, *Sharovipteryx* glides beneath a canopy of *Podozamites* (conifer) and *Vittaephyllum* (seed fern) leaves.

Plate 6.2. After falling from the branches of a cycad, the rotting carcass of a *Longisquama* attracts the attention of a variety of insect carrion feeders, including two beetles, *Notocupoides* and *Hadeocoleus*. The dipteran *Axioxyela* buzzes over the carcass. The Madygen Formation is one of the few deposits in the world that is currently known for abundant Triassic insect remains. As well as *Longisquama*, Madygen is also home to another equally controversial tetrapod, *Sharovipteryx*. One of the reasons that so few insect assemblages are known for the Triassic is not necessarily the poor preservation potential of insects and the consequent real absence of Triassic insect assemblages. It seems just as likely—in fact more probable—that additional rich insect assemblages do exist, but that the fossils are readily overlooked in the field.

number of different taxa, including *Prototitan primitivus*, *Nanotitan extentatus*, and *N. magnificus*, but *Gigatitan vulgaris*, with a wingspan of up to a foot, was undoubtedly the most spectacular. The impressively large stridulatory organs must have been capable of producing incredible sound. As well as the flies and titanopterans an array of beetles, cockroaches and bugs are also known. Other deposits in the same part of the world have also yielded abundant insect remains. The Tologoi Formation in southeast Kyrgyzstan is known for a wealth of water bugs as well as isolated dipteran remains. Despite their abundance, it is worth remembering that for the most part, the insect remains from the Madygen Formation are isolated wings rather than complete individuals. As a consequence, there are some limitations to the extent they can be referred to different taxa.

In addition to the insects, the invertebrates discovered include various bivalves and crustaceans. Fishes include the dipnoan *Asioceratodus* and the ubiquitous *Saurichthys*. *Longisquama* and *Sharovipteryx* are not the only tetrapods. Ivakhnenko (1978) described *Triassurus*, a single poorly preserved specimen, as the earliest urodele, and Tatarinov (1980, 1994) described the cynodont *Madygenia*. Yet for many, the importance of the Madygen rests almost entirely on the small blocks of sediment that bear *Longisquama* and *Sharovipteryx*.

Plate 6.3. Among the *Ginkgoites* leaves of the Madygen woodland rest a *Gigantitan* and a *Zeunerophlebia*.

Discerning the details of both taxa is difficult. This is in part due to rather poor preservation of certain areas of the specimens, but perhaps more importantly, both have been coated
with a preservative that obscures details of articulations between individual elements. Despite this, or perhaps because of it, both taxa continue to be the subject of intense debate. In
addition to speculation about their aerial capablilities, they have been the focus of debate in
connection with the two most popular vertebrate groups of the Mesozoic. *Longisquama* has
recently been indirectly involved in the debate on theropod dinosaurs and the origin of
birds, whereas *Sharovipteryx* has been cited in a novel theory on the origin of pterosaurs.

Sharovipteryx

The sole specimen of *Sharovipteryx* comprises part and counterpart slabs. Although it is
difficult to discern details of the various cranial and postcranial elements, the membra-

Figure 6.1. Only known specimen of *Sharovipteryx*, showing the remarkably preserved
gliding membrane attached
to the hindlimbs.

nous structures, or uropatagium, supported by the disproportionately elongate hindlimbs are strikingly clear. The skull is slender and pointed, with an elongate snout, and the neck is rather long, being equal in length to the short trunk. There appear to be at least seven cervical vertebrae, with vertebrae 3 through 7 being noticeably elongated (fig. 6.1). With few clear details of the osteology, it has proved difficult to assess the phylogenetic relationships of *Sharovipteryx*, but most opinion favors close links to the prolacertiforms (e.g., see Unwin et al. 2000). The most stunning features of the specimen are the superbly preserved gliding membranes. There are also impressions of skin around various parts of the body that bear scales. Sharov (1970) argued that *Sharovipteryx* was an arboreal glider that flew from branch to branch by using the head and body as a kind of rudder and the tail as a counterweight. Gans et al. (1987) suggested a slightly different reconstruction of the gliding apparatus, but essentially agreed with Sharov's original proposal. They also indicated camouflage and display as possible functions of the membranes.

Longisquama

Longisquama (fig. 6.2) was first described by Sharov in 1970. He considered it to be an arboreal pseudosuchian, which he reconstructed with a single row of very long, featherlike appendages running down the length of the back. Sharov believed that these structures could have functioned as a kind of parachute and facilitated the animal's movement through the trees as it jumped from branch to branch. However, with a single row of these structures, it is difficult to envisage just how this was achieved. Perhaps they moved independently of each other and extended out to either side in an alternate fashion. More recently, Haubold and Buffetaut (1987) reexamined the material and, noting the apparent presence of two temporal arches and an antorbital fenestra, concurred with Sharov's referral of the animal to the "pseudosuchians." Like Sharov, Haubold and Buffetaut considered it highly probable that the dorsal appendages were derived from elongated scales. On the other hand, they found it difficult to accept that a single row of appendages could function in any gliding or parachuting fashion. Instead, they proposed that in life, *Longisquama* possessed a double row of the featherlike dorsal structures. It would appear that these appendages became shorter moving posteriorly: in the holotype, the most anterior structure is 150 millimeters long, whereas the most posterior one is only 100 millimeters long. When spread out to the sides, the appendages would collectively have formed a gliding membrane consistent in shape with the gliding membranes supported by the ribs of the kuehneosaurs (see chapter 10). Haubold and Buffetaut proposed that at rest the two rows were held together above the body. This is certainly a much more plausible reconstruction, but it must remain rather equivocal for a variety of reasons, principal among the fact that the appendages are not actually attached to the skeleton in the holotype. Additional specimens that have been referred to *Longisquama* consist of isolated appendages and no skeletal material, so the pattern of attachment to the body must remain conjectural.

The featherlike appendages of *Longisquama* are certainly difficult to interpret, and perhaps because they are not particularly clear, they have generated a good deal of heated debate. They are represented by impressions that have a central shaft from which radiates a continuous series of pinnae. As a consequence, Jones et al. (2000) argued that parts of the central axis were hollow (similar to the central shaft of a bird feather), and that the distal ends of successive pinnae formed a ribbonlike margin. Indeed, they consider the structures to be protofeathers, and *Longisquama* therefore to represent a crucial stumbling block to the conventional view that birds arose from theropod dinosaurs. However, Reisz and Sues (2000) argue strongly against the idea that the *Longisquama* appendages represent any type of feather. Feathers tend to fray at the edges, but the appendages on *Longisquama* seem to have a solid rim around the edge. They look more like ribbed

Figure 6.2. Extraordinary specimen of *Longisquama*.

membranes, and Sues and Reisz prefer to recognize them as peculiarly elongate scales, a view that I share.

Indeed, it is not unreasonable to question whether these featherlike structures actually belong with the skeleton. At least it does seem strange that there is only one skeleton known while there are several partial sets of the featherlike structures (Haubold and Buffetaut 1987). It would seem likely that the skeletal material would be more likely to preserve than the appendages, which are supposedly soft tissue structures. Could it be possible that the appendages are actually foliage of an unknown plant? Admittedly, if these impressions are considered to be representative of some kind of foliage, then it has a peculiar venation pattern that is inconsistent with any known Triassic foliage types. When dealing with radically different new fossils, one must be prepared to consider every option.

The Islands of Italy

Some time after *Sharovipteryx* and *Longisquama* were apparently gliding in the Fergana Valley, a little further to the south and west, early members of a more widespread group of winged tetrapods had taken to the air. On the ground were some equally bizarre animals called drepanosaurs.

Lombardy is a region of Europe noted for its culinary delights. Even when collecting fossils in the remotest mountain regions, the paleontologist is always guaranteed a gastronomic reward at a neighboring hostelry tucked away on a steep hillside. Once the arduous work of collecting and preparation has been accomplished, the visiting scientist can marvel at the specimens, sample the ancient architecture and art, and enjoy the exquisite cuisine at the likes of the Ristorante Bernabo, a stone's throw from the Museo Civico di Scienze Naturali "Enrico Caffi," in Bergamo. There is even the opportunity to enjoy authentic Triassic water, for the world-famous San Pellegrino mineral water comes directly from springs in the Norian dolomite. But the spectacular alpine scenery and cuisine of northern Italy are surely matched by its Triassic fossils and by the attention that the regional museums pay to their conservation and care. I suspect that there is no place in the world that the needs of the visiting paleontologist are better rewarded!

Many of the thick fossiliferous units that extend into southern Switzerland around the Lake of Lugano have for many years produced exquisitely preserved Middle Triassic marine vertebrate fossils, particularly actinopterygian fishes (e.g., Bürgin et al. 1989) and nothosaurs (e.g., Sander 1989) (see chapter 4). However, as we have seen, the so-called Grenzbitumenzone of Switzerland and the Besano Formation, its equivalent in Italy, are also known to contain occasional terrestrial forms such as *Ticinosuchus* (Krebs 1965). In more recent times, the Upper Triassic sediments have attracted much-deserved attention. These too are rich in marine vertebrates, but a reputation for the bizarre has been added to the bill of fare after the discovery of several unusual terrestrial vertebrates. Tetrapods such as *Megalancosaurus* have stimulated a great deal of discussion and heated debate. This region is also home to the earliest known pterosaurs—the group that was to dominate the Mesozoic skies.

There are two principal areas for Upper Triassic fossils in northern Italy. The first occurs to the north of Milan and extends from Bergamo northwestward as far as Lake Lugano. The other is centered on the Preone Valley north of Udine. Both areas have yielded similar faunal assemblages, and it is tempting to discuss the two together as if they represented one contiguous unit. However, to do this might tend to hide subtle differences in fauna and environments between the two.

In the vicinity of Bergamo are two Norian units (Revueltian LVF equivalents), the Calcare di Zorzino (Zorzino Limestone) Formation and the Argillite di Riva di Solto shales. The Argillite di Riva di Solto is the slightly younger of the two formations. The Zorzino Limestone is coeval with the Dolomia di Forni (Forni Dolostone) of the Udine area of Friuli.

The Udine Region

In the region of Udine (northeastern Italy) Triassic sediments include the early Carnian, Raibler Schichten, and the middle Norian Dolomia di Forni. The Raibler Schichten consists of thinly bedded black limestones that contain fish such as *Saurichthys*, *Peltopleurus*, *Pholidopleurus*, *Ptycholepis*, and *Birgeria*. It is also known for its crustaceans, terrestrial plants, and ammonoids, but only a single fragment of an aquatic reptile has been found.

By contrast, the middle Norian Dolomia di Forni has a much more diverse faunal and floral assemblage. Although there is not a single site producing all the different taxa, all come from a restricted area in the upper valley of the Tagliamento River and originate in the same type of sediment. Invertebrates include a variety of decapod crustaceans (e.g., Garassino and Teruzzi 1993), comprising at least six genera. Additional crustaceans described from the Dolomia di Forni include the thylacocephalans, *Microcaris* and *Atropicaris*. This extinct group of crustaceans is typically considered to have been scavengers living in the benthos (Alessandrello et al. 1989). Their phylogenetic position remains obscure. These crustaceans, flitting around on the bottom of the shallow seas, would have provided a rich source of food for many of the fish and the marine reptiles. But crustaceans were not the only invertebrates: coleoid cephalopods, ophiuroids, gastropods, and bivalves are also known to occur in the deposits, albeit in much fewer numbers.

Parts of Europe remained covered by shallow seas into Late Triassic times, and many of the same genera of fish that were present in the Middle Triassic continued to dominate the waters. In the Dolomia di Forni they include *Saurichthys*, *Sargodon*, *Birgeria*, *Eopholidophorus*, coelacanths, and the ubiquitous pycnodonts. However, the reptile faunas took on a much different flavor. Overhead, some of the earliest pterosaurs swooped down upon fish as they shoaled near the surface. Two genera, *Eudimorphodon* and *Preondactylus*, have been recorded from the Udine area. Like many modern seabirds, did the

Triassic pterosaurs form raucous nesting colonies on emergent landmasses? Were they principally confined to the air spaces over land feeding on insects, or did they spend much of their day over the shallow seas plucking fish from the surface waters? Such questions cannot be answered with complete conviction. *Eudimorphodon* (fig. 6.3) possessed multicusped posterior teeth and procumbent, acutely conical anterior teeth. Judging from the gut contents of one specimen, *Eudimorphodon* was at least partly piscivorous. Other taxa may have been insectivorous. *Preondactylus* has simple conical teeth like *Dorygnathus* and was probably insectivorous. Despite the general paucity of Triassic insect assemblages worldwide, a glimpse of the insects that may have inhabited the adjacent islands and landmasses can be gained from the Argillite di Riva di Solto Shales. These are discussed a little later on.

One rather unusual tetrapod from the Udine area is a prolacertiform called *Langobardisaurus* (Renesto 1994). It was apparently an aquatic form, and it shows a number of similarities to the aquatic prolacertiform *Tanytrachelos* from the Carnian of the United States (see chapter 10). The pronounced elongation of the first phalanx on the fifth digit of the hindfoot is a unique character that *Langobardisaurus* shares with *Tanytrachelos* and *Tanystropheus*. It has been postulated that both these animals may have possessed webbed feet. However, the simple conical teeth of *Tanytrachelos* are consistent with an insectivorous diet and are unlike the differentiated teeth of *Langobardisaurus*. That animal possessed a most unusual dentition comprising acutely conical teeth at the front of the jaws, followed by a series of multicusped teeth. These in turn were followed by a single crushing molariform-like tooth at the posterior end of each jaw quadrant. Perhaps *Langobardisaurus* fed on crustaceans, or even small mollusks.

Without question, the most notorious and controversial tetrapod from Udine is *Megalancosaurus*. It was first described on the basis of a single tantalizing specimen that preserved the lightly built skull, but only the anterior portion of the postcranial skeleton, including long slender forelimbs. The skull preserved only scant details of the articulations between the individual cranial elements. At the time of its discovery and description, Wild (1978) speculated on possible affinities of *Megalancosaurus* with pterosaurs. Since then, ad-

Figure 6.3. Specimen of *Eudimorphodon.*

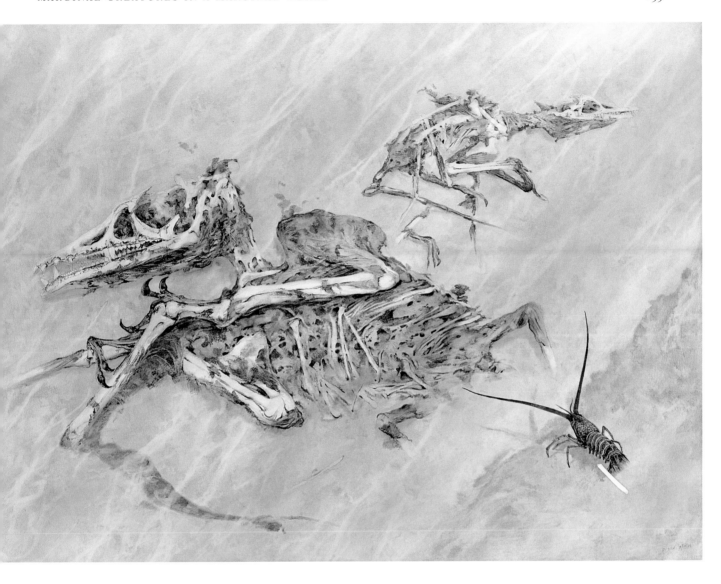

ditional specimens have been found that together provide a very different perspective on this animal. We now consider it to be a close relative of another strange creature that is known from the neighboring Zorzino Limestone. I shall discuss both these forms in more detail below, but first let's look at the fauna of the Zorzino Limestone in more general terms.

Plate 6.4. The decaying carcasses of two pterosaurs, *Preondactylus* and *Eudimorphodon* lie in shallow water near a Tethyan beach.

The Zorzino Limestone (Norian, Late Triassic) of the Bergamo Region

The vertebrate component of the Calcare di Zorzino fauna consists mostly of fishes, but there are also more than 10 genera of tetrapods. In addition, a wonderful invertebrate fauna is known. As with the Forni Dolostone fauna, several discrete localities have been particularly productive. The invertebrates include echinoderms (echinoids, asteroids, ophiuroids, and crinoids), corals, brachiopods, mollusks (bivalves and gastropods), and crustaceans. Because the seafloor was apparently anoxic at the time of deposition, these benthic invertebrate fossils must be considered to be allochthonous—that is, the animals lived in a different environment from the one in which they became buried and fossilized. Reptiles, and possibly also the fishes, likely swam above the anoxic bottom waters, and in their case, the fossils might be autochthonous—that is, spending at least part of their time in the environment in which they were ultimately buried.

Marine reptiles found here include the thalattosaur *Endennasaurus* and the placodont *Psephoderma*. The placodonts were superficially like turtles, with a rigid carapace-like structure protecting the body. They typically possess large molariform cheek teeth that seem to have been perfectly adapted for crushing mollusk shells. *Psephoderma* was no exception in this respect. The largest specimen is nearly 2 meters long, but isolated teeth indicate that some individuals attained much greater size. The flat and broad head terminates in a relatively narrow rostrum. Renesto (1995) postulated that *Psephoderma* frequented the basin margins. Here it would poke around the seafloor and use its rostrum to pull up byssate mollusks, including *Isognomon* and *Modiolus*. The distal enlargement of the femur and humerus and poorly ossified tarsal elements are indicative of a fully marine lifestyle. In addition to *Psephoderma*, there were a number of durophagous fish that also fed on mollusks.

Endennasaurus was approximately 1 meter long with an elongate skull, but its jaws were toothless. The limbs seem to be reasonably well developed for walking on land but not adapted for providing a power stroke in water. Consequently, the long, laterally compressed tail would have provided the main propulsive force in swimming (fig. 6.4). Periodically *Endennasaurus* came out onto land, perhaps during the breeding season to lay eggs. *Endennasaurus* has gastralia that are particularly robust, and these would have acted as a form of ballast, allowing the animal to freely swim

Figure 6.4. Thalattosaur *Endennasaurus*.

down to the bottom waters. It is likely that it fed on soft benthic organisms, probably using its long snout to stir up the mud and silt on the bottom, and then quickly snapping at any of the benthic animals that it disturbed in this way. The best-known thalattosaur, *Askeptosaurus*, also originates from this part of the world, but from the Middle Triassic Genzbitumenzone (see chapter 4).

When first discovered, an isolated skull of a phytosaur (with affinities to *Mystriosuchus*) was considerd to be a freshwater form because phytosaurs are typically associated with continental sediments. However, Buffetaut (1993) noted that the Austrian Dachsteinkalk—a shallow marine deposit—was known to contain *Mystriosuchus*. More recently, Renesto and Lombardo (1999) described a second phytosaur specimen from the Calacare di Zorzino that they again compared with *Mystriosuchus*. The second specimen also preserved a significant proportion of the postcranial skeleton, and the tail is suggestive of enhanced swimming capabilities. It is exceptionally long and laterally compressed, and the hemal arches are expanded distally. In addition, the limbs are relatively short. Combined, these features are indicative of axial swimming in the sense used by Braun and Reif (1985). It seems highly probable that this voracious predator frequently strayed into coastal marine waters. So maybe it was the Triassic equivalent of the saltwater crocodile, *Crocodylus porosus*.

As in the Udine area, the Zorzino Limestone has yielded remains of *Eudimorphodon.* Although *Preondactylus* has not been recorded from the Bergamo region, a third genus, *Peteinosaurus,* has been described (Wild 1978, 1984).

A handful of specimens of *Langobardisaurus* are also known from the Zorzino Limestone. Although *Langobardisaurus* probably had an amphibious habit, there are certainly indications of some fully terrestrial members of the Zorzino limestone assemblage. These include a single specimen of a sphenodontian (with affinities to *Diphydontosaurus*) and isolated fragments of dorsal armor that are diagnostic of the stagonolepidid, *Aetosaurus.* That brings us back to the enigmatic *Megalancosaurus,* which has also been documented from the Bergamo region, and another apparently related form, *Drepanosaurus,* which to date has only been found in the vicinity of Bergamo. Although some would argue for an aquatic habit for these animals, an equally convincing case has been made for an arboreal lifestyle. But before we talk about their lifestyle, let's take a look at their anatomy.

Plate 6.5. The slender-snouted phytosaur *Mystrio-suchus* is mocked by flocks of pterosaurs (*Eudimorphodon* and *Peteinosaurus*) as it prowls along the beach.

The Drepanosaurs

Without doubt, *Drepanosaurus* (fig. 6.5) is one of the most unusual of Triassic vertebrates, *Sharovipteryx* and *Longisquama* not withstanding. Pinna (1980, 1984) gave a detailed account of this animal on the basis of a single articulated specimen, complete except for the

Figure 6.5. *Drepanosaurus.*
(A) Almost complete single
specimen showing its enormous claws and peculiar
forelimb with a large extra
bone in the region of the
"elbow." (B) Skeleton seen
from above, as restored by
Pinna (1980). He recognized
the huge claw on the end of
the second digit in the hand
but was conservative in his
reconstruction of the forelimb, presumably doubting
that any animal could have
such a peculiar extra element
between the humerus and
the ulna. The skull is completely unknown. (C) The
tail possesses a strange clawlike element at its end.

A

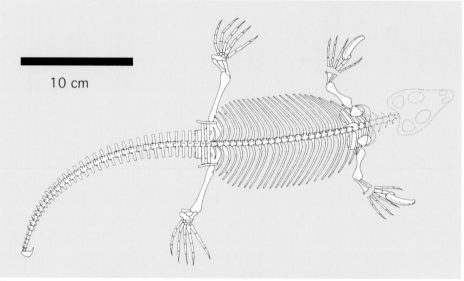

B

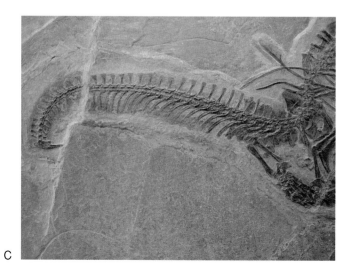

C

head and part of the neck. The tail possessed greatly elongated neural and hemal spines. Pinna interpreted *Drepanosaurus* as a lepidosaur and reconstructed it with a curious spike at the end of the tail and with the second digit of each hand bearing a massive ungual phalanx. However, the striking similarity of the spike on the end of the tail to a typical ungual phalanx certainly raised doubts concerning its original position, and for many at the time, it seemed more appropriate to consider it as a displaced phalanx. The size and form of the putative aberrant ungual phalanges on the second digits of the hands exhibit a remarkable resemblance to elements of the pectoral girdle (coracoids or sternal plates). In the holotype, the enlarged phalanx on the left side is also positioned in the proximity of the shoulder girdle. A more conservative view would be to regard these curious elements as part of the shoulder girdle.

That we have a tendency to shoehorn new forms into already known and accepted categories is well illustrated and documented (Gould 1989) by Walcott's original work on the Burgess Shale fossils. For example, Walcott placed into conventional taxa (e.g., Trilobita)

Plate 6.6. A small group of *Drepanosaurus* scurry between their burrows dug into mudbanks along the water's edge.

many of the bizarre arthropods for which more recent researchers, such as Whittington (1975) and Conway-Morris (1985), have erected completely new, higher-order taxa. There is an inherent reluctance to recognize novelties when we discover them. Thus with *Drepanosaurus* there has been an attempt to explain away some of the unique and bizarre features by reinterpreting the bone identities. However, that Pinna was at least partly correct in his original interpretations has been recently borne out by the discovery of exciting new specimens of the same taxon. These new specimens, together with comparison with some other equally unusual taxa, suggest that *Drepanosaurus* may have been even more aberrant than even Pinna had thought.

First, the new specimens have the spike positioned at the end of the completely articulated tail, thus countering any suggestion that it could be a displaced terminal phalanx. Second, a closer examination of the elements of the forelimb reveal what appears to be a peculiar arrangement of bones distal to the humerus. It is hard to get away from the fact that there are three distinctly separate bones between the humerus and the carpal elements. Two of these bones are relatively slender and typical of epipodial elements, but the third has a flat, platelike form that is uncharacteristic for an epipodial. Assuming that the bone is in natural articulation (and there is nothing to indicate that it is not), it would seem to be a neomorph.

Enter *Megalancosaurus* (fig 6.6). In many ways *Megalancosaurus* stands as an icon of the Upper Triassic: enigmatic, bizarre, controversial, yet all the while a wonderful inspiration to the imagination. *Megalancosaurus* was actually discovered a few years before *Drepanosaurus* and was originally described by Calzavara et al. (1981) on the basis of a single specimen from the Udine area that preserved the skull, neck, and part of the forelimb. However, because of the high degree of dolomitization, details of the cranial sutures could not be determined. Nevertheless, the rather lightly built and superficially birdlike form of the skull, coupled with the enlarged forelimb, particularly attracted the attention of those workers who remained unconvinced by an increasing body of evidence supporting the theory that birds arose from theropod dinosaurs.

For instance, Ruben (1998) suggested that *Megalancosaurus* might have been an effi-

Figure 6.6. Superficially bird-like head of *Megalanco-saurus* as preserved in the holotype.

cient glider and that a patagium (flight membrane) may well have been supported by the long, partially extendable forelimbs. Moreover, he considered such morphology as a plausible model for the ancestor of birds. But as Renesto (2000) pointed out, climbing abilities are typically a prerequisite for many gliding forms, and it is important not to jump to conclusions regarding gliding (or especially flying) abilities. The characters cited by Ruben may merely reflect adaptations to an arboreal lifestyle and do not by themselves indicate adaptations for flight. Unlike *Sharovipteryx*, there is certainly no evidence in the preserved specimens of *Megalancosaurus* for the presence of gliding membranes.

Since the first discovery, new specimens of *Megalancosaurus* have continued to show up on an irregular basis from both the Dolomia di Forni and the Zorzino Limestone. The additional specimens have greatly added to our picture of this animal, although at the same time they have also raised intriguing questions. Renesto (2000) did not deny an at least superficial similarity of the skull of *Megalancosaurus* to birds, but at the same time, he was quick to point out that no detailed characteristics of its osteology were in keeping with an avian association. In fact, the link to *Drepanosaurus* is almost undeniable.

The structure of the *Megalancosaurus* carpus may shed some light on the identity of the enigmatic elements described in the forelimb of *Drepanosaurus*. Although these large, rounded elements could be dismissed as part of the pectoral girdle and sternum, close examination of the specimens suggests that they do form part of the forelimb, and that they do indeed articulate with the proximal carpals and the ulna. Renesto (2000) describes an elongate ulnare and intermedium in *Megalancosaurus* that are aligned parallel to each other. It is thus conceivable that in *Drepanosaurus*, the extra elements represent the elongated ulnare and intermedium that have been much exaggerated. In accordance with this identification, the ulna could then be assumed to have developed into the bizarre rounded and flattened element. Presumably such a condition of the forelimb is correlated with the tremendous development of the ungual phalanx on digit II and its possible fossorial (digging) habit (although Renesto considered *Drepanosaurus* to be an arboreal animal that used the large claw for ripping open bark while in the pursuit of insects).

One of the interesting and unusual features of the *Megalancosaurus* skeleton is the occurrence of exceptionally tall and well-developed neural spines on the anterior dorsal vertebrae. They are expanded distally and fused with the adjacent neural spines so that in effect they form the equivalent of a notarium. (In pterosaurs, the first few trunk vertebrae fuse, thereby stiffening and anchoring the shoulder girdle.) Indeed, Renesto (2000) has suggested that it may have functioned in much the same manner as the notarium of pterosaurs.

The elongate neck vertebrae are very like those of early pterosaurs, including *Eudimorphodon*. Moreover, they are both closely comparable to isolated cervical vertebrae described from the fissure deposits at Cromhall Quarry (Fraser 1988a; Renesto and Fraser 2003). Renesto (2000) also noted the similarity of the cervical vertebrae of *Megalancosaurus* to those of *Protoavis* (see chapter 7 for more details of this controversial taxon).

A survey of all the available *Megalancosaurus* specimens indicates that there is some variation in the arrangements and form of the digits on the foot. In one, the ungual phalanx of the first toe takes on the form of a small, rounded element with a blunt distal end and not the usual claw. Renesto (2000) suggested that these differences might very well reflect the presence of two species of *Megalancosaurus* within the Italian deposits. However, it is also equally likely that they could merely represent sexual dimorphism. Unfortunately, the foot is not preserved in the holotype.

Together, the specimens of *Megalancosaurus* present an intriguing picture of small reptiles hanging from trees something in the manner of a pangolin. Grasping hind feet and a prehensile tail would free the head and forelimbs for the business of catching supper. Certainly the grasping digits are reminiscent of those of a chameleon. Likewise, judging from the form and preserved coil of the tail, it does not seem unreasonable to suggest that *Megalancosaurus* had a prehensile tail. Perhaps, like the chameleon, *Megalanco-*

saurus was able to dangle from branches unsupported by anything but its tail. This is just one interpretation, and more detailed study of these and other new specimens could just as easily suggest a quite different habit as provide further support for an arboreal one.

The Argillite di Riva di Solto

The slightly younger Argillite di Riva di Solto shales are generally less fossiliferous than the Zorzino Limestone. The fossiliferous rock is bituminous shale and contains fish such as the nektonic pholidopleurids, and predacious forms like *Saurichthys* and the aptly named flying fish *Thoracopterus magnificus.* As previously mentioned, the former genus had a worldwide distribution and ranged from the Lower Triassic through the Lower Jurassic. They were slender fish characterized by a long snout bristling with teeth. Undoubtedly they would have caused chaos as they darted into the midst of shoals of fish such as the ubiquitous pholidophorids.

Thoracopterus, with its greatly enlarged pectoral fins, was also a fast and elusive swimmer that could flit over the water. Although a predator itself, it seems likely that the flying habit was an escape response initiated as a result of attack from much bigger predators. The modern flying fish, *Exocoetus*, is capable of being airborne for as long as 20 to 30 seconds and traveling up to 150 meters. Interestingly, *Exocoetus* and its kin, once out of the water, are prone to predation by seabirds: it's a bit like flying out of the frying pan and into the fire! Whether juvenile *Thoracopterus* faced a similar dilemma from pterosaurs circling overhead is difficult to say, but remains of *Eudimorphodon* are known from the Argillite di Riva di Solto.

Of the invertebrate remains, crustaceans are the most abundant, but lamellibranch and gastropod mollusks are not uncommon, with occasional isopods and echinoderms, and a single annelid. However, the insect fossils make the Argillite di Riva di Solto unique. So far, very few taxa have been described, but the collections in the Bergamo museum comprise a variety of forms, including dipterans, beetles, and dragonflies. Whalley (1986) described a well-preserved odonate as *Italophlebia gervasuttii.* Five specimens of this dragonfly have so far been collected, on the basis of which Whalley concluded that *Italophlebi* belonged in a new family. The largest specimen has a wingspan of approximately 60 millimeters. Preserved on the second segment of the abdomen are structures that have been interpreted as part of the male copulatory apparatus. Modern mating dragonflies adopt a characteristic ring or wheel posture, which the males assume to transfer sperm to the female. The presence of this special copulatory structure in the males of *Italophlebia* is strong evidence that this unique mating method in odonates goes back at least to the Triassic.

On the same block as the holotype of *Italophlebia* is a single elytron of a coleopteran. As mentioned later (see chapter 9), the elytra of beetles often have distinctive patterns, but unfortunately, this does not automatically permit definitive referral to a particular family. The distribution of elytron sculpture patterns among the different families of Coleoptera has not been well studied, and it will always remain difficult or impossible to assign isolated elytra to a particular family. The legs and reproductive structures are required in order to make an accurate identification. Nevertheless, by examining the variation in sculpturing patterns, it is sometimes possible to get a picture of beetle diversity levels at a given insect locality. What is so interesting about the Argillite di Riva di Solto elytron is the distinctive pattern of parallel lines running down its length. Similar patterns have been described in Upper Triassic and Lower Jurassic elytra from Britain that have been referred to the genus *Holcoptera.* Of course, we have not been able to satisfactorily refer this genus to a known family of beetles—it is simply a form genus, but one that is now known from Italy, Britain, and the United States (Virginia Solite Quarry). The most productive area for finding these insects is approximately 20 kilometers from Bergamo in the Imagna Valley, where active quarrying is still going on. It seems likely that in future the importance of these deposits will only grow. The presence of complete insects, coupled with the occasional terrestrial vertebrate skeleton, indicates that the Argillite di Riva di Solto was deposited relatively close to a body of land.

The Italian assemblages represent a mixture of paleoenvironments, including terrestrial situations, marine reefs, and marine basins, and there is the added challenge of determining the habitat of the different constituents of the assemblage. Was *Drepanosaurus* definitively an arboreal form, or could it have been an aquatic animal, as others (Berman and Reisz 1992; Colbert and Olsen 2001) have suggested for related forms from North America (see chapters 7 and 10)?

With the exception of palynomorphs (spores and pollen), there is no evidence of the vegetation cover that grew on the exposed landmasses. Nevertheless, it is likely that it was similar to other neighboring areas of Europe and was probably dominated by conifers and cycads. Despite the proximity of land, remains of terrestrial vertebrates are sparse, and those that do occur are typically arboreal forms that exhibit some unique features. Furthermore, with the exception of a single sphenodontian specimen, there are no traces of modern terrestrial groups. But this is not the case with many other Late Triassic vertebrate assemblages.

chapter seven

The Enigmas of Texas and the U.S. Southwest

On the western side of North America, the great swathes of the Chinle and Dockum Groups document a period similar to that of the Newark Supergroup in the east. Although the Newark Supergroup has been considered rather unfossiliferous (which, as we shall see later, is something of a fallacy), the Chinle conjures up images of vast forests and abundant vertebrate fossils. Perhaps because of the great exposures in the American Southwest, the Chinle was extensively excavated for the better part of the twentieth century, and it has revealed a wealth of vertebrate and plant fossils that provide a vivid picture of life in the area from the Carnian through the end of the Norian. By contrast, the eastern part of North America is a tangled mass of vegetation and concrete—not an area likely to inspire an "Indiana Jones" to go exploring. It is probably this rather daunting landscape rather than any intrinsic lack of fossils that has been in large part responsible for the notion that the east has few Triassic fossils.

As a result of the apparently unending exposures of Triassic sediments in the Southwest, it is possible to find a treasure trove of fossils almost anywhere, and indeed that has been the experience of generations of fossil hunters. The discoveries are endless: Holbrook in Arizona to Moab and Grand Staircase-Escalante National Monument in Utah; Lubbock, Texas, to Tucumcari, New Mexico. Clearly in a book of this sort it is impossible to even begin to provide adequate coverage of all the fossiliferous localities in the Chinle and Dockum. For more details, the reader is referred to Long and Murry (1995). Here, in order to get a flavor of the life that teemed in the forests and valleys of this corner of the world during Triassic times, we shall look at a handful of specific areas.

The whole Carnian-Norian interlude is not represented at any one particular locality. Some of the oldest fossiliferous sequences of the Chinle group crop out in Howard County, Texas, near what is now the ghost town of Otis Chalk. A rich and diverse tetrapod fauna originates from this region that includes the phytosaurs *Paleorhinus* and *Angistorhinus*, the aetosaur *Longosuchus*, the rhynchosaur *Otischalkia*, and the temnospondyls *Latiscopus*, *Buettneria*, and *Apachesaurus* together with a variety of other animals such as the strange archosaur *Doswellia*, the poposaurid *Poposaurus*, and the protorosaur *Malerisaurus* (Chatterjee 1981). Spencer Lucas and coworkers (e.g., Lucas and Hunt 1993; Lucas et al. 1993; Lucas 1997) have recognized that *Paleorhinus*, *Angistorhinus*, and *Longisuchus* are apparently common elsewhere, but are also restricted in time and therefore useful in recognizing early late Carnian-age sediments.

Perhaps even more importantly, small remnants of the skeletons of these forms can be

Plate 7.1. The early morning mist rises from a swamp deep in the *Araucarioxylon* forest where a large group of metoposaurs congregate, mouths open to breathe the fetid air. Surrounding the pool are stands of horsetails.

distinctive and therefore readily identifiable. For instance, in *Longosuchus*, the lateral dermal plates that protected the body are developed as prominent hornlike structures. *Paleorhinus* had a long and narrow garial-like snout, and the nostrils (external nares) are situated relatively far forward in front of the eye socket (orbits), and even in front of the antorbital fenestrae. Furthermore, the skull is dorsoventrally flattened, and in life, the eyes would have pointed upward to a great extent. *Angistorhinus*, by contrast, had a deeper skull with the nostrils positioned slightly further back on the rostrum and level with the antorbital fenestra.

Some 1000 kilometers to the north of Otis Chalk, in Fremont County, Wyoming, the Popo Agie Formation yields similar animals to the Otis Chalk locality, including *Buettneria*, *Paleorhinus*, and *Angistorhinus*. Even further afield (over 9000 kilometers today, but more like 4500 kilometers in Triassic times), the Schilfsandstein and Kieselsandstein in Germany also yield *Paleorhinus* and *Metoposaurus*, thereby directly linking them with the Otis Chalk localities. Even further away, in India, the lower Maleri fauna has also yielded *Paleorhinus* and another species of *Malerisaurus* (Chatterjee 1974). Lucas and Hunt (1993) proposed the name Otischalkian to describe this particular unit of time, so even some of the most obscure places on the globe have achieved international recognition because of their fossils.

Petrified Forest National Park

Today, Petrified Forest National Park is an awe-inspiring but bleak and arid monument to the past. During Triassic times it was an even more impressive "National Park," one that teemed with life in a hot and wet environment. Situated just to the north of the equator and at the same latitude as the more southern parts of the Newark basins, it might be imagined that it was an area equivalent to the tropical rain forests of today. Certainly the huge trunks of the conifer *Araucarioxylon* that are common in parts of the Chinle are ample evidence of significant vegetation, but whether they formed dense forests or were distributed as isolated stands of trees along riverbanks is unclear. At the same time, it must be remembered that the deposits of the Chinle represent a few million years of sedimentation, and that at different times through that period, vegetation cover may well have varied significantly. Likewise, as mentioned in chapter 5, the "Chinle fauna" does not represent a single community of animals living in close proximity at the same time. Instead, a fairly extensive duration of time is represented between the bottom and the top of the formation, which is divided into five members. The basal Shinarump Member of the Petrified Formation is typically regarded as upper Carnian in age (roughly equivalent to the Otischalkian), whereas the upper Owl Rock Member is middle Norian (Alaunian, approximately equivalent to the Revueltian land vertebrate faunachrons [LVF]). In between the Shinarump and the Owl Rock members, the Sonsela Sandstone is sandwiched between the lower and upper Petrified Forest members.

At the park itself, the vertebrate finds can be broadly divided into three distinct assemblages. The oldest sediments can be found at the Teepees area of the Park. The Blue Mesa area has produced some of the more important lower Chinle vertebrates (Adamanian equivalents), whereas up to the north in the Painted Desert region, somewhat younger sediments have yielded a plethora of upper Chinle vertebrates, including a variety of dinosaur taxa. The Chinle also covers an area extending from northeast Arizona into southeast Utah and New Mexico. Over such a large area there are naturally various facies changes, and correlating the different units is not without its difficulties and, inevitably, its controversies.

That is not to say that the Chinle cannot be described in broad terms. Typically the environment is described as one of floodplains and braided streams, and the macrovertebrate tetrapod fauna is dominated by phytosaurs and metoposaurs—forms that are usually regarded as semiaquatic. On the other hand, there are significant departures from the norm. The famed *Placerias* Quarry (Adamanian) is one such instance. Microvertebrates from this quarry support a depositional setting consisting of a quiet pond or marshland. Surprisingly,

though, the macrovertebrates found here are a little more terrestrially oriented than the phytosaurs and metoposaurs that are common elsewhere. In addition to *Placerias* itself, aetosaurs and poposaurs are also common at the quarry, and phytosaur remains are unusually rare. Kaye and Padian (1994, 193) postulated that the "large, hippopotamus-like *Placerias* may have habitually wallowed in the marsh-like muds of the site."

When depicted in books, the Mesozoic is frequently shown with volcanoes steaming in the background. The volcanic ashes of the Chinle are clearly evidence of volcanic activity at certain times, but such activity was probably not the norm. Today's active volcanoes only rarely erupt, and a much more typical scene would have been quiet and dormant volcanoes that from a distance would have looked little different from many of the volcanic hills and mountains that we see today to the west of Albuquerque. That is not to say that the eruption of such volcanoes was not instrumental in the formation of rich fossiliferous deposits.

The Teepees

At the Teepees, the mudstones and lenses of sandstone are approximately 225 million years old. The sediments are thought to represent an extensive, muddy tropical floodplain. Numerous small rivers and streams meandered across this vast plain, heading to the north-

Plate 7.2. A group of the dicynodont *Placerias* plod dolefully through conifer and seed fern woodland.

west and a sea whose coastline was somewhere in the vicinity of today's Utah-Nevada border. The large herbivorous dicynodont *Placerias* was one of the common animals browsing on the luxuriant vegetation that undoubtedly surrounded the riverbanks.

The Blue Mesa

Looking today at the dry, harsh landscape of Petrified Forest National Park, it is hard to imagine that it once teemed with amphibians. But as we have seen, not only was the Triassic world a very different place from that of the present day, but many of the amphibians were also nonconformists. What we typically understand by the term *amphibian* is an animal reliant on water to breathe and reproduce, but such creatures do not form a natural assemblage of vertebrates. Many of those animals that relied on water for reproduction in the Triassic, like many of their Paleozoic predecessors, represent completely separate lineages from the modern frogs and salamanders. Some of them even seem to have had only the most tenuous of links to water, being perfectly adapted to the terrestrial realm. The large metoposaur *Buettneria* is particularly well represented in the lower Chinle deposits of the Petrified Forest. The eyes of these animals were positioned on top of their broad, flat heads, and it is probable that they congregated in swamps, lurking just beneath the surface of the water or resting on half-submerged logs, perhaps listening for the tread of a careless procolophonid or juvenile aetosaur coming down to drink at the water's edge. Around such swamps, dense stands of horsetails provided additional cover. Beyond them, *Araucarioxylon* trees soared into the humid, misty air. Moisture dripped from mosses that coated the trunks and rotting logs, and the dank air was filled with the eerie chirps of myriad insects.

If the region in which Chinle deposition took place was indeed subjected to marked seasonality as a result of the exaggerated monsoon, then the dry season would have inflicted a heavy toll on life in the swamps. In the same way that watering holes in the savannas of Africa periodically dry up and wreak havoc on the living communities, so too would competition for waning resources have intensified during the Triassic droughts. Drying pools would bring about increased intraspecific conflicts. Perhaps the bite marks that are sometimes found on the snouts of large phytosaurs resulted from battles over depleting supplies of food and water. During particularly harsh seasons, finally deprived of water, even the strongest individuals would have succumbed. When the rains returned, torrents of water swept ash and debris over their mass graveyards, where the skeletons became buried and ultimately fossilized.

However, it is not certain that such marked seasonality was prevalent in Chinle times. As we have seen, the *Araucarioxylon* trunks lack true growth rings. If the region remained humid year round, then we have to invoke other causes for the mass death assemblages—perhaps increased levels of water toxicity. Ash (1986b) noted that the common occurrence of water-loving plants such as horsetails, lycopods, and ferns is highly suggestive of humid conditions, although we don't know whether such plants were restricted to the areas immediately adjacent to the watercourses. Frequently cycads and ginkgoes are found together with these hydrophilic plants. They probably grew nearby, but in slightly higher and drier habitats, and still not far removed from the watercourses. It is interesting to note that remains of more xeric plants, such as conifers, often form part of these floral assemblages. But such fossils are rather fragmentary, and Ash (1986a) suggested that in all probability, they were transported some distance from drier upland regions. Significantly, the cuticle (the waxy coating that restricts water loss from the outer surface of plant parts) of such fossils shows no adaptations for life in an arid climate. The stomata are not sunken or protected, there are no stomatal plugs, and the cuticle is not especially thick. It therefore seems highly likely that even in those areas somewhat removed from the rivers and streams, the climate was still rather humid and encouraged fairly extensive vegetation cover.

Demko et al. (1998) suggested that the ferns found in Chinle sediments were restricted entirely to the margins of watercourses and lakes in the basin, and that the surrounding

Plate 7.3. Deep in the Chinle forests, metoposaurs share the ponds and pools with a variety of other residents, including freshwater sharks (*Lissodus*), coelacanths (*Chinlea*), and phytosaurs.

Plate 7.4. A large phytosaur (*Rutiodon*) scatters a small group of the sphenosuchian crocodile *Hesperosuchus*.

land was quite arid and rather sparsely vegetated. On the other hand, some authors (Ash and Creber 1992; Ash 2001) have suggested that the moisture regimes along modern rivers are not so very different from those of the general regions through which they flow. Thus where rivers such as the Rio Grande and Colorado flow through semidesert regions, hydrophilic plants like ferns do not survive. Nor do very humid climates exist in narrow corridors around rivers and lakes in an otherwise arid or markedly seasonal environment, as proposed by Demko et al. (1998). Even so, where rivers, like the Nile, flow through desert regions, the banks of the river are well vegetated.

The most common vertebrate remains in the Chinle belong to phytosaurs. Most phytosaurs likely had a very similar way of life to crocodiles, which they resembled. They had long, slender snouts, and their jaws were lined with an array of acutely conical teeth. Rows of bony scutes covered the back of their bodies, a characteristic of other pseudosuchians. Although such protective measures might seem unwarranted in these fearsome predators, deep puncture holes that have been found in some bones, including snouts, testify to probable attacks from members of their own kind.

Other carnivorous archosaurs with these parallel rows of scutes extending the length of the back include the sphenosuchian crocodiles. Hans Sues (personal communication) has suggested that the scutes acted as a rigid support to the back during locomotion. There is no doubt that sphenosuchians such as *Hesperosuchus* and *Terrestrisuchus* were agile tetrapods. Any additional support of the backbone would help to alleviate extra stress imposed on it as a result of the active lifestyle. It is equally plausible that phytosaurs frequently adopted the same "high walk" that living crocodiles use when they move quickly. In this case, the dorsal scutes of phytosaurs would have provided much-needed extra support to the backbone.

The slender-snouted phytosaur *Rutiodon* is the dominant form in the Petrified Forest, and as Lucas (1998) and coworkers (principally Adrian Hunt) have repeatedly suggested, this form would appear to be restricted to the latest Carnian and is potentially an excellent biostratigraphic indicator (Adamanian LVF). More recently, however, the biostratigraphic use of *Rutiodon* has been questioned. Hungerbühler and Sues (2001) referred a phytosaur specimen from the lower Dockum to *Rutiodon carolinensis*. They also noted that *Rutiodon carolinensis* occurred in both the eastern (Newark Supergroup) and southwestern United States and concluded that the animal is synonymous with *Angistorhinus*. Certainly *Angistorhinus* and *Rutiodon* are sufficiently alike for them to be grouped in the same family. However, whereas *Angistorhinus* is considered to be one of the more primitive groups of phytosaurs, *Rutiodon* has a depressed supratemporal fenestra, which is considered to be a feature of more derived phytosaurs.

Hunt (1989) broadly divided the shape of phytosaur skulls into three main feeding types that he considered to represent ecological vicars of three morphologies of modern crocodiles: piscivorous (gavials), generalist feeders (caiman), and predatory (Nile crocodile). The piscivores are characterized by an elongate narrow snout bearing a homodont dentition consisting of cylindrical nonserrate teeth. The generalists also have rather slender snouts, but the rostrum is more robust than that of the piscivores and usually surmounted by a crest. The dentition is more heterodont and includes slicing, bladelike teeth. Finally, the predatory type is characterized by massive skulls that are proportionally short with a heterdont dentition. Hunt showed that all three forms of phytosaur co-occurred within different stratigraphic sequences. Thus in the lower Dockum the long-snouted *Paleorhinus* is recorded with the generalist *Angistorhinus* (but see above and the suggested synonymy with *Rutiodon*) and the heavy-set *Brachysuchus*. In the lower Chinle, Hunt identified three different species of *Rutiodon*, *R. zunni*, *R. adamensis*, and *R. gregorii* as the piscivore, generalist, and predator, respectively. In the upper Chinle, he considered the long-snouted *Pseudopalatus pristinus* as the piscivore, *Belodon buceros* as the generalist, and an unnamed taxon from Revuelto Creek, New Mexico, as the predator. This latter from was initially collected by J. T. Gregory and referred to *Rutiodon gregorii* by most authors, presumably because of the massive proportions of the skull. However, Hunt noted that the temporal region was more akin to *Pseudopalatus*.

Sexual dimorphism has also been reported in at least one phytosaur, *Pseudopalatus*. A fossil assemblage in the Canjilon Quarry, near Ghost Ranch, high in the Chinle succession, has produced a number of phytosaur skulls. Charles Camp collected 11 skulls from the site in the years 1928 and 1930. A detailed examination of some of these specimens (Zeigler et al. 2001) has revealed two distinct morphotypes. In one, the septomaxillae, which define the shape of the rostral crest, are short and produce an abrupt, volcano-like narial crest. In the second form, the septomaxillae are much longer, resulting in a longer and taller crest. It seems more plausible to accept that these are variants of the same species, and that

Figure 7.1. Skeletal restoration of the rauisuchian *Postosuchus* (after Long and Murry 1995).

the morph with the longer crest may be representative of the male. In this scenario, the more prominent crest could be regarded as a display feature. It is even conceivable that it was brightly colored in life.

Another characteristic carnivore of this time period is the rauisuchian *Postosuchus kirkpatricki* (fig. 7.1). This large predator was first described by Sankar

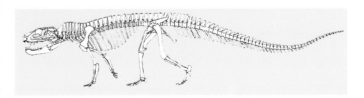

Plate 7.5. (A) *Postosuchus* is at first bewildered by the array of spines and bony plates that protect the aetosaur *Desmatosuchus*. (B) However, after some worrying, the *Desmatosuchus* finally makes a wrong move and is then tossed into the air by its powerful adversary.

A

Chatterjee from the Dockum of Garza County, Texas (see below), but its remains are also known in the Petrified Forest. Throughout lower Chinle times, *Postosuchus* was the undoubted top carnivore on land. Chatterjee (1985), when he first described *Postosuchus*, considered it to be an early relative of the carnosaurs, including *Tyrannosaurus rex*. Although this view is no longer tenable, it does speak to the overall nature and appearance of *Postosuchus* as a formidable Triassic carnivore. Whether it actively pursued prey is an-

B

other matter. It may have been an ambush predator, lurking in the wooded areas near the riverbanks, and, like all large carnivores, it was almost certainly something of a bully, scavenging on the hard-earned kills of other, smaller predators.

An equally impressive rauisuchian is *Chatterjeea* (Long and Murry 1995). At first glance, many features of the postcranial skeleton would suggest that *Chatterjeea* is actually a dinosaur, but a closer look reveals subtle differences, and, most importantly, it does have a pronounced crocodile-normal type ankle joint.

With all these carnivores around, the herbivores needed to protect themselves in some way. The aetosaurs were not only large animals with stout claws, but they were also heavily protected by their dermal armor. In fact, they bore some resemblance to enormous armadillos. At least three different species of aetosaur have been described from the Blue Mesa Member of the Petrified Forest. These are *Desmatosuchus haplocercus*,

Plate 7.6. Chancing across a dead *Typothorax*, a large *Postosuchus* tries to ward off a competitor by baring its teeth and hissing a warning. Such scenes would have been just as common an occurrence among Triassic predators as they are today for big cats in the savanna grasslands of Africa. Although *Postosuchus* undoubtedly harried down prey from time to time, scavenging a carcass would have been even more acceptable. Any opportunity for stealing the hard-fought prize of another would be welcomed.

Stagonolepis wellesi, and *Paratypothorax andressi*. *Paratypothorax* was a particularly wide-bodied animal with greatly elongated dermal scutes covering its back. When viewed from above, it had a particularly dumpy appearance. *Stagonolepis* and *Desmatosuchus* were of a much narrower build, but they were still protected by bony armor covering the entire back. The armor of *Desmatosuchus* was further enhanced by large spikes projecting out from the side of the neck (Small 1985). Although these small-headed animals may have appeared benign, they would certainly have had the ability to give an enemy a most unappetizing mouthful!

During the deposition of the Sonsela Sandstone (about 220 million years ago), the Petrified Forest region was crisscrossed by large rivers with gravel beds. Stands of the large *Araucarioxylon* trees towered along the riverbanks, and forest and woods probably extended up into the higher elevations. The aetosaurs, such as *Desmatosuchus*, foraged among the leaf litter on the forest floor.

The Painted Desert

About 215 million years ago, Petrified Forest National Park reverted once again to a vast muddy floodplain, but this time, the rivers flowing across it were much larger. Enormous phytosaurs, some reaching 7 meters long, lurked along the banks.

Chindesaurus bryansmalli is a herrerasaurid dinosaur that is largely known on the basis of a single incomplete skeleton collected from the north end of the park. Despite the lack of any cranial material, there seems little doubt that it was indeed a relative of the Argentinian *Herrerasaurus* (chapter 9), and currently it is regarded as the youngest member of the Herrerasauridae.

As Jacobs and Murry (1980) state, the Chinle is without doubt "one of the major sources of Late Triassic vertebrate fossils in the western hemisphere." Abundant phytosaurs and metoposaurs are typical of the faunal assemblages, but in certain areas, there seem to be rich pockets of diverse microvertebrate assemblages. Sometimes these microvertebrate assemblages are also found together with the larger forms. Two of these rich pockets are situated near the town of St. Johns, Arizona, just to the southeast of Petrified Forest National Park. These are the famed *Placerias* and Downs Quarries.

Plate 7.7. From high in the branches of a *Ctenophyllum* the spiny outlines of two *Longosuchus* (aetosaurs) stand out among a small clump of the fern *Phlebopteris*.

©DOUG HENDERSON 6/20/93

The Placerias Quarry

The *Placerias* Quarry (Adamanian LVF) was originally worked by C. L. Camp and crews from the University of California between 1930 and 1934. More than 1600 disarticulated bones of the mammal-like reptile *Placerias* were recovered from the site during that time, together with bones of a number of other large reptiles. These include abundant scutes of *Typothorax* and *Desmatosuchus*. The phytosaurs and metoposaurs that are common elsewhere in the Chinle were not well represented. Camp and Welles (1956) also noted that no *Ceratodus* teeth or *Unio* shells were found at the site, despite being common elsewhere in the vicinity. These authors considered heavy concentrations of lime to be indicative of a pool, marsh, or spring bog depositional environment. Concentrations of small masses of carbonized vegetation were frequent. Bones were sometimes encased in crystalline gypsum. They attributed patches of free sulfur and abundant gypsum to

Plate 7.9. The heavily tusked *Placerias* are no less a formidable opponent.

Plate 7.8. *(opposite page)* For a fleeting moment the moonlight betrays the presence of two herrerasaurids as they furtively scuttle among the forest's undergrowth.

Plate 7.10. Two small tanystropheids explore a sunken
tree trunk. No complete
skeletons of tanystropheids
have been found in the
Chinle, but fragmentary remains including elongate
neck vertebrae hint at the
presence of at least smaller
members of the same family.

chemosynthetic bacterial action in highly organic sediments. Camp and Welles noted that the upper surfaces of some bones had rotted while the undersurfaces were better preserved, where they had apparently been pressed into the mud.

The bones were scattered and formed a pavement on the irregular floor of the quarry. In addition, coprolites were numerous. This suggested to Camp and Welles (1956) that animals congregated around the pond or bog to feed on the vegetation and each other. It is easy to picture scenes not unlike those that take place around an east African watering hole today, where herbivores (antelope and gazelles) come to take the water and may be preyed upon by aquatic carnivores (crocodiles), and where semiaquatic herbivores (hippos) lie partially submerged, cooling in the water. Unfortunately, the analogy fails because of the lack of phytosaurs in the *Placerias* Quarry sediments.

The *Placerias* Quarry was again worked in the late 1970s, but this time by crews from the Museum of Northern Arizona, who concentrated their efforts on the microvertebrates through wet-screening of the sediments. Later in 1989 and 1990, crews from the University of California at Berkeley under the direction of Kevin Padian reopened the site and greatly supplemented the faunal list. This combined work has resulted in the identification of an extensive fish and tetrapod fauna. A number of small sharks have been found, including *Lissodus*, *Acrodus*, the ctenacanth shark *Phoebodus*, and *Xenacanthus*, the last of which is particularly common. Scales and teeth of redfieldiids such as *Cionichthys* are also common (Jacobs and Murry 1980). Palaeoniscids are represented by *Turseodus* and the pholidopleurid *Australosomus* can also be recognized from its distinctive fused vertebral elements. In addition, semionotids, lungfish, and coelacanths can be found. The latter appear to include *Moenkopia* and *Diplurus* (Kaye and Padian 1994).

Ghosts of New Mexico

No discussion of the Triassic would be complete without a mention of the famous Ghost Ranch Quarry from north central New Mexico (Apachean LVF). Here the dense assemblage of *Coelophysis* skeletons is bewildering. How such a massive assemblage of predators came to be buried together in the one place is a mystery. But equally mysterious is that the only other remains so far recovered from the Ghost Ranch also are almost entirely those of small predators. One is the beautiful skeleton of the sphenosuchian crocodile *Hesperosuchus* (Clark et al. 2001), and the other is a bizarre creature that has yet to be named. The one possible exception to the carnivore prevalence is the sparse remains of a drepanosaur (Harris and Downs 2002).

The skeleton of the new form is notable for its much-reduced limbs and the hundreds of small osteoderms that appear to have covered the entire body, giving it an appearance something akin to a pangolin. The tail is long and also covered with osteoderms. The neural spines are tall, and along the top of the tail, the osteoderms stand upright in a most unusual fashion. Small and Downs (2002) postulated that the animal was aquatic and used the deep tail to scull. The small limbs are consistent with this idea. The skull is equally remarkable in that it has a huge lower temporal fenestra, but no upper one. The large, laterally compressed, and serrate teeth (with a thecodont implantation) are consistent with an archosauriform, as is the presence of a laterosphenoid. However, in addition to lacking an upper temporal fenestra, the reptile's skull also lacks any antorbital and mandibular fenestrae. Quite simply, it is difficult to establish any relationships for the animal, and it is just one more example of the weird and wonderful tetrapods of the Late Triassic. But let's return to the initial discovery of the Ghost Ranch site.

In 1947 Ned Colbert led a team from the American Museum of Natural History on a reconnaissance trip to study the area around Ghost Ranch. The discovery of a complete phytosaur skull on the first day was ultimately to pale into insignificance compared with the incredible density of small theropod bones that awaited them. On the fourth day of

Plate 7.11. Chattering like small birds, a vast "flock" of *Coelophysis* run nervously into the forests.

the field season, having removed the phytosaur skull, the field party resumed their prospecting and discovered the first *Coelophysis* bones weathering out at the surface. After locating the horizon from which the bones were originating, it soon became apparent that this was part of one of the most remarkable fossil assemblages ever discovered. Not only was this a particularly dense accumulation, but there were also numerous articulated skeletons, and the bone was exceptionally well preserved. Because many of the skeletons were interlocked, the most effective way to collect the material was to expose the fossiliferous horizon of a large area, and then to arbitrarily divide it into a series of huge blocks. A field crew returned in 1948, and a considerable amount of additional material was recovered. In 1949, 1951, and 1953 Colbert returned to the area, but most of the work in these years centered on stratigraphic and sedimentary studies. In total, the American Museum

collected 13 blocks. In 1981 a joint field crew from the Carnegie Institute, Museum of Northern Arizona, and Yale Peabody Museum reopened the quarry and removed an additional 15 blocks. These 28 blocks are now distributed in a number of institutions in North America, where they are still being prepared.

   Coelophysis was first described by Cope on the basis of fragmentary bones collected by David Baldwin for Edward Drinker Cope in 1881. As we shall see, the precise location of the discovery of these bones has proved to be somewhat of a contentious issue. Cope originally considered the material to belong to the genus *Coelurus*, and he believed that the rather fragmentary material was a mixture of two different forms. He therefore assigned the material that Baldwin had brought him to two different species of this Late Jurassic theropod (Cope 1887a). However, he soon realized that they were actually different from

Plate 7.12. As the sun rises higher in the sky, some members of the group roll in the mud, while a small sphenosuchian passes by.

Plate 7.13. A lightning strike during a violent thunderstorm scatters the flock.

Coelurus, and so later that year, he transferred the material to the genus *Tanystropheus* while at the same time taking the opportunity to split the material even further in describing a third species, *Tanystropheus willistoni* (Cope 1887b). Triassic dinosaurs were suddenly becoming more numerous. However, in 1889, the material was once again transferred to a new genus, this time *Coelophysis*, the name it holds today, even if it has been somewhat tenuous at times.

Friedrich von Huene also examined this material and reassigned specimens to different species. At the same time, he noted that the principal difference between the three forms was size. In 1930 Hay designated *C. bauri* the type species of *Coelophysis*.

Colbert concluded that Cope's original material was all found not too far from the Ghost Ranch site, perhaps as little as 3 miles away. He also concluded that the original

material that was assigned by Cope to three separate taxa should all be assigned to a single species, *Coelophysis bauri*.

In recent years, there has been some serious question that the original *Coelophysis* material actually came from the same stratigraphic horizon as Ghost Ranch and that there are two different species. Indeed, Hunt and Lucas (1991) went so far as to erect a new generic name ("*Rioarribasaurus*") for the Ghost Ranch material. This upset the apple cart in the paleontological world for a time. However, partly because the name *Coelophysis* for the Ghost Ranch theropod was so ingrained in the literature, the name "*Rioarribasaurus*" was thrown out by the ICZN (International Code for Zoological Nomenclature). Perhaps just as importantly, at least one authority (Padian 1986) expressed concern that *Coelophysis bauri* was named on scrappy material that was never adequately diagnosed or compared with other taxa.

Baldwin collected at two localities in north-central New Mexico. One of these, the Arroyo Seco, is the locality for at least some of the dinosaur fossils that were given the names

Plate 7.14. As the shadows lengthen at the end of the day, two *Coelophysis* set out on one last hunting foray.

Coelophysis longicollis, *C. bauri*, and *C. willistoni* by Cope (1889). Much more recently, additional dinosaurian material was collected from a locality called Orphan Mesa that has since been shown to be the same as the Arroyo Seco locality. Significantly, the Orphan Mesa site is situated in the Petrified Forest Formation, while the Whitaker (Ghost Ranch) Quarry occurs in the stratigraphically higher Rock Point Formation. More significantly, the new material, together with some of Baldwin's original material, has been shown to be clearly different from the classic *Coelophysis* material from the Ghost Ranch Quarry. To underscore the point that the Ghost Ranch theropod should have been assigned a new name, Sullivan and Lucas (1999) chose the name *Eucoelophysis*, or "true *Coelophysis*," for the Orphan Mesa theropod.

Typically, the Ghost Ranch bones exhibit no sign of damage, and this, together with the fact that completely articulated skeletons predominate, points to rapid burial after transportation over very short distances. The most likely cause of death is mass drowning in floodwaters. But at least one more spectacular—and harrowing—demise has been proposed: that the animals were the victims of volcanic eruptions and that they were overcome by hot ash. However, this seems to be a most improbable scenario because silts adjacent to the quarry exhibit no evidence whatsoever of volcanic ash.

What would bring together such a large number of animals in life? Some kind of migration or breeding behavior would seem to be the most likely explanation. If the enormous congregation of *Coelophysis* was linked to some aspect of breeding, then it might be expected that there would be some more direct evidence of breeding activity, such as preserved egg shell fragments or even nests. Of course, that this is not the case does not automatically rule this theory out. It is even possible that *Coelophysis* gave birth to live young. In fact, the discovery of the remains of small individuals within the body cavities of two of the larger specimens tend to bolster this theory. However, it would seem that the bone is too well ossified to be representative of embryonic material, and a far more likely explanation is that the large adult individuals were indulging in cannibalism or perhaps feeding on carrion. Cannibalism is certainly practiced by certain reptiles today, including monitor lizards and crocodiles. Interestingly, in the Komodo dragon, intraspecific aggression seems to increase when small feeding aggregations are formed. And, of course, as one might expect, the bullying instinct results in most of the aggression being directed toward the smaller members of the group. In the Nile crocodile, Hugh Cott noted that individuals tended to segregate into age classes on the basking grounds. This is probably a direct result of their cannibalistic tendencies. That the *Coelophysis* assemblage represents a complete mixture of size ranges tends to speak against this type of situation, and more in favor of the feeding group.

One other feature of the *Coelophysis* mass graveyard is that the skeletons tend to fall into one of three clear size classes. A robust form is probably indicative of one of the sexes in the mature animal; a gracile morph is suggestive of the other sex; and the third grouping appears to represent juveniles. Interestingly, of a sample examined by Rinehart et al. (2002), the robust and gracile morphs each constituted 30 percent of the sample, and the remaining 40 percent appeared to be juveniles. Such a finding strongly supports the notion that these animals were sexually dimorphic.

Not far from Ghost Ranch in Abiquiu is another intriguing locality, the Snyder Quarry (Zeigler et al. 2003). Worked by crews from the New Mexico Museum of Natural History, this site yields elements of the usual Chinle phytosaurs, aetosaurs, dinosaurs, and metoposaurs. They are indicative of a mid-Norian age. More interesting is the taphonomic setting of the quarry. Zeigler (2003) showed that the bones accumulated as the result of a catastrophic event and were transported by water. The fact that bones are only occasionally articulated, but often associated, indicates that they were in a state of decay before transport and deposition. There are a high percentage of subadults represented in the material. The bones exhibit no evidence of scavenging. What is particularly striking is the presence of a significant amount of charcoal within

the deposit. All the evidence points to a wildfire raging through the area, overwhelming animals that could not escape. Soon thereafter, heavy rains produced flash floods that washed unimpeded through the burned-out forests before eventually dumping their load of sediment, charcoal, and bones.

Another high point of the New Mexico Triassic is a fossil called *Dolabrosaurus*. Berman and Reisz (1992) described *Dolabrosaurus* on the basis of portions of the vertebral column and partial fore- and hindlimbs of a single individual found in Rio Arriba County. The greatly elongated neural spines and hemal arches of the caudal vertebrae suggested to Berman and Reisz that the tail formed a sculling organ. They reasoned that this, coupled with a supposed paddlelike structure of the manus and pes, indicated an aquatic or amphibious habit. In fact, as Berman and Reisz pointed out, *Dolabrosaurus* shares a number of similarities to *Hypuronector*, a form described originally as "the deep-tailed swimmer" from the Upper Triassic Lockatong Formation of Connecticut and New Jersey (Colbert and Olsen 2001) (see chapter 11). *Hypuronector* (only about 15 centimeters long) occurs in lacustrine deposits and has exceptionally elongate neural arches and hemal spines, with the latter fused to the centrum. This certainly seems to be evidence of an aquatic habit. The pectoral girdle of the deep-tailed swimmer has the same curious structure, with a narrow rodlike scapula and a broad, platelike coracoid (Colbert and Olsen 2001). Both these animals are now regarded as members of the Drepanosauria, the group that was erected for those Italian mavericks *Drepanosaurus* and *Megalancosaurus*. Inter-

Plate 7.15. On one of their treks to the lake to drink, the group avoids a *Redondasaurus* basking near the shoreline.

Plate 7.16. Startling a small sphenodontian out of a crevice in the rocks, an adult *Coelophysis* prepares to strike at its prey.

estingly, Renesto and Paganoni (1995) cast some doubt on the assumption of an aquatic habit, preferring an arboreal way of life for the drepanosaurs as a whole. They pointed out that the presence of long preungual phalanges, long, sharp, and narrow claws, and slender limbs are more consistent with an arboreal lifestyle. More importantly, they felt that the extent of the caudal zygapophyses precluded much lateral movement of the tail, although this did permit free vertical movements.

In *Megalancosaurus*, there are also opposable digits on the manus and pes. In *Drepanosaurus*, the large ungual phalanx on the second digit of the hand could have been an equally effective digging implement in an aquatic or arboreal animal. Renesto (2000) described a change in the orientation of the hemal arches in the tail of *Megalancosaurus*, which he considered to be more consistent with prehensile, not swimming, ability. It may be that different members of the group occupied different niches.

The most recent drepanosaur finds come from the British fissure deposits (see chapter 10) (Renesto and Fraser 2003). It will be interesting to see whether future discoveries of drepanosaurs from the Chinle, Dockum, or elsewhere exhibit the same development of

Plate 7.17. In an effort to escape a predator, the "flock" swarms into a creek.

Plate 7.18. Underwater, they disturb a baby phytosaur and small fish.

bones in the forearm that are observed in the Italian drepanosaurs. The fact that many of these curious forms are only known by one or two specimens (e.g., "Vallesaurus"), and that the interpretation of certain atypical features is equivocal, open them to all kinds of speculation about their mode of life. Although at least one member has been implicated by some authors in the origins of birds, there is no substantive evidence for such hypotheses, and I would argue that this speculation merely reflects our current poor understanding of these intriguing tetrapods.

Denizens of the Dockum

Further to the east of the Ghost Ranch, the Late Triassic strata are referred to the Dockum Group. The sediments of the Dockum Group, which range in thickness from 70 to 700 meters, were probably deposited in a fluvial-deltaic-lacustrine basin. They comprise continental red bed sequences, and they stretch over vast areas of southeastern New Mexico and west Texas. But naturally enough, it is still not possible to follow the Chinle strata continuously across the continent, and so in Texas, we have a completely different strati-

Plate 7.19. Two
Trilophosaurus bask in a
clearing of cycads (*Zamites*)
and the ginkgo *Eoginkgoites*.

graphic nomenclature. Nevertheless, the Tecovas Formation is roughly time equivalent to the Shinarump Member and lower Petrified Forest Member of the Chinle; the Trujillo Formation is approximately equivalent to the Sonsela Sandstone and part of the upper Petrified Forest Member; and the Cooper Formation is the lateral equivalent of the up-permost part of the upper Petrified Forest Member.

Like the Chinle, the Dockum has a long history of research. Edward Drinker Cope was certainly no stranger to the Dockum. He described vertebrate fossils from this unit as early as 1893. But despite Cope's pioneering work, he did miss one or two exciting out-crops in his travels. One of these is situated near the settlement of Post, Garza County. It was discovered in 1980 by a field crew from Texas Tech under the guidance of Sankar Chatterjee and has turned into one of the jewels of the Dockum. Aquatic vertebrates dominate the assemblages and include freshwater sharks, coelacanths (*Chinlea*), dip-noans (*Ceratodus*), and redfieldiids, as well as aquatic, or at least semiaquatic, tetrapods such as a brachyopid, metoposaurs (*Apachesaurus*), and phytosaurs. More terrestrial ani-mals include the sphenodontian *Clevosaurus*, a possible sphenosuchian, *Placerias*, the or-nithischian *Technosaurus*, and some possible mammal remains. In addition, the range of herbivores includes aetosaurs (*Desmatosuchus* and *Typothorax*), an ictidosaur (*Pachy-*

genelus), and perhaps rhynchosaurs. (Chatterjee [1980] felt that rhynchosaurs may have been more partial to mollusks than plants. He noted the abundance of unionids [freshwater mussels] in the Dockum, and considered the tooth batteries of the rhynchosaurs to have been ideally suited to crushing their shells.) The top carnivore of the Post Quarry was *Postosuchus*. This rauisuchid was first described by Chatterjee (1980) on the basis of a variety of skeletal elements. These included partially associated material of at least 12 individuals.

The wealth and diversity of the Post Quarry material are often overshadowed by two controversial inhabitants. The first is *Shuvosaurus*. Chatterjee (1993) described it as an ornithomimosaur, but if he is correct, this would greatly extend the temporal range of these toothless theropods back from the Lower Cretaceous (Aptian-Albian). Although there is no question that the skull of *Shuvosaurus* parallels that of the ornithomimosaurs, there are also marked differences that cast considerable doubt on its ornithomimosaur affinities. Until a detailed study of *Shuvosaurus* is undertaken, it is just as likely that the animal will turn out to be a strange pseudosuchian. Indeed, Long and Murry (1995) have suggested that there may be a close link between *Shuvosaurus* and *Chatterjeea*. The rauisuchian *Chatterjeea* was first described by Long and Murry (1995) solely on the basis of postcranial material from the Dockum of Texas. (The generic name was given in honor of Sankar Chatterjee's work.) Probable *Shuvosaurus* remains have been found in New Mexico and Texas in close association with *Chatterjeea*, and the size and preservation of the two forms are consistent with each other. If they do ultimately turn out to be the same animal, *Shuvosaurus* will have priority, and the name honoring Sankar Chatterjee will unfortunately disappear into obscurity.

But if *Shuvosaurus* created a stir, it was but a tempest in a teacup alongside *Protoavis*! Known on the basis of two fragmentary skeletons, *Protoavis* was described by Chatterjee (1991) as the oldest bird. Somewhat surprisingly, it has been recorded from two different horizons: besides the specimens from the Post Quarry (Cooper Canyon Formation), *Protoavis* is also reported from the Kirkpatrick Quarry, which lies in the older Tecovas Formation.

There is no denying that there is at least a superficial similarity between some bird elements and their counterparts in *Protoavis*, but it must be said that the material is fragmentary and difficult to interpret. Although Chatterjee described scars on the ulna that he considered to demonstrate the presence of wing feathers, others have strongly disagreed. To add to the difficulties, the material was not clearly associated, let alone articulated. There are therefore no guarantees that we are dealing with the remains of one taxon, and unless a relatively complete articulated specimen is found, it is likely that *Protoavis* will remain an enigma. However, there remain vast tracts of unexplored Dockum and Chinle sediments in Arizona, New Mexico, and Texas, and I would venture that the chances remain high that additional material will show up one day. This will either settle the dispute—or, more than likely, open up a completely new one.

chapter eight

Life in the Sand Dunes?

The Elgin Reptiles

Elgin lies on the periphery of Speyside, home to some of the finest whisky distilleries in Scotland. It is also home to a handful of rather poorly preserved vertebrate fossils that are found in moderately coarse sandstones. Although not exactly the Glenlivets of the pale-ontological world, they are nevertheless the center of a great deal of international scientific scrutiny, and the story of the Elgin reptiles is a key component in our understanding of the Triassic world.

To gain an accurate perspective on the Elgin reptiles, it is worth reflecting briefly on typical sandy environments in the world today. It must be remembered that a fossil deposit preserves just a small sample of life in a given area at a given time. For example, when you walk among the sand dunes behind a beach, it is common to see nothing but sand piled high on either side. Yet you know that a few feet away is an ocean teeming with a variety of marine life ranging from plankton to seaweed and large fish. Surrounding you, the extensive dune fields are home to a much more limited fauna adapted to the specialized conditions of a dry landscape that is constantly shifting. A little further away in a back bay, the fauna and flora change once again, and you see animals that are adapted to brackish water. Perhaps there are freshwater bogs and swamps nearby too.

All of this diversity can occur in the space of a couple of miles. Because of the proximity of these very different ecosystems, components of one are easily introduced into another; for example, predators may live in one area and feed in another. Each of these ecosystems is interconnected to some degree. So what happens during fossilization processes? When some catastrophic event such as a flood occurs, parts of more than one of these different ecosystems can become mixed together. If this is the case, then an accurate record of any one individual ecosystem is instantly lost. As a result, it could easily appear that the different components of the fauna and flora that had accumulated together in a deposit were all living together in one relatively uniform environment. However, on closer inspection, the great diversity of forms and the variable state of preservation might indicate that some fossils were transported further than others, and that the fossil assemblage represented a mixture of paleoecosystems. Alternatively, if only part of one of these ecosystems enters the fossil record—suppose, for example, the sand dunes shift and bury the adjacent area—then once again it becomes difficult to assess what the ancient environment was really like. Would we be able to tell, for instance how extensive the sand dunes were? Were they narrow coastal dunes, or did they extend in all directions for miles? Is it possible that we are even looking at an extensive sandy desert such as parts of

Plate 8.1. A lone *Ornitho-suchus* prowls along the ridge of a sand dune.

the Sahara today? Did the animals that were buried, and ultimately became fossilized, actually live in the dunes? With such uncertainties in mind, we shall examine a variety of possible scenarios for this important Carnian sequence from northeast Scotland: the Lossiemouth Sandstones.

Early Discoveries and the Age of the Sandstones

As with so many early discoveries, the history of research on the Elgin reptiles is closely tied in with the local quarrying operations. The sandstone was used as a building stone and was worked for many years. The commercial value of new and rare fossils was no less in the 1800s than it is today, and local quarrymen on occasions could get a good bonus by selling their finds to the local naturalists. Although earnings for such sales were undoubtedly good, they were nothing compared to the prices that are unfortunately part of the commercial trade in fossil vertebrates today.

As found, the fossils typically consisted of soft bone in a hard sandstone matrix, and they were difficult to prepare and study. Frequently the bone crumbled away to leave partial natural molds. It certainly took a well-trained eye to differentiate between a mold caused by differential weathering of the sediment and one that resulted from weathering away of the bony material. Working in the mid-1900s, Alick Walker turned the crumbly, soft nature of the bone to great advantage and developed a technique that involved creating complete natural molds of the fossil bones. He deliberately dissolved away all bone using acid and then took rubber and plastic casts from the natural mold that he had thereby created. The result was a faithful cast of the original specimen.

Originally, all the sandstones in the Moray Firth area were considered to be of equivalent age and thought to part of the Old Red Sandstone series, approximately 370 million years old. This was in large part due to the occurrence of a number of fish specimens, the first remains being discovered as early as 1836. In 1844, one of the quarrymen at Scaat Craig discovered a block of sandstone that contained several rows of scales that were somewhat larger than the typical finds. A local geologist, Patrick Duff (the Elgin town clerk), pronounced that these were the scales of a large fish, a view that was "confirmed" by the eminent anatomist and fish expert, Louis Agassiz. For many academics of the time, this merely served to reinforce the opinion that the sandstones were indeed all Old Red, and therefore Devonian in age. Agassiz named the new form *Stagonolepis robertsoni*. Additional specimens were found a few years later at other quarries in the area, most notably Findrassie and Lossiemouth, but surprisingly, limb bones were also recovered—clearly this was no fish! The new finds were passed on to T. H. Huxley, who correctly determined that they belonged to a tetrapod, not a fish (Huxley 1875). In 1877 Huxley published a detailed description that identified *Stagonolepis* as an ancient crocodile.

Perhaps more significantly, in 1851, eight years before the realization that *Stagonolepis* was a reptile, the remains of a different tetrapod had been described from the sandstones, and this discovery had sown seeds of doubt in the minds of some geologists about the Devonian age of the sediments. The history of the discovery and description of the small reptile *Leptopleuron* is a fascinating one involving two of the giants of vertebrate paleontology, and is reminiscent of the feuds that later developed between Cope and Marsh in North America. The first specimen of *Leptopleuron* was discovered in the Spynie Quarry by Patrick Duff in 1851. He sent the specimen to his brother, Dr. George Duff, in London, who first allowed Charles Lyell to examine it.

Lyell was naturally intrigued by the specimen because of its putative Devonian age. This specimen of a procolophonid, which Lyell thought to be an amphibian (Lyell called it a batrachian), seemed to be a critical piece of evidence supporting the antiprogressionist views that he held at that time. Lyell urged Gideon Mantell to describe the fossil in detail. At the same time, Lyell had plates prepared of the specimen so that he could illustrate it in his new edition of *Manual in Elementary Geology*. However, George Duff also permitted Sir Richard Owen to examine the specimen. Mantell prepared a paper together with Lambert Brickenden, one of Mantell's collectors, describing the specimen as a new batrachian under the name *Telerpeton*. The paper was supposed to be presented at the meeting of the Geological Society on December 17, 1851, but the talk was postponed because of a lengthy discussion of an earlier presentation. As a result, Owen beat him to the punch. On December 20, Owen published a short description of the specimen in the *Literary Gazette* as a lacertian (lizard), with the name *Leptopleuron*. Thus Mantell was denied the honor of naming the new beast by a cruel twist of fate.

Although Lyell may not have been swayed from his convictions regarding the age of the sandstones, it was becoming apparent that the remains of fish and reptiles were actually never found together in the same beds. This certainly made many geologists suspicious, and J. W. Judd even went so far as to say that he would not accept that the reptile-bearing sandstones were Devonian unless he was shown a fossil fish in the mouth of a reptile (Benton 1977)!

The weight of evidence against the Devonian attribution continued to pile up. The ubiquitous rhynchosaurs turned up in the Elgin area in the form of *Hyperodapedon*. The first specimens were collected in 1858 by Dr. George Gordon, and these also were passed onto T. H. Huxley. With the subsequent collection of even more material, Huxley was able to describe most of the skeleton in 1869. By that time, other rhynchosaurs were known from England and India in sediments that were well dated as Triassic. Thus the description of *Hyperodapedon* was an important landmark in the debate on the age of the Lossiemouth Sandstone, named after the town on the Moray Firth where some of the prominent specimens originated.

Over the next few decades, a handful of additional Elgin reptiles were discovered that were also remarkably similar to those from Upper Triassic strata elsewhere in the world, and a Late Triassic age assignment became even more firmly established. However, a new twist to the age debate began in 1885 with the discovery of a dicynodont, an animal known widely from the Permian and Lower Triassic. Subsequent discoveries of an additional dicynodont and a pareiasaur indicated that at least some of the sandstones were more likely older. It is now recognized that in addition to the Old Red Sandstones, there are also Permian or Lower Triassic sequences in the Cutties Hillock area, even though the majority of the tetrapod yielding localities in the area are of Upper Triassic age (equivalent to the Adamanian land vertebrate faunachron [LVF]).

In 1877 Huxley was the first to describe a carnivore from the Lossiemouth Sandstone. On the basis of a fragment of maxilla, he believed that it might be representative of a dinosaur, and he called it *Dasygnathus longidens*. Later, Newton (1894) described additional material of a carnivore from Spynie and named it *Ornithosuchus* ("bird-crocodile") *woodwardi*, believing that it was ancestral to both crocodiles and dinosaurs. Further remains were discovered in the early 1900s that indicated an animal twice as large as *Ornithosuchus*. Consequently it was designated by Broom as a different species, *O. taylori*. Much later, Walker (1964) reexamined all this material and came to the conclusion that together they really only represented different growth stages of the same species, which he called *Ornithosuchus longidens*.

The widespread practice in the early 1900s of describing new, but not directly comparable, fragmentary specimens of fossil vertebrates as new species often led to an image of much greater diversity in the fossil record than is justified on the evidence. Although assemblages such as those represented by the Elgin fossils do not even begin to approach the full diversity of the once living community, we must still exercise some caution and not create too many taxa.

Two intriguing small archosaurs, *Erpetosuchus* and *Scleromochlus*, were described in 1894 and 1907, respectively. As we shall see, *Scleromochlus* has since turned into one of the most problematic Triassic reptiles known. Another small reptile from the Elgin area was discovered by William Taylor and described by Huene in 1910a under the name

Brachyrhinodon. With a lizardlike skeleton that was very different from the large barrel-shaped body of *Hyperodapedon*, *Brachyrhinodon*, together with its living relative, the tuatara (*Sphenodon*), was for a long time considered to be related to the rhynchosaurs because of its beaklike premaxillary region. Only recently have the sphenodontians been separated completely from the rhynchosaurs and recognized as the sister group to the Squamata.

The last reptile to be discovered in the Lossiemouth Sandstones was also described by Huene in 1910b. It was a single specimen found by William Taylor and represented a small animal that was only about 2 feet long. Huene referred it to the Dinosauria and called it *Saltopus*, or "hopping foot," in reference to the long hind legs and appressed metarsals, which Huene considered to be adaptations to a saltatorial habit. The specimen is poorly preserved and extremely difficult to interpret. Benton and Walker (1985) suggested that it was a primitive theropod with three sacrals and a short ilium. However, as Norman (1990) points out, it might represent the remains of a dinosauromorph, such as *Marasuchus*, or even an early pterosaur.

Today it is still possible to search through the old quarries in the vicinity of Elgin, including Spynie and Findrassie, and imagine the natural mold of some ancient beast in the weathered fragments. Sculptured sandstone surfaces with intriguing cavities abound, but they are almost always nothing more than weathered surfaces, merely generated by Mother Nature to tease and taunt us. Most of the outcrops are largely overgrown, and the chances of finding a new specimen are slim. Occasionally a walk along the edge of the Moray Firth at Lossiemouth will yield a fragment of bone adhering to a poorly defined depression on a small sandstone block. A visit to the local museum in Elgin or the Royal Scottish Museum in Edinburgh is likely to be a more satisfying venture.

The Fauna and Paleoenvironment

A dry wind howls across the sand dunes as a small *Scleromochlus* struggles across the open wilderness, desperately searching for some shelter. Huddled together in the lee of a rock, a small group of rhynchosaurs, weak from lack of food and water, await death. On any given day, this could have been the harsh reality of life 220 million years ago in what today is northeast Scotland. But was this a typical scene? Was there a dry, sandy desert stretching for miles, or was this just one of many different ecosystems within fairly close proximity?

The Lossiemouth Sandstone is thought by many paleontologists to be representative of a fairly barren dune environment, similar perhaps to the Cretaceous Djadochta Formation of the Gobi Desert, or perhaps even parts of the Sahara Desert today. The sandstone exhibits characteristic dune bedding, and it is clear that the sediments were deposited by wind rather than water. However, given the variety of different fossil reptiles that have been found in the sandstone, it is reasonable to assume that nearby was a relatively well-vegetated area providing the primary food source. Just how close this vegetated area might have been is anyone's guess. There is no evidence of plant fossils, and reconstructions of the vegetation cover that appear in texts are based on deposits of similar age elsewhere (for example, see Benton 1999). Such reconstructions are, of course, highly speculative.

Paradoxically, the earliest discoveries of reptiles in the Lossiemouth Sandstone are all of herbivorous forms, and this is almost certainly a reflection of the fact that herbivores predominate in the assemblage. The numbers of individuals are also quite high; for example, there are well over 30 individuals of *Hyperodapedon* represented in collections in various museums. In turn, the abundance of herbivores could be a hint that the depositional environment (eolian sand dunes) does not represent an accurate record of the environment in which these animals actually lived. Not only are there no plant fossils, but neither are there any invertebrate remains to offer any additional clues. Plant fossils, and in particular their preserved cuticle, might have given us some indication of possible adaptations to a xeric climate. But in their absence, it is impossible to say anything about the levels of humidity that these animals were typically accustomed to. Certainly there is nothing about the vertebrate fossils to indicate that the animals were in any way adapted to a marginal environment.

As we have seen, there are many explanations for the association of a diverse tetrapod assemblage with an apparently harsh depositional environment. Although we cannot completely rule out the possibility that these animals lived entirely in arid dune fields, it seems much more likely that the large- to moderate-sized herbivores, *Stagonolepis* and *Hyperodapedon*, would have required vegetative cover to fulfill their dietary requirements. Just exactly where this cover was located and what it consisted of is difficult to say. Benton (1983b) felt that both *Stagonolepis* and *Hyperodapedon* must have sought plants in adjacent floodplains or around interdune pools, and it is conceivable that extensive dune fields were broken up by isolated oases. The shifting sands would from time to time bury the oases and the herbivores living in them. The carcasses of animals that perished may have attracted scavengers across the sands. However, the fact that most of the specimens are well articulated indicates that the cadavers were rapidly buried, and among the known specimens, there are no unequivocal signs of extensive scavenging. In many specimens, the limbs are positioned in a natural resting pose, with the forelimbs flexed and pointing forward and the hindlimbs directed either forward or backward, although Benton (1983b) did note some broken bones in two specimens of *Hyperodapedon* that could have been caused by a large carnivore such as *Ornithosuchus*. In both instances, well-preserved skulls, including sclerotic plates in orbits that are only slightly displaced, show some clean breaks in the bone that were probably caused at or after death. Walker (1964) also noted similar limited crushing in a specimen of *Ornithosuchus* that he thought was almost certainly the result of scavenging or even predator action. Nevertheless, there are no definitive tooth or claw marks preserved.

It is also possible that many of the animals spent only parts of their life cycle in the dunes. Perhaps the inhospitable sand dunes afforded some measure of security from predators and a place in which to breed and lay eggs. Another possibility is that the sand dunes periodically migrated over a neighboring, well-vegetated area and inundated the unfortunate reptiles living there. Whatever the cause, we have to accept that the Lossiemouth Sandstones do not by themselves provide us with a particularly clear image of day-to-day life in the Triassic. As we shall see, neither are the specimens stunning. However, what the fossils may lack in completeness and beauty, they certainly make up for by their uniqueness and their importance to our understanding of key issues in phylogeny.

Whatever the exact nature of the terrestrial environment, it supported a mix of ancient and modern tetrapods (a theme that is recurrent among Late Triassic fossil assemblages). Among the ancients are the rhynchosaur *Hyperodapedon*, the aetosaur *Stagonolepis*, the ornithosuchian *Ornithosuchus*, the erpetosuchid *Erpetosuchus*, and the procolophonid *Leptopleuron*. The ornithodiran *Scleromochlus* might be considered as one of the ancient forms. However, if it is the sister group to pterosaurs and dinosaurs, then it is part of the clade that gave rise to bird. The sphenodontian *Brachyrhinodon* is certainly best considered to be a modern representative of the fauna because descendants of this lineage are alive today, if only just! (See chapter 9 for a general discussion of the sphenodontians.)

The Plant Eaters

Three herbivores dominate the Lossiemouth Sandstone assemblages, and they range in size from the small but squat *Leptopleuron* through the pig-sized *Hyperodapedon* to the fairly large *Stagonolepis*.

Like other Late Triassic rhynchosaurs, *Hyperodapedon* was a midsized quadruped (fig. 8.1). It was similar to *Scaphonyx* from South America and may be congeneric with *Paradapedon* from India (Chatterjee 1974). It reached a length of about 5 feet, and it had a distinct barrel-shaped body. Presumably this accommodated a large gut that facilitated

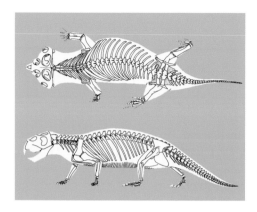

Figure 8.1. Restoration of the skeleton of the rhynchosaur *Hyperodapedon* (after Benton 1983b).

the slow digestion of plant matter. Another distinctive feature was the beaklike snout and the expanded cheeks that bore a battery of teeth. The "beak" was an ideal cropping apparatus, and with the sophisticated array of cheek teeth, *Hyperodapedon* would have at least partially processed its food orally with its precision shear bite. *Hyperodapedon* possessed no protective armor. Nevertheless, when backed into a corner, an adult would have been quite capable of delivering a nasty bite with its powerful jaws.

In his detailed account of *Hyperodapedon*, Benton (1983b) suggested that it had large eyes with good powers of accommodation and had a diurnal habit. He argued that the well-developed hyoid apparatus would have provided for the firm attachment of a large tongue, as well as aiding in the detection of groundborne sounds. The large tongue would have manipulated food in the mouth and passed it back from the premaxillary beak to the crushing tooth plates. On the basis of the large nasal capsules, Benton believed that *Hyperodapedon* had a good sense of smell. Huene (1939) suggested that the rhynchosaur hindlimb was capable of being used as a powerful digging tool, scratching backward so that it could expose roots. Benton supported this interpretation, commenting on the nearly parasagittal movement of the femur during the normal stride that could be used for controlled digging. He also noted the relatively large claws and their flattened shape, which are consistent with a dorsoventral action of the pes. Although Huene (1939) compared the digging habit of rhynchosaurs with that of moles, Benton preferred the scratch-digger model (Hildebrand 1974), as exemplified by some turtles and ground squirrels. In moles, digging is accomplished by rotation of the humerus and a thrusting back of the powerful hand, whereas in the scratch-diggers, the claws are used to break up and loosen the soil and dirt, which is then pushed back with the sole of the foot. Benton also commented upon the pointed daggerlike premaxillary "beak," citing its possible function as another digging implement.

Stagonolepis (fig. 8.2) was an armored quadruped about twice the length of *Hyperodapedon*. Its head was relatively much smaller, and it seems unlikely that it used a great deal of oral processing of its food. Its leaf-shaped teeth would have served to nip off plant matter, but the form of the teeth is not consistent with sustained grinding of coarse plant fibers. The edentulous snout was somewhat blunt and in life was possibly piglike. Perhaps on occasions *Stagonolepis* rooted in the ground for some of its food. Specimens tend to fall into two discrete size groups, with an approximately equal number of individuals in each group. Because morphological variation between each group is slight, the two groups have been considered to reflect sexual dimorphism. Other fossil vertebrates have also been shown to have similar dimorphism—for instance, the well-known synapsid *Dimetrodon*. *Stagonolepis* is one of the best-known aetosaurs (or stagonolepids) prevalent in Late Triassic strata throughout the world.

The smallest herbivores known from the Lossiemouth Sandstone are procolophonids. Although the procolophonids were restricted to the Triassic, they were widespread and were clearly an important component of many terrestrial vertebrate communities. The Late Triassic forms are particularly striking, often showing the development of fairly extensive spikes in the cheek region (on the quadratojugal) (e.g., *Hypsognathus*, the Cromhall taxon, and the Owl Rock procolophonid). These presumably afforded the animal some protection against attack from predators. Some forms known from the Newark Supergroup, and probably others from the fissure deposits and Germany, were further protected by bony armor that covered the neck region and maybe the rest of the body surface. The taxon known from the Lossiemouth Sandstones, *Leptopleuron*, was typical of other Late Triassic procolophonids in possessing pronounced spiny processes on the lateral corners of the skull. There is, however, no evidence of bony dermal plates covering any part of the body.

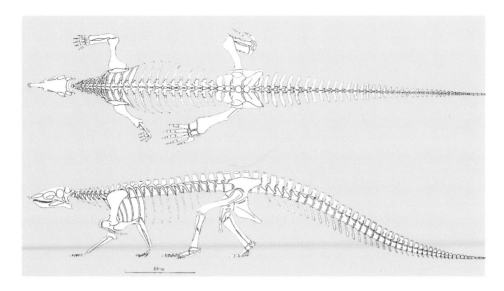

Figure 8.2. Reconstruction of the skeleton of the aetosaur *Stagonolepis* (after Walker 1961).

The Top Predator

The largest carnivore in the Elgin assemblages is *Ornithosuchus*. At up to 10 feet long, and armed with jaws bearing teeth like a steak knife, it would certainly have been a fearsome predator (Fig. 2.13). Nevertheless, taking on the well-armored aetosaurs and the rhynchosaurs would have been no easy task. A rhynchosaur backed into a corner would have been a formidable opponent, with a powerful bite of its own. Juveniles and animals weakened by previous injuries and poor health would likely have been much more attractive options. *Ornithosuchus* had two rows of dermal scutes running the length of its back, which would have provided some protection from a retaliatory strike from an intended victim or from a larger aggressor of its own kind.

For a long time *Ornithosuchus* was regarded as a dinosaur (Walker 1964). Although Romer, for example, recognized that it was "primitive" in many respects, he was impressed by the basic similarity of the skull to that of a theropod dinosaur. There was therefore little doubt in Romer's mind that *Ornithosuchus* was an early dinosaur. Chatterjee (1985) followed a similar line of reasoning in his description of the rauisuchid *Postosuchus* (chapter 7) when he chose to ally it with the carnosaurs.

A closely related form is *Riojasuchus* from the Los Colorados Formation (Norian of Argentina), which, together with the poorly preserved *Venaticosaurus*, is placed in the family Ornithosuchidae. The taxonomic position of Ornithosuchidae has been controversial. Although they have traditionally been grouped together with the pterosaurs and dinosaurs (Ornithodira) as the Ornithosuchia, Sereno (1991) cast doubt on this split and suggested that too much significance had been paid to the nature of the ankle joint. As a consequence, other potential synapomorphies were ignored, and Ornithosuchidae are probably more closely related to the Suchia and Parasuchia.

The Vermin of the Triassic? Small Insectivores

It is unfortunate that many of the smaller vertebrates of the Mesozoic tend to be left in the shadows of their larger contemporaries. It is almost as if they are an afterthought, something akin to vermin—merely troublesome animals that got in the way of the dinosaurs!

But, of course, that was not the case, and there was a great diversity of small amphibians, reptiles, and mammals that formed critical components of Mesozoic communities. In the Lossiemouth Sandstone, *Brachyrhinodon* is one such small vertebrate. This lizardlike reptile (fig. 8.3) is actually one of the earliest members of the Sphenodontia.

Remains of *Brachyrhinodon* are relatively common, although it was one of the later reptiles to be discovered (Huene 1910a). A redescription by Fraser and Benton (1989) showed that *Brachyrhinodon*, although one of the oldest known sphenodontians, was more derived than such basal members of the Sphenodontia as *Diphydontosaurus* and *Planocephalosaurus* (from the British fissure deposits; see chapter 9). Wu (1994) concurred with this view, considering *Brachyrhinodon* to be the sister group of *Clevosaurus.*

Compared with many lizards of equal size, these sphenodontians had relatively deep and robust jaws, and I have suggested (Fraser 1988b) that they were ideally suited to capturing and dealing with the tough exoskeleton of large insects such as cockroaches, orthopterans, and certain coleopterans. Farlow (1976) observed the living *Sphenodon* eating insects, and described how the powerful jaw muscles actually bulged through the temporal openings as it crunched through its prey.

A Critical Fossil or a Paleontological Red Herring?

The little archosaur *Scleromochlus* was a lightly built animal with a relatively short body and long hindlimbs (fig. 8.4). Different authors have postulated different modes of life. Huene (1914) considered *Scleromochlus* to have been arboreal, climbing into trees, where, grasping twigs and branches with the hand and prehensile fifth toe, it fed on insects. Huene thought *Scleromochlus* to be capable of jumping from branches of one tree to another and went on to postulate that it was a parachuting animal with "skin duplications on the forelimbs, perhaps also in other places." Although this is speculation—such fine details were not preserved even if they had been present—the scenario is strangely reminiscent of one proposed for *Sharovipteryx.* Nevertheless, it must be observed that neither the hand nor the foot display modifications for grasping. Hildebrand (1974) noted that in typical climbers, the penultimate phalanx is longer than the more proximal phalanges, but this is not true for *Scleromochlus.*

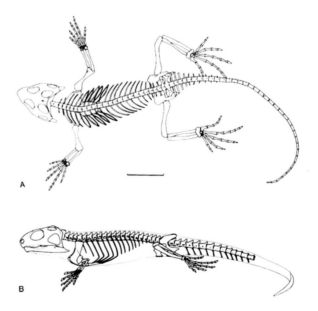

Figure 8.3. Skeleton of the sphenodontian *Brachyrhinodon* in (A) dorsal and (B) lateral views (after Fraser and Benton 1989).

Huene was not alone in considering *Scleromochlus* to have been saltatorial, but others have envisaged it as a ground saltator. C. W. Andrews, in a discussion of Woodward's (1907) paper, first suggested that *Scleromochlus* was a sand hopper. The basis for this argument was the similarity between the hindlimb of the jerboa and *Scleromochlus*. Alick Walker (1961) agreed with this interpretation. Certainly there are a number of features that are suggestive of a jumping habit, including the fused and long metatarsal bones.

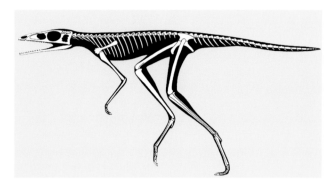

Figure 8.4. Restoration of the skeleton of *Scleromochlus* (after Sereno 1991).

At the same time, many modifications associated with jumping are also consistent with a cursorial mode of life. These include the elongate hind legs, and the relatively elongate distal elements of the leg (tibia and metatarsals). The reduced forelimb is consistent with a bipedal cursor. On the other hand, most cursorial animals with these adaptations are moderate- to large-sized animals, and the fact that *Scleromochlus* is so much smaller tends to at least hint at something more than a bipedal cursor. Thus I favor an agile animal capable of running and jumping in the sand dunes or nearby undergrowth. Carcasses left partly buried by sandstorms would attract a variety of scavenging insects, which in turn could form a feast for the little *Scleromochlus* cavorting around the sand.

It is not just the mode of life of *Scleromochlus* that has been the subject of so much debate; its phylogenetic position is equally controversial. When it was first described, Woodward (1907) placed *Scleromochlus* in the Dinosauria. But first Broom (1913) and then Huene (1914) reassigned it to the Pseudosuchia. Presently it is typically placed within its own family, the Scleromochlidae (Benton 1999), but this family floats among a number of positions relative to other archosaurs. Walker argued that *Scleromochlus* fell within the crocodilian lineage of archosaurs (the Pseudosuchia). He was particularly impressed with an apparent similarity between *Scleromochlus* and stagonolepidids (aetosaurs) in the configuration of their temporal region. Other authors tended to follow this lead (e.g., Krebs 1976; Benton and Walker 1985). In sharp contrast, Padian (1984, 1997) considered *Scleromochlus* to be an ornithodiran and in particular recognized it as the basal outgroup of pterosaurs. Likewise, Sereno (1991) regarded *Scleromochlus* as an ornithodiran, but he was less sure of its position relative to pterosaurs and dinosaurs. He argued that it was equally likely that it could be placed as the sister group to either Dinosauria or Pterosauria, or even basal to both. In fact, Benton (1999) considered the latter to be the most strongly supported phylogeny.

Why would competent paleontologists place *Scleromochlus* in such fundamentally different groups as the Pseudosuchia and Ornithodira? After all, it might be argued that a close examination of the ankle joint should immediately point you in the right direction. The answer is that there is no agreement on the precise identity of the different tarsal elements. And therein lies the answer to most of the issues concerning the small tetrapods from the Lossiemouth Sandstone. The coarse sandstone does not preserve really fine details, and it can be difficult indeed to perceive which features of the molds might be real, which are artifacts of the sediment, or which are merely figments of the imagination! Huene (1914) says that he used "strongly one-sided, artistic illumination." This certainly implies something other than rigorous scientific observation. Needless to say, a slight shift in the angle that the light falls on the casts can potentially produce a quite different impression. Thus one should always adopt an open mind and take a cautious approach when interpreting the molds.

In the case of *Scleromochlus*, one of the specimens, BMNH R3557, appears to have a well-preserved ankle. Two of the elements seem to correspond closely to the proximal tarsals of a crurotarsan type ankle. Thus both Woodward (1907) and Huene (1914) recognized one of the elements as the calcaneum because it possessed a calcaneal spur, a feature that is diagnostic of the crurotarsan ankle. The other element they believed to be the astragalus. However, Padian (1983, 1984) questioned this interpretation, suggesting instead that the so-called proximal tarsals were really the distal tarsal elements. He argued that the putative calcaneum was really the lateral distal tarsal, and the supposed astragalus the medial distal tarsal. Furthermore, he made a compelling argument that as distal tarsals, they were consistent with those elements in pterosaurs. He assumed that the proximal tarsals had fused to the epipodials. If so, the ornithodiran interpretation becomes more tenable.

Sereno (1991) accepted Padian's interpretation of these elements as distal tarsals, but he did note that the best-exposed tarsal bone on BMNH R3557 was remarkably similar to the left calcaneum of a crurotarsan. However, because these bones are clearly associated with a right limb, Sereno felt that this interpretation must be misleading and that the tarsal bones were definitely distal elements. Another option available is that the small bones were indeed proximal tarsals, but the bone that Woodward and Huene presumably considered to be the calcaneum (because of the spur) is in fact the astragalus. This is the approach that Benton (1999) has adopted, and this interpretation is certainly not without its merit. This view is also consistent with inclusion of *Scleromochlus* within the Ornithodira.

The second issue is the possible close relationship of *Scleromochlus* to the pterosaurs. This was an idea that was first proposed by Huene (1914), a view that he reiterated in 1956, but one that apparently went ignored by most people, including the eminent Alfred Sherwood Romer, who apparently considered *Scleromochlus*, along with *Ornithosuchus*, to be a dinosaur.

Padian (1984) revisited Huene's proposal and reexamined all the casts of *Scleromochlus* in this context. He concluded that *Scleromochlus* probably was the sister group to pterosaurs. Padian listed four principal features that he thought allied *Scleromochlus* with pterosaurs. These are (1) the lightly constructed skull with large fenestrae, (2) long straplike scapula, (3) short, deep, trapezoidal pelvis, and (4) the four elongated, closely appressed metatarsals. Padian also felt that *Scleromochlus* possessed distal tarsal elements that were identical to those of pterosaurs such as *Dimorphodon*.

Interestingly, although Sereno (1991) accepted the possibility of a pterosaur-*Scleromochlus* sister-group relationship, he actually dismissed the particular arguments of Padian, stating that there was nothing unusually large about the cranial fenestrations, that the scapula was remarkably short, and that the pelvis was triradiate and not trapezoid. Instead, he offered four potential synapomorphies of his own: (1) the skull more than 50 percent the presacral vertebral column length, (2) the length of the scapula less than 75 percent of the humeral length, (3) the absence of a fourth trochanter on the femur, and (4) the length of metatarsal I is at least 85 percent of the length of metatarsal III. At the same time, Sereno acknowledged that there was an equally convincing case to be made for a sister-group relationship between *Scleromochlus* and the Dinosauria.

Certainly exacerbating the debate is the poor preservation that gives the observer a tantalizing glimpse of a wonderful little animal—a glimpse that is sufficient to make the imagination run wild, but fails to keep in mind the hard realities and limitations of the fossils themselves. A second problem lies in the use of qualitative rather than quantitative descriptions of characteristics. Thus we have to contend with features such as a "long, narrow, strap-like" scapula. What is long and thin to one author may be ordinary and unremarkable to another. What we really need are hard-and-fast definitions. Certainly precise ratios are needed to enable subsequent authors to make meaningful comparisons. But perhaps the most critical aspect is what other parameter is chosen as a comparison. Is the bone of concern equal in length to the humerus or the equivalent length of the first five cervical vertebrae? Padian and Sereno have adopted different approaches here, and this is at least one reason for their disagreement. Of course, if the hypothesis that pterosaurs are part of the prolacertiform radiation ever attains a level of some respectability as a viable alternative, then *Scleromochlus* will be examined in a new light.

A Crocodylomorph?

Erpetosuchus is another poorly known small archosaur that is difficult to pigeonhole taxo-nomically. It was a small animal (skull under 9 centimeters long) with gracile front limbs (the posterior half of the animal is unknown). Superficially, the skull bears a resemblance to sphenosuchian crocodiles, with the quadrate angled forward and a deeply emarginated otic region. However, a closer examination of the sutures (although poorly defined in the sandstone molds) reveals that the quadrate does not hook back under the squamosal, and it is unlikely that there was a quadrate-prootic contact (a character widely distributed among crocodylians). Neither is the quadratojugal positioned anterior rather than lateral to the quadrate. Furthermore, the radiale and ulnare are not elongated, and the coracoid lacks an attenuated posterior process.

The most unusual feature of *Erpetosuchus* is the distribution of the marginal teeth. In the upper jaw, they are restricted to the premaxilla and the very anterior portion of the maxilla. In the lower jaw, there are at least two or three acutely conical teeth situated more posteriorly. Another erpetosuchid from the Newark Supergroup (see chapter 10) was first

Plate 8.2. In the middle of a sandstorm, a small *Sclero-mochlus* evades any possible capture attempts by an *Er-petosuchus.* The lack of teeth in the cheek region of *Er-petosuchus* is curious and makes its dietary habits un-clear. It may have preyed upon small reptiles, or alter-natively, it could have fed on insects.

recognized on the basis of this peculiar dentition. Just how this dentition functioned is unclear, but it seems plausible that the acutely conical teeth in the front of the jaw captured
large insects such as cockroaches, piercing the exoskeleton and preventing them from es-
caping. Thus immobilized, the tough exoskeleton could be pierced by the teeth in the
lower jaw and crushed against the broad edentulous region of the maxilla and palate.

The North American erpetosuchid originates from the lower part of the New Haven
Formation and is therefore of Norian age. Although it reveals little additional information,
it nevertheless gives renewed hope for the discovery of more complete material. As with
the controversy surrounding *Scleromochlus*, resolving the relationships of *Erpetosuchus*
probably depends on more conventional specimens.

Unless the Elgin area is opened up again, it seems unlikely that better specimens of
scleromochlids or erpetosuchids will surface. Our best chance for resolving the phylogenetic position of these beasts lies in finding representatives from some of the other Triassic sites dotted around the globe. Like the shadows that play across the latex casts of the
small Elgin tetrapods, we get only a misty glimpse of the animals as they once existed in
what is today northeast Scotland. The next chapter in their story may await us in the rocks
on the other side of the North Atlantic.

chapter nine

Life in the Deep South

The Late Triassic of Gondwana

The separation of Pangaea into Gondwana in the south and Laurasia in the north has always been considered to have greatly affected the terrestrial faunas and floras in the two halves of the supercontinent. As we saw in chapter 1, provinciality between the northern and southern Late Triassic floras is well established, with *Dicroidium* dominating the Gondwanan plant assemblages and cycadophytes and conifers holding sway in Laurasia. A distinction in the faunas has also been indicated, with phytosaurs being largely limited to the northern hemisphere and tetrapods such as traversodont cynodonts often predominating in southern hemisphere vertebrate assemblages. The great embayment of ocean that was the Tethys Sea presumably acted as a major barrier to dispersal throughout Pangaea. However, as we are about to see, it is not that simple. As with interpretation of the Triassic climate, traditional views of faunal distribution have been challenged in recent years.

The Maleri Faunas of India

Although the Indian plate is today part of the Asian continent, during Triassic times, it was firmly attached to Africa and Antarctica, and therefore, we should expect to see close ties among their Triassic fossil assemblages. To some extent, that is what we do see. For example, Indian floral macrofossil assemblages are typically dominated by the foliage of *Dicroidium* (see chapters 1 and 2). But the same cannot be said for some of the faunal elements. In our look at the array of forms known from the Chinle and Dockum (chapter 7), we saw that there are many examples of species represented by complete skeletons, and these are inevitably what we tend to concentrate on. Nevertheless, we also saw that microvertebrate assemblages composed of hundreds of disassociated elements provide us with additional insights into diversity levels of these ancient communities.

As a result, some of the animals listed from the Chinle and Dockum are based on fragmentary material. For instance, fragments of elongated cervical vertebrae and ribs in the Petrified Forest National Park have been referred to the prolacertiform *Tanytrachelos*. As we shall see (chapter 11), this tetrapod is well known from complete skeletons from a couple of locations in the eastern United States. On the basis of just fragmentary material, it is difficult to be sure that *Tanytrachelos* did indeed occur in the American Southwest, but if not, something similar lived in the Chinle ponds and lakes. Another of these unusual long-necked prolacertiforms that occurs in the Dockum is *Malerisaurus*. This occurrence is particularly pertinent to the present discussion because this prolacertiform genus was first

Plate 9.1. A shaft of light catches the outline of a *Pisanosaurus* in the Ischigualasto forest. *Pisanosaurus* is based on fragmentary material but is currently the oldest known ornithischian dinosaur.

described for material from the Triassic of India. Once again, it is hard to be completely confident that the same animal occurred in such widely separated areas of the world, but the possibility does at least serve to draw attention to some similarities in faunas.

The Maleri Formation crops out in central India in the region of the Pranhita-Godavari Valley. It apparently contains two successive assemblages of tetrapods. It is within the lower of the two assemblages that *Malerisaurus* occurs (Chatterjee 1980). However, this is only one component of a rich fauna. In actual fact, the specimens of *Malerisaurus* were found within the body cavity of a specimen of the phytosaur *Paleorhinus* (=*Parasuchus*). Obviously the unfortunate *Paleorhinus* died before it could completely digest its equally unfortunate last meal! Additional members of the lower Maleri assemblage are the temnospondyl *Metoposaurus*, the rhynchosaur *Paradapedon*, and the cynodont *Exaeretodon*. This fauna therefore shows a remarkable similarity to tetrapod assemblages in the lower part of the Chinle Group (Otischalkian land vertebrate faunachron [LVF]). It is not unreasonable to suppose that the environment of the two regions was also similar—low-lying flood plains that were fairly well vegetated. The occurrence of unionids as well as a variety of fishes certainly points to a well-watered region.

The upper part of the Maleri has also yielded phytosaurs, but they show a much closer resemblance to *Rutiodon* and *Leptosuchus*, two taxa that are characteristic of the upper Carnian (Adamanian LVF). The long, slender-snouted *Rutiodon* is the phytosaur so abundant in the Shinarump Member of Petrified Forest National Park. Another possible connection with other Carnian localities is *Alwalkeria maleriensis*. This poorly preserved dinosaur has been referred to the Herrerasauridae (see below), but there is nothing in the material that is diagnostic of this family, and its relationships to other Triassic dinosaurs remains unclear.

It is remarkable that India, although a part of the Gondwanan (southern hemisphere) region of Pangaea in Triassic times, already exhibited affinities in its vertebrate faunas to the northern continents of which it would ultimately become a part. However, although much has been made in the past of marked differences between the north and the south, it is becoming increasingly apparent that these differences are fading. For instance, phytosaur remains have recently been reported in the upper Carnian of South America (Kischlat et al. 2002).

The Molteno Formation (Carnian)—A Myriad of Insects

Today we must cross a vast expanse of ocean to get to the African continent from India, but in Triassic times, the trip was simply a long trek across land. The vertebrate fossil riches of Africa have been particularly well known since the German expeditions of the early 1900s to Tendaguru in Tanzania (Maier 1997, 2003) and the Bahariya sands of Egypt. In 1907 an ailing Eberhard Fraas from the Natural History Museum in Berlin began work on Jurassic sediments cropping out near the village of Tendaguru. Werner Janensch then took over this work, and for the next few years, he used large native crews to excavate the spectacular remains of *Brachiosaurus*, *Kentrosaurus*, and their kin that are still housed in Berlin. At around the same time, Ernst Stromer from the museum in Münich began explorations further north and discovered the theropods *Carcharodontosaurus* and *Spinosaurus*. Unfortunately these discoveries did not survive the bombing of Münich during the World War II. In more recent years, Philippe Taquet, Angela Milner, Paul Sereno, and Josh Smith have heightened our appreciation for the Mesozoic of Africa. They have continued in the footsteps of Janensch and Stromer and gone on to explore countries such as Niger, Morocco, and Egypt, bringing to light spectacular fossils of *Ouranosaurus*, *Baryonyx*, *Sarchosuchus*, *Afrovenator*, and *Paralititan*, to name a few. However, these are all Jurassic or Cretaceous beasts, and they have eclipsed what are equally important sedimentary units further to the south.

We examined the Early Triassic sequences of the Karoo Basin in chapter 3 and saw the role that this area of the world has played in influencing ideas on Triassic life. However, ongoing work is proving that southern Africa also has important caches of Upper Triassic fossils. The Molteno Formation of southern Africa is particularly notable for a huge array of plant and insect fossils in addition to one spider (Selden et al. 1999). According to Anderson and Anderson (1993a, 1993b), it documents a period in Gondwanan Triassic history when *Dicroidium* and other seed ferns reached their greatest diversity levels. The sediments of the Molteno were deposited in an extensive floodplain that spread over not only southern Africa, but also into parts of what is today South America.

Although the plains were probably quite uniform, a number of different habitats are thought to be documented by the faunae and florae. First there were riparian forests of *Dicroidium*, together with *Dicroidium* woodland regions of the floodplain. Anderson and Anderson (1993a, 1993b) felt that the different species of *Dicroidium* ranged from shrublike plants to tall trees. Around the margins of lakes were woodlands primarily formed of the ginkgo *Sphaenobaiera*. The conifer *Heidiphyllum* is a ubiquitous component of Molteno sediments, and according to Anderson et al. (1996), it probably grew in the floodplain or on braid-river bars where the water table was high. Finally, ferns and herbaceous gymnosperms grew on the sandbars and banks forming meadows.

The broader-leafed species of *Sphenobaiera* appear to have been less common elements of the *Dicroidium* forests, and the sclerophyllous-leaved species of this genus grew in areas of open bush. The latter habitats yield high-diversity insect faunas dominated by cockroaches and, to a lesser extent, beetles. In the riparian forests, coackroaches and beetles are equally abundant. However, just as with the Madygen Formation (chapter 6), it should be remembered that the insect remains from the Molteno comprise largely isolated wings and elytra. Differences between the ornamentation on the elytra are sometimes small. It is possible that the diversity levels are somewhat exaggerated, as many are really just form taxa, and it is difficult to factor in the extent of individual variation. However, if Anderson et al. (1996) are correct, then the diversity levels of insects and plants were extremely high, perhaps even approaching today's diversity levels.

Although the Molteno may contain no vertebrates represented by skeletal remains, the occurrence of tracks and trackways does testify to the presence of a variety of different tetrapods (Ellenberger 1970, 1972, 1974; Haubold 1986). Tracks have their own special features that allow us to recognize different types for which we can assign taxonomic names. However, at best, we can only make a general referral of a track to a particular group of animals. Thus three-toed tracks with well-developed claw impressions are indicative of theropod dinosaurs, but we can typically go very little further with the identification. One of the most widespread three-toed tracks from the Triassic is *Grallator*. Another larger form has been named *Eubrontes*, but there are no clear-cut size distinctions. The result is that it can sometimes be difficult to distinguish between the two. It is almost impossible to attribute a fossil footprint to a taxon that is based on skeletal material.

The nature of a footprint depends not only on the structure of the foot, but also on the condition and type of substrate that it is made in. For example, the footprints we produce at the beach depend on the consistency and nature of the sediment we walk through. Furthermore, no two individuals walk exactly alike or produce identical footprints. On one prospecting trip to the Lossiemouth Sandstone near Elgin (see chapter 8), I was walking along the sandy beach at Lossiemouth with Alan Charig. On our return trip, Alan noted the different foot patterns that we produced, with mine being much deeper along the outside edge than his. We must therefore be aware when dealing with fossil trackways that slight differences in footprints may not reflect taxonomic differences in the track makers, but differences in the substrate. With these caveats in mind, the Molteno tracks indicate the presence of a number of different theropod dinosaurs (*Tritotrisauropus*, *Masitisisauropus*, *Trisaurodactylus*, etc.), and an assortment of quadrupeds (*Pseudotetrasauropus*, *Tetrasauropus*, *Deuterosauropodopus*, etc.)—presumably archosaurs such as rauisuchians, phytosaurs, and aetosaurs.

Much is still to be done in the Molteno, and we can expect future fieldwork to yield many more surprises. However, the rough sketch that we currently have again paints a very different picture of the Triassic from the traditional image of sun-cracked, barren red beds.

South America

Like Africa, South America has figured prominently in recent spectacular dinosaur finds. The enormous meat-eating *Giganotosaurus*, rivaling *Tyrannosaurus* and *Carcharodontosaurus* in size, and the vast sauropod nesting grounds of Maheuva, are just two examples. However, unlike Africa, the Triassic of South America has caught the public's interest, largely as a result of the discovery of the nearly complete remains of two animals touted as the world's oldest dinosaurs.

The Ischigualasto

Sediments of the Ischigualasto Formation were deposited in small rift basins that, like those of the Newark Supergroup, were formed before the breakup of Pangaea. ^{40}Ar/^{39}Ar radiometric dates of bentonites indicate that the oldest sediments at the base of the Ischigualasto are about 228 million years old. The upper boundary of the formation is unconstrained but is thought to be not much younger than 224 million years old (Rogers et al. 1993). On the other hand, Lucas and colleagues (Hunt and Lucas 1991; Lucas et al. 1992; Lucas and Hunt 1993; Heckert and Lucas 1996) argue for a somewhat younger age (equivalent to the Adamanian LVF) on the basis of the vertebrate fossils.

The Ischigualasto Formation of San Juan Province, Argentina, is well known for its early dinosaurs or dinosauromorphs, *Herrerasaurus* and *Eoraptor*. The primitive ornithischian *Pisanosaurus* also originates from these strata, but it is not known from such complete material and is often left in the shadows of its more complete distant relatives. However, amid the brouhaha over dinosaurs in general, and the earliest dinosaurs in particular, other inhabitants of the Ischigualasto rocks are almost totally neglected. For instance, the Ischigualasto also includes other predatory reptiles such as the large rauisuchid *Saurosuchus* (Reig 1959; Sill 1974; Alcober 2000). With its narrow 2-foot-long skull bearing sharp, serrate teeth, this was undoubtedly the top predator of the area. There are also prestosuchids, ornithosuchids, and carnivorous cynodonts. The herbivores include gomphodonts, aetosaurs, kannemeyeriid dicynodonts (*Ischigualastia*), and rhynchosaurs. In fact, the most abundant animals of the Ischigualasto are the rhynchosaur *Scaphonyx* and the cynodont *Exaeretodon*.

The aetosaurs were originally assigned to the genera *Aetosauroides* and *Argentinosuchus*, but Heckert and Lucas (2002) have shown that this material can all be referred to *Stagonolepis*. Indeed, Heckert and Lucas even suggested that much of the aetosaur material from the Ischigualasto could be referred to *Stagonolepis robertsoni*, the same species that occurs in the Lossiemouth Sandstone, and thereby intimated that the Ischigualasto Formation and Lossiemouth Sandstone are coeval deposits.

One interesting feature of the Ischigualasto is that *Scaphonyx* becomes noticeably less common toward the upper third of the formation and eventually disappears altogether. At the same time *Exaeretodon* continues right to the top and even continues into the overlying Los Colorados Formation. It thus appears that there was a local extinction of *Scaphonyx* in this region. What was responsible for the extinction is unclear. It has been suggested that the demise of the rhynchosaurs in general might be linked with the decline of the *Dicroidium* flora worldwide, but in this particular instance, *Dicroidium* floras are prevalent throughout the Ischigalasto and also occur in the Los Colorados, and so the disappearance of *Scaphonyx* cannot be associated with this particular floral change.

But to return to the dinosaurs, and the two most famed taxa of the Ischigualasto: *Her-*

Plate 9.2. A single *Eoraptor* pauses to listen by the trunk of a large *Araucarioxylon* tree.

Plate 9.3. Slowly moving through the *Araucarioxylon* forest, a small family herd of *Rio-jasaurus* migrates along the margins of a swampy clearing. *Riojasaurus* was a large prosauro-pod dinosaur that was almost certainly a quadruped.

(c) Doug Henderson 1/7/93

rerasaurus and *Eoraptor. Herrerasaurus*, at 3 or 4 meters long, is the larger of the two and was first described by Reig (1963). It remained something of an enigma until additional remains were discovered by crews under the leadership of Paul Sereno (Sereno and Novas 1992; Novas 1994; Sereno and Novas 1994; Sereno 1994). With these additional discoveries, it became apparent that *Herrerasaurus* exhibits a strange mixture of derived and plesiomorphic characters. For instance, it has only two sacral vertebrae and lacks a brevis shelf on the ilium. On the other hand, it has a prominent pubic "foot" that is reminiscent of tetanuran theropods. Because of this mix of characters, the phylogenetic position of *Herrerasaurus* is murky, and it has been vigorously debated whether this form is a basal dinosaur close to the ancestor of all dinosaurs, a member of the theropod lineage, or possibly not even a true dinosaur. Whatever its real pedigree, there is no doubting that it was one of the top carnivores of its day: one specimen was recovered with the remains of a young *Scaphonyx* in the confines of its ribcage. Only the contemporaneous rauisuchians might have been more formidable hunters.

Eoraptor was also a predatory form, but only reached a length of 1 meter. First described by Sereno and colleagues (1993) on the basis of a single fairly complete specimen, it is rather distinct from *Herrerasaurus* and its kin. It is a more gracile form with a slender skull. Although it has been reported to possess three sacral vertebrae and a brevis fossa on the ilium (synapomorphies of Dinosauria), in other respects, it would seem to be more primitive than *Herrerasaurus*. Nevertheless, most authorities tend to recognize it as a basal theropod dinosaur (e.g., Novas 1996). Sereno et al. (1993) regard the specimen as an adult, although Bonaparte (1996) expressed some doubts.

Finally, the small early ornithischian *Pisanosaurus* has also been described (Casamiquela 1967) from the Ischigualasto. Although the material is incomplete, it is apparent that this was a basal ornithischian possessing a battery of closely packed teeth. It may well have foraged on low-growing ground cover plants.

Santa Maria Formation

The contemporaneous Santa Maria Formation has also yielded a herrerasaur in the form of *Staurikosaurus*. Although we have not been fortunate enough to find complete skeletons, there is no doubt that the structure of the hindlimb and pelvic girdle is similar to that of *Herrerasaurus*. Another important member of the dinosaur fauna is the sauropodomorph *Saturnalia*. Found in the red beds of the Alemoa Member (Langer et al. 1999), *Saturnalia* was a gracile plant eater that was approximately 1.5 meters long. It originates in the same beds as *Staurikosaurus* and is thought to be from the same locality as the cynodont *Gomphodontosuchus* and the archosaur *Hoplitosuchus*.

Although the sacrum of *Saturnalia* has three sacrals (the third thought by Langer et al. to have been coopted from the caudal series), the acetabulum is not fully open. Because the open acetabulum is regarded as a key feature of dinosaurs, Langer et al. suggested that the complete opening of the acetabulum may have taken place more than once during dinosaur evolution. After all, *Herrerasaurus*, with just two sacrals, has a perforated acetabulum. In fact, the overall shape of the ilium in *Saturnalia* is not unlike that of *Agnostiphys* from the British fissure deposits (see chapter 10 for details), although the humeri of these two taxa are very unlike each other. It is worth noting that Langer et al. also noted that the astragalus of *Saturnalia* was similar to that of *Herrerasaurus* but was more derived in that the ascending process was separated from the anterior margin of the bone by a platform. Like Fraser et al. (2002), Langer et al. considered *Herrerasaurus* and *Staurikosaurus* either to form a monophyletic Herrerasauridae that is the sister group to the Dinosauria, or to be successive sister taxa of Theropoda plus Sauropodomorpha. This conclusion disagrees with the recent phylogenetic propositions of Sereno and Novas (1992) and Sereno et al. (1993).

Plate 9.4. Deep in a conifer forest dripping with mosses, two *Herrerasaurus* pass by the rhynchosaur *Scaphonyx* sheltering behind a fallen log. A variety of seed ferns also grow on the forest floor.

Hyperodapedon-like rhynchosaurs also occur in the Santa Maria type. As we have already seen, other formations that are rich in *Hyperodapedon*-like rhynchosaurs have a worldwide distribution; along with the Ischigualasto and lower Maleri, these include the Lossiemouth Sandstone (Scotland) far to the north in Laurasia. Other components of the Santa Maria tetrapod fauna include the stagonolepid *Stagonolepis* (*Aetosauroides*), and traversodontid and tritheledontid cynodonts.

For the most part, known assemblages from the Upper Triassic of Gondwana exhibit few hints of the wave of change that would sweep the world at the close of the period. With the exception of the dinosaurs, the major groups are largely representative of lineages destined for extinction. On the other hand, we know of one or two Late Triassic vertebrate assemblages dominated by representatives of groups that were about to flourish. In the next chapter, we shall examine one corner of the world that is particularly rich in assemblages bearing the definitive stamp of the modern world.

Part Four

THE BIRTH OF MODERN TERRESTRIAL ECOSYSTEMS

chapter ten

Pitfalls in Paleontology

The British Fissure Deposits

An Ancient Archipelago

Today the Cotswolds and Mendip hills of southeast England conjure up a bucolic land-scape dotted with small villages, each with its own church and local pub. But 220 million years ago, what is today southern Britain was a flat area of land dotted with playa lakes and mudflats interspersed by low limestone hills. This was apparently an ever-changing land-scape. Lakes came and went, and over the next 20 million years, the area gradually be-came inundated by a shallow sea until only islands in archipelagos remained. These is-lands were later submerged by the advancing sea. In many ways, the landscape may have been similar to that of the European alpine region at the same time, and indeed at least one of the terrestrial vertebrates, the sphenodontian *Diphydontosaurus*, seems to have ex-isted in both places at approximately the same time. The remains of small pterosaurs are also common to both regions (although none of the species identified occurs in both). But for the most part, terrestrial vertebrates are lacking from the northern Italian sequences, and those cave deposits in the Tethyan realm that include a significant terrestrial compo-nent preserve sediments from the Middle Triassic, so the two areas are not directly com-parable. Where the Italian sequences contain a strong element of the bizarre and im-probable, in the British deposits, there is at least a whiff of some of the groups familiar in today's terrestrial vertebrate faunas.

Part of the story of life in this marginal environment has been captured in the so-called fissure deposits of England and Wales (fig. 10.1). The remains of many of the vertebrates were washed into cave and joint systems that fissured the ancient limestone surface. These vertebrates include exquisitely preserved bones (albeit completely disassociated) of the some of the world's first mammals. For this reason, the fissure sediments have attracted a great deal of attention among paleontologists. They are also the source of a great diversity of reptiles, including pterosaurs, dinosauromorphs, sphenodontians, crocodiles, and rauisuchians. Here once more in the Late Triassic world is the mixture of ancient and modern tetrapods.

The picture of playa lakes, mudflats, and island archipelagos has not always been the prevailing interpretation. The tetrapods of this area were long regarded as unique in representing an upland fauna, and it is only in the last decade that the notion of an upland fauna has been abandoned. The history of research into these Triassic fissures is a fascinating story that documents constantly changing ideas and opinions, with the

Plate 10.1. Seeking a quiet place to consume the freshly killed sphenodontian *Planocephalosaurus*, the early crocodile *Terrestrisuchus* crosses a small stream on one of the limestone islands in what is now southern En-gland. These islands teemed with a variety of small sphen-odontians that would scuttle away under rocks at the first signs of danger. However, if caught in the open, *Planocephalosaurus* and *Di-phydontosaurus* would have been no match for the agile *Terrestrisuchus*. Even the larger *Clevosaurus* seen in the foreground would need to be wary of attack. The conifers are based on the fo-liage form taxa *Pagiophyllum* and *Brachyphyllum*.

Figure 10.1. One of the Triassic fissure fills at Cromhall Quarry. The reddish sediments are the Triassic sands and mudstones filling in an ancient cavern that was etched out of the Carboniferous Limestone.

question of sediment ages being central to most of the debates. The reason is that the fissure and cave systems are completely isolated, both from each other and from other, more typical continuous sequences of Early Mesozoic sediments. Determining the ages of these sequences is fraught with difficulties.

What's in a Cave?

Karst scenery is a term that refers to the extensive development of caves, underground water channels, and sinkhole (doline) systems in landscape predominated by limestones. The name comes from the Karst region of the eastern Adriatic, where these natural features are common. It actually derives from the Slavic word *kras*, which means "a waterless place" (because all the water drains underground from the surface). Limestone is readily dissolved by water, consequently, below the water table, the groundwater seeks out and dissolves cracks and lines of weakness in the bedding planes of the limestone (fig. 10.2). Over time, these underground chambers develop and enlarge as dissolution of the limestone continues. The next phase of cave development occurs when the water table drops, leaving the upper part of the chambers and channels dry. Dissolution still continues in the lower part of the caverns, and may extend deeper to open up new chambers. Eventually openings extend up to the surface, and at this stage, sediments begin to be washed in from outside by rainwater. As contact with the surface increases, fluctuations in the temperature, humidity, and carbon dioxide content all begin to increase, bringing about instability in the walls and roof of the caves, which start to collapse.

During Triassic times, the ancient land surface of present-day southwest Britain was mostly limestone. The presence of well-developed karst features with massive underground channels and huge rounded caverns is strongly suggestive of high humidity (rainfall) around that time. As these features developed and matured, sediments washed into them from the outside, bringing with them remains of animals living up at the surface (fig. 10.3). They are thus perfect traps to entomb the remains of life at the surface.

Although all limestone caves and fissures can be attributed in some part to dissolution of the limestone by water, tectonic influences can also play a role. Movement of the earth's crust also tends to open up cracks in the limestone. These are typically narrow, parallel-sided vertical slots running through the limestone. Water passing through will dissolve more limestone and thus enhance the feature.

The history of research into the Triassic fissure and cave deposits goes back to the nineteenth century, but when the first work was carried out greatly depends on one's interpretation of a cave. Halstead and Nicoll (1971) regarded Riley and Stutchbury's (1840) account of dinosaur remains from Durdham Down as the earliest record of such work. The latter authors considered the bone-bearing breccia to be the result of a collapsed cave roof. On the other hand, some workers, including Pamela Robinson (1957), chose not to recognize the Durdham Down sediments as part of a cave system. For her, the earliest paleontological work on the fissure deposits was conducted by Charles Moore, who in the mid-1860s undertook a survey of the Mesozoic deposits in the Mendip region. Moore's major find was at Holwell Quarry, where he discovered teeth of haramyid mammals and reptiles. Unfortunately, the Durdham Down locality is no longer accessible, having been completely buried under the city of Bristol and its parklands. Additional studies of Mesozoic sediments in southwest Britain were intermittent until the 1930s, when renewed prospecting in the local quarries began a sustained period of fieldwork that continues to this day.

A number of different localities are recognized at Carboniferous Limestone Quarries in the Mendips, Cotswolds, and southeast Wales (Fraser 1985). Over the years, here in the tranquil English countryside of stone bridges and thatched cottages, many quarries have exploited the limestone for road construction and for fluxing stone for blast furnaces. In the middle part of the twentieth century, the limestone was worked by firing small explosive charges in the quarry faces, and the stone was loaded by small mechanical shovels or even by hand. This slow, methodical method of working meant that any fissure fillings that were encountered tended to be either left in place and worked around, or slowly removed and dumped in an abandoned part of the quarry. What was a downright hindrance to the quarry operators ultimately became a treasure trove for paleontologists. The fact that any exposed fissures, or at least the sediment infills, tended to be abandoned turned into an added bonus for paleontologists, because there was no real urgency to complete the fieldwork! It was possible to make repeated return trips to spoil piles to continue searching the weathered surfaces.

In the late 1930s, Walter Kühne took to examining a variety of quarries in the area, starting with Moore's Holwell locality. Here Kühne found additional microvertebrate teeth, including two triconodont specimens. During this time Kühne spent many days walking and cycling the hedge-lined highways and byways in his quest for early mammal remains. Soon after his success at Holwell, he discovered a new site in Somerset. At a quarry in Windsor Hill, he found sediments containing numerous bones of a new tritylodont. This locality is still notable for having yielded over 2000 tetrapod bones, all of a single genus (*Oligokyphus*). Although as a German national Kühne was interned during the war, upon his release, he resumed prospecting work with his wife, Charlotte. In 1947 he discovered early mammal remains in Duchy Quarry, Glamorgan. This material was located in one of the spoil heaps that the quarry had put to one side. Kühne returned to his native Germany in 1951, but Pamela Robinson and Kenneth Kermack of University College, London, continued the work that Kühne had begun. Robinson concentrated her activities in the Mendips while Kermack worked principally in Wales. Other quarries in South Wales were soon found to contain the remains of small vertebrates, and in the period between 1952 and 1984, research groups from University College found fossiliferous fissure infills in 16 different quarries in Wales (Evans and Kermack 1994). Most of the material collected by these groups is now housed in the Natural History Museum. Paleontologists from Cambridge University, under the direction of Rex Parrington, also became involved in the research, and there are consequently significant collections in the Museum of Zoology at Cambridge, England.

With the development of the motorway system in Britain, the demand for stone aggregate grew, and new and bigger quarries opened up. At the same time, quarrying procedures became more and more sophisticated, and increasingly bigger areas could be blasted and removed in ever-decreasing time frames. Prospecting and collecting for microvertebrates thus became much more difficult. Quarries had to be regularly visited, but it was still possible for whole fissure systems to be removed with one or two blasts and be lost forever. In the late 1900s new localities were becoming increasingly harder to find, although on occasion, a new Mesozoic exposure would appear in an active quarry. The collecting effort from numerous fissure systems in Tytherington Quarry by crews from the University of Bristol in the 1970s and 1980s was particularly impressive. The most recent work has been undertaken by researchers at Bristol, Cambridge, Aberdeen, and London.

Over the last 10 years, new fissures have been encountered, along the fresh faces of active quarries, but with the exception of Tytherington (fig. 10.4), only sparse quantities of bone have been recovered in this way. The greatest successes have resulted from visiting the abandoned quarry faces of the sites first worked by the University College and Cambridge University teams. Together with Gordon Walkden of Aberdeen University, I worked the old west wall of Cromhall Quarry intermittently from 1979 to

A

B

Figure 10.2. Modern karst features. (A) Large underground channel that has been dissolved out of limestone. (B) Scalloped walls of an underground passageway.

Figure 10.3. Water washing through cave systems can bring in tremendous amounts of sediment. Sediments transported into ancient cave systems would have also carried in animals and plants from the world above.

Figure 10.4. Ancient fissure choked with dark gray mudstone at Tytherington Quarry.

1993. This work resulted in the recovery of hundreds of thousands of isolated elements and numerous blocks of unprocessed bone-bearing matrix. Ruthin Quarry in Glamorgan has expanded enormously since Robinson collected from the locality in the 1950s. Nevertheless, despite the vast pit that was exposed in the 1980s, it was still the original fissure situated above the quarry offices that yielded almost all the fossiliferous matrix that Gordon and I collected from this locality.

The Age of the Sediments

Early in the research on the fissure phenomenon, a broad two-part division was recognized in the assemblages. On the one hand, there were infills containing mammal remains, and these were also typically found to contain palynomorphs of ferns, cycads, and conifers. On the other hand, there were infills apparently lacking any mammal remains whatsoever, and these were also seemingly devoid of palynomorphs. The former were considered to be Latest Triassic (Rhaetian) to Early Jurassic in age, whereas the latter were regarded as being somewhat older and were typically referred to the Norian. Robinson (1957) believed that the fissure systems lacking mammal remains were mature, solution-etched features, whereas those that contained mammals tended to be immature, slotlike fissures. Certainly there appeared to be two distinct types, and Shubin and Sues (1991) referred to these as the Complex A (lacking mammals) and Complex B (containing mammals) type assemblages.

Nevertheless, the absolute distinction was muddied by two different pieces of evidence. Fraser et al. (1985) described two therian mammal teeth from a cave system at Emborough quarry, a locality that until then fell fair and square into the Complex A type. It was a large, open, mature system that contained abundant reptile remains, in particular the gliding reptile *Kuehneosaurus*. Despite systematic sampling, no palynomorphs had ever been recovered from this locality. On the basis of a topographic study of the immediate vicinity and the relationship of the cave sediments to patches of Rhaetic and Jurassic sediments, Robinson argued convincingly for their Norian age. Indeed, of all the fissure sediments, the strongest case for a Norian age was undoubtedly that for the Emborough sediments. Fraser et al. (1985) did not dispute the age assignment but instead suggested that the presence or absence of mammal remains was not a foolproof method of distinguishing between the two types of assemblage. However, a few years earlier, Marshall and Whiteside (1980) did contest the age assignments. Discoveries by Bristol University crews at Tytherington Quarry of a number of mature solution features containing reptile-only assemblages were indicative of typical Complex A faunas and sediments. However, in one such fill, Marshall and Whiteside recovered palynomorphs consistent with those from the European Rhaetian—thus seemingly younger than previously supposed. These authors suggested that none of the fissure tetrapod assemblages was older than Rhaetian. Even more significant were the associated acritarchs and dinocysts. These offered strong evidence of marine influences, which in turn quashed the idea that the assemblages were representative of upland communities.

Although it is difficult to envisage a scenario where the marine components might have been introduced into an upland environment, the evidence that Marshall and Whiteside cited to argue that all Complex A assemblages were no older than Rhaetian is weaker. In fact, there is now a school of thought that suggests that the Complex A assemblages could even extend down into the Carnian. Simms (1990) first made the observation that periods of high humidity provided the optimum conditions for the formation of the extensive solution features, and that this could well be coincident with the middle Carnian pluvial event. That does not mean that the sediments necessarily began to accumulate at that time. Benton (1991) and Benton and Spencer (1995) noted that certain components of the German Stubendsandstein (lower to middle Norian) and the Lossiemouth Sandstone (upper Carnian) are closely comparable with forms from the assemblages at

Cromhall Quarry. In particular the procolophonids, *Leptopleuron*, and an unnamed form from Cromhall, as well as certain sphenodontian taxa, seem to be particularly close. Nevertheless, there has been no suggestion that any of these forms are congeneric, and thus the closely comparable forms cannot be regarded as strong evidence for the age of the fissure sediments.

Perhaps more significant is the record of a single *Aetosaurus ferrox* scute from one of the fills at Cromhall Quarry (Lucas et al. 1999). Aetosaurs have a well-established value in correlating nonmarine Upper Triassic units. Their scutes are highly diagnostic, and thus isolated fragments of these aetosaurs can be confidently identified to genus and species. The scute from Cromhall is a match for scutes of *Aetosaurus ferrox* from the mass death assemblage from Stuttgart-Heslach, Germany, albeit somewhat smaller. The temporal range of *Aetosaurus* has been established as early middle Norian, and therefore its presence in an infill at Cromhall is highly suggestive of a similar age for this particular unit (equivalent to the Revueltian land vertebrate faunachron [LVF]).

It is important to remember that the age of other Triassic fossils from Cromhall Quarry could be substantially older or younger. This is because the quarry contains several separate karstic fissure infills, and as we shall see, they were evidently subject to many episodes of filling during the Late Triassic. Indeed, in one part of the quarry, a normally bedded sequence of Triassic deposits overlies many immature slotlike fissures. In addition to isolated sphenodontian bones, the cover sediments have been found to contain fish remains that are highly suggestive of a marginal marine habitat. This in turn implies a Latest Triassic (Rhaetian) or even Early Jurassic age assignment for this section in the quarry.

Most authors would now agree that the region that is now southwest Britain lacked extensive upland areas during Late Triassic times. Although the precise details are still a matter of some conjecture, during early Late Triassic times desert plains and giant playa lakes (Tucker 1977, 1978; Jeans 1978) stretched across the entire region. These playa lakes were really just a distal expression of the extensive carbonate platforms and marine basins that extended over vast areas of southern and northwest Europe. Mudflats, ponds, lakes, and extensive algal mats seem to have been a major feature. At times there may have been vast stands of horsetails stretching across the wide-open landscape. This would have been an ideal habitat for a variety of insects that in turn would have attracted small insectivorous tetrapods, including early mammals that scuttled in the undergrowth, and small cursorial sphenosuchian crocodiles and their allies that lurked by the water's edge. It is not hard to imagine small pterosaurs swooping across this landscape, taking small insects over the surface waters, before returning to their roosts at night on small rocky outcrops. As the Triassic came to a close, the seas advanced so that marine influences began to dominate, with a shallow sea encroaching over the land, leaving only small islands as refuges for terrestrial animals.

Although the fissure systems are still divided into two broad categories (and they will be discussed separately in this chapter), we have seen that they cannot be unequivocally distinguished purely on the basis of their faunas. More importantly, it is not possible to say that all the animals from one locality inhabited the same environment or were even alive at the same time. Typically there are a number of Mesozoic fissure and cave systems at each quarry. Although there are similarities among the different vertebrate assemblages, there are typically subtle variations in faunas from one fissure to the next, and it would appear that the age of the individual fissure deposits is likely to be one variable responsible for these differences. However, it is unlikely to be the sole factor; other variables include the role of predators as agents of accumulation and sampling of slightly different paleoenvironments.

Preservation

Many of the Triassic fossils referred to in this book instantly provide the observer with a glimpse, or at least a strong clue, of the animal's original habitat. For example, associated plants and insects on the same bedding planes as *Tanytrachelos* at the Solite Quarry elicit an image of a watery pond with aquatic vegetation (chapter 11). The wonderfully complete skeletons of cynodonts and archosaurs from the Chañares (chapter 4) provide ample evidence of small carnivores, insectivores, and herbivores scurrying around a fully terrestrial environment. One look at the battery of cheek teeth in the gomphodont, *Massetognathus*, testifies to its herbivorous habit. But the secrets of the fissure assemblages are not so readily unraveled. Isolated elements of many different taxa all jumbled together provide no clear picture of the animals from which they came. Having freed the bones from their limestone matrix by dissolution in acetic acid baths, one is left with the painstaking task of building up three-dimentional jigsaw puzzles, but without a picture of the real thing to guide you. As if this were not enough of a brainteaser, numerous jigsaw puzzles have been mixed in together. Furthermore, none is 100 percent complete, and most are far less than 50 percent complete!

In places, the sediments lodged within the fissures are packed full of isolated bones that are usually completely disarticulated but frequently exquisitely preserved, so that all the details of articulation facets and muscle scars are readily apparent. These particular assemblages may well be attributable to rainwater running off the ancient land surface and carrying the remains of animals that had died at the surface. In other instances, articulated material is found, and it is possible that these were the skeletons of animals that sought shelter in the cave systems and ultimately died there. It is also possible that some of the assemblages are the result of floodwaters and surface runoff washing through the middens of predators that were located near the entrances of these cave systems. Signs that small terrestrial crocodiles may have been responsible for creating such middens include some of their isolated teeth among these bone assemblages lacking roots and exhibiting resorption pits at the bases of the crowns. Clearly these teeth could have been shed while eating prey. On the other hand, some sediments bear isolated teeth of the same crocodiles with the entire root still preserved. Furthermore, a variety of crocodile bones also occur in many of the deposits, indicating that it was not just the washings of their middens that contributed to the bone assemblages. Further discussion of the possible role of predators is undertaken later in this chapter.

Although vertebrates are the principal fossiliferous components of the fissures, rare invertebrates are sometimes encountered (for example, the conchostracan *Euestheria*, and remains of beetles and millipedes). Palynomorphs in the B suite provide us with some indication of the type of vegetation cover, and the presence of charcoal in many of these fills is indicative of extensive forest fires—perhaps started by lightning. If these were indeed island communities, there would be little chance of escape for many of the tetrapods, and many would have perished in the wildfires.

The Holy Grail

Upland regions of the world are rarely the sites of sediment deposition; instead, they are the areas of greatest erosion. I am certainly forced to admit that the chances for representatives of upland faunas and floras becoming entombed in sediments in their preferred habitat are on a par with my beloved Scotland winning the World Cup! Thus for many, upland faunas constitute the "holy grail" of paleontology. As mentioned earlier, the fissure deposits were long seen as the answer to that search. Robinson (1957) thought that plotting the fissure localities on a relief map of the area during Upper Triassic times provided ample evidence of this fact. She suggested that by comparison with other Late Triassic vertebrate faunas, the fissure assemblages were markedly different. She considered forms such as rhynchosaurs, phytosaurs, and metoposaurs to be restricted to lowland areas such as river valleys. Their conspicuous absence from the fissure fills and their domination by relatively small forms (such as sphenodontians) she attributed to the fact that these represented upland faunas.

Robinson envisaged the fissure animals living in a denuded semiarid to arid landscape on top of hills formed by Carboniferous limestone. However, a reexamination of the geology against modern-day topography gives no reason to believe that the fissure openings were significantly elevated during Norian times. The lack of large- to moderate-sized tetrapods that typify Late Triassic terrestrial assemblages in many areas, including the Chinle, Dockum, and Godvari Valley, might be attributable to the fissure systems representing a specialized paleoenvironment. Perhaps the combination of limestone rocks and paleolatitude resulted in a less common vegetation type that supported a less conventional fauna. Alternatively, perhaps the larger forms did inhabit the area, but selection biases peculiar to fissure and cave systems (for further discussion, see below) resulted in their exclusion from the fossil assemblages. It is therefore interesting to note that in Norian exposures at Bendrick Rock and elsewhere in southwest Britain, dinosaur tracks do hint at the presence of larger animals in close proximity.

In short, just as the hopes of a World Cup for the Tartan Army were dashed in 1978, so too has the concept of some of the fissure assemblages representing an upland community been quashed. Although the concept of a complete division of the fissure assemblages into two separate groups has now been abandoned, it cannot be denied that the assemblages represent deposition over a considerable time range that spans the Triassic-Jurassic boundary. On one hand, the opposite ends of this time spectrum are characterized by different faunal assemblages, and it is still sometimes useful to discuss the Complex A and B faunas separately. On the other hand, the two assemblage complexes are so intimately interrelated that it can be difficult to separate them. For this reason, although the subject matter of this book does not strictly include the Early Jurassic, the Complex B faunas are also discussed.

Playa Lakes and Mudflats—Complex A Assemblages

These faunal assemblages comprise almost exclusively nontherapsid reptiles, and in most cases, over half of these are sphenodontians. The remaining component typically consists of small archosaurs, although in a couple of instances, namely at Emborough and Batscombe Quarries, the predominant tetrapod was found to be one of the gliding kuhneosaurs. In addition to Emborough and Batscombe, Tytherington and Cromhall Quarries in England and Ruthin and Pant-y-ffynon in Wales are the principal localities yielding Complex A type assemblages (i.e., those lacking mammals).

The Sphenodontians

Today represented solely by the tuatara (*Sphenodon*) living on a few isolated islands off New Zealand, the sphenodontians (or rhynchocephalians) were a diverse group during the Early Mesozoic. The tuatara gets its common name (although tuatara is anything but common!) from the Maori word meaning "spiny back." *Sphenodon* is still curiously encumbered with the "living fossil" label. Although those sphenodontians living in southwest Britain during the Late Triassic might also have been island inhabitors, there the close similarity ends. *Sphenodon* is no more like its Triassic and Jurassic predecessors than Jurassic lizards are like modern-day forms, yet lizards are never considered to be living fossils. Today we can see the great diversity of habitats that lizards occupy, and we know that there are over 2500 species. Despite a similar body plan, each family of lizards has its own special characteristics that have evolved over millions of years. By contrast, there is but a single living genus of sphenodontian, and it is difficult to picture the Mesozoic forms as living taxa, each having its own distinct adaptations for a particular way of life. Rather than regarding *Sphenodon* as a specialized animal with a body plan superficially similar to certain Mesozoic forms, many authors have chosen to consider the Mesozoic forms as a uniform group that occupied the same niche as the modern survivor.

Our ignorance of the group and an apparent inability to imagine the fossil forms as living animals has unfortunately perpetuated the concept that *Sphenodon* is a reptile that has hardly changed from the time of the first dinosaurs. It is an oft-repeated phrase that fossils are testimony to species that did not adapt to the environment and are now extinct. In that case, although Triassic and Jurassic sphenodontian genera such as *Clevosaurus* and *Homoeosaurus* did not adapt to changing environments, it is clear that the lineage from which *Sphenodon* is descended was successful in this. Obviously we do not know the behavior patterns of the fossil forms, nor do we know any of the details of their physiology. However, by comparison with other living reptiles, the ability of *Sphenodon* to be active at relatively low temperatures (52°F), among other unusual physiological characters, is perhaps indicative of an animal that has altered quite a bit from Mesozoic members of the group.

If we look closely at the sphenodontians in the Complex A fissure deposits, we can see a diverse assemblage that clearly occupied different niches in order to avoid competition with each other. First of all, there is a size difference: the smallest taxon is *Diphydontosaurus*, in which an adult probably measured not much more than 20 centimeters from snout to the tip of its long tail. It bore acutely conical marginal teeth that, although posteriorly exhibiting the characteristic acrodont tooth implantation of sphenodontians (teeth fused to the summit of the jaw margins), anteriorly are pleurodont (set in a groove along the apex of the jaw). The small and relatively delicate skull with slender mandibles was clearly not well adapted to crushing large and tough insect exoskeletons. It seems more likely that *Diphydontosaurus* used more of a snap-and-gulp feeding action on small insects crawling over the vegetation.

Probably one of the most abundant genera is *Planocephalosaurus* (fig. 10.5). This was a slightly larger animal than *Diphydontosaurus*, and one that possessed a wholly acrodont dentition. The jaws were robustly built and capable of exerting a more powerful crushing action than *Diphydontosaurus*. It seems likely that *Planocephalosaurus* fed upon somewhat larger insects, possibly ground beetles and small cockroaches scuttling in the leaf litter.

Clevosaurus is perhaps the most ubiquitous sphenodontian (Robinson 1973). Although first described on the basis of remains from fissure sediments at Cromhall Quarry in Avon, sphenodontian bones attributed to this genus have been found in North America, China, and South Africa. At least three species have been referred to the genus *Clevosaurus* from Cromhall Quarry alone. *C. hudsoni* was the first to be described and is certainly the best known, with a handful of articulated skulls and postcrania in addition to numerous disassociated elements (Fraser 1988b). A smaller species, *C. minor*, is known from several infills at Cromhall but has never been found in the same deposits as *C. hudsoni*. This suggests that they were nonoverlapping in age. On the other hand, a form bearing distinct, transversely broadened teeth, *C. robusta*, does occur together with *C. minor*. It seems likely that *C. robusta* may have at least been a facultative herbivore, and even if an insectivore, it would have had very different dietary requirements than *C. minor*.

Other sphenodontian taxa described from the fissure fills are much rarer, but some are sufficiently large and have sufficiently deep jaws and a robust dentition that also hint at herbivory. In fact, the dentitions of *Sigmala* and *Pelecymala* (Fraser 1986) are reminiscent of the living agamid lizard, *Uromastix*, which is considered an obligate herbivore. Certainly herbivory is strongly favored for at least two Late Mesozoic sphenodontians from North America: *Eilenodon* from the Late Jurassic, and *Toxolophosaurus* from the Early Cretaceous. On the other hand, it is difficult to judge just how restricted were the diets of these Triassic sphenodontians. Even supposedly obligate herbivores will take a variety of insects, perhaps accidentally, as they munch through foliage. Many insectivorous lizards such as anoles will

Figure 10.5. Skeletal restoration of the sphenodontian *Planocephalosaurus* (after Fraser and Walkden 1984).

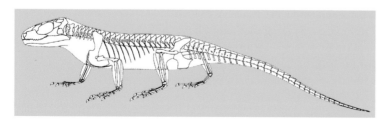

contentedly ingest flower petals and pollen as they search for insects, and some reptiles may be insectivores when young and herbivores as adults (Pough 1973).

Diphydontosaurus (Whiteside 1986) has been recorded from several different localities in the Bristol Channel region; however, whether there were many different species remains an open question. The fact that there are consistent, albeit minor, differences among the fossils from some of the different quarries hints at a number of different species. For instance, at Tytherington, from which the genus was first described, the mandible, as illustrated by the holotype, has in excess of 30 teeth. *Diphydontosaurus* is also abundant at nearby Cromhall Quarry, but here the mandible typically has in the region of 20 or so teeth. At just what level these differences represent biologically distinct taxa is impossible to say.

In the past there was a tendency among vertebrate paleontologists to oversplit taxa. The discovery of a new specimen exhibiting relatively minor differences from previously described specimens apparently warranted the erection of a new species. Thus, for example, we had a plethora of dicynodont species from the Karoo that were each known from but a single specimen (e.g., Broom 1913). Recognizing that there is often considerable variation within species, including such characteristics as the configuration of cranial sutures, there has been a general move in recent years toward a more conservative approach.

Today, many genera that once held many species have reverted to monospecific genera. At the same time, we must remember that certain extant species are recognized purely on features such as color variation, behavior patterns, and geographic distribution and exhibit minimal osteological variation among closely related species. Estimating the diversity levels of fossil assemblages is therefore a complex undertaking, and one that is always open to considerable skepticism.

Turning back to our question of the variability observed in *Diphydontosaurus* elements, if there were indeed several distinct species, we then have to ask whether they were separated spatially or temporally. This question is almost impossible to answer. Correlating the precise ages of the sediments lodged within each of the individual fissures has not proved possible because they are mostly isolated from overlying and continuous well-bedded sequences.

A cursory inspection of habitats where anole lizards are abundant might indicate that somehow or other a variety of similar species are defying accepted concepts in ecology by living together in completely overlapping ranges and with similar feeding habits. However, closer study indicates that the species are frequently separated by perch height: some live in the grasses, others on twigs and woody surfaces, and others high up in the tree canopy. Detailed studies also reveal recurrent differences in morphotype. Those species that have narrow perches, such as the distal branches of trees, have shorter limbs. This specific compartmentalization of feeding niches permits these otherwise similar anoles to live together without too much direct competition.

Such behavior patterns leave no trace in the fossil record, and it is doubtful whether it would be possible to separate out all the individual taxa solely on the basis of skeletal data. In all likelihood, the diversity of Triassic sphenodontians was greater than a literal reading of the fossil record implies. Similar species may have fed at different times of the day or lived in different microhabitats (like the species of *Anolis*) or had specific dietary tastes, with some feeding on just one species of insect and others having more catholic tastes.

Kuehneosaurs

The kuehneosaurs constitute one of the more enigmatic groups of tetrapod in the fissure fills. Growing up to a meter in length, these diapsids are characterized by extremely elongate and straight ribs that were attached to long transverse processes on the vertebrae. Just like the modern-day gliding lizard, *Draco*, kuehneosaurs almost certainly possessed a membrane that was supported by the ribs. By extending the ribs out to the sides, the membrane would be stretched to produce an aerofoil suitable for gliding.

Two sites, the Emborough and Batscombe Quarries, have yielded abundant kuehneosaur remains. At first, Robinson (1962) felt that the two sites yielded different species, *Kuehneosaurus lattissimus* and *Kuehneosaurus latus*. But after the discovery of *Icarosaurus,* a similar animal from the Newark Supergroup (see chapter 11), she considered them to be two different genera, naming the Batscombe form *Kuehneosuchus.* However, the supposed differences between the Emborough and Batscombe forms have not since been confirmed.

Although the kuhneosaurs were first thought to have affinities with lizards, their systematic position is now less certain, and they may even have closer affinities with basal archosauromorphs.

The Earliest Crocodiles

For much of the Mesozoic, crocodylians were largely marine animals, inhabiting coastal waters or even the open seas. Today's crocodylians are also mostly aquatic, or at least semi-aquatic, although many of them inhabit freshwater environments. A first glance at these scaly creatures may well be deceptive. Alligators in the Florida Everglades will typically appear sluggish and slow-moving as they lie basking in the sun or partly submerged in swampy areas, with snowy egrets peacefully perched on their backs. This calm belies more sinister behavior: alligators are capable of sudden, explosive bursts of activity as they attack prey. Many a family pet has fallen unsuspecting victim to an unexpectedly not-so-sluggish gator. And this type of predatory behavior is widespread among the group. Crocodiles on the Nile tend to congregate according to size, and for good reason: they are fierce predators, and potential prey tends to be anything smaller, or at least less ferocious, than them, including other crocodiles. Overtly sedate animals can suddenly become agile and quick-moving predators. There was, however, nothing deceptive about Triassic crocodiles—they definitely looked the part of active hunters.

The sphenosuchian crocodile, *Terrestrisuchus,* is one of the most widespread predators in the fissure assemblages. Just like the sphenodontians, there appear to have been several different species that may well have been restricted in time and space. First described on the basis of partially articulated and associated material from Pant-y-ffynon Quarry, the proportions of *Terrestrisuchus gracilis* (Crush 1984) are not in doubt. Completely disassociated material of *Terrestrisuchus* was recorded from Cromhall Quarry, and close examination of certain of the cranial elements revealed subtle differences from those of *T. gracilis* from Pant-y-ffynon—in particular in the shape of the squamosal. Cromhall has produced an articulated postcranium from a sphenosuchian crocodile (most probably *Terrestrisuchus*) that has different limb proportions to that of *T. gracilis.* This latter individual has markedly shorter hind legs and is quite clearly a different taxon.

The sphenosuchians were fast-moving, entirely terrestrial predators. In general, sphenosuchians were much smaller than today's representatives of the group, ranging in size from 2 to 4 or 5 feet.

Proportions of the axial skeleton and the relatively short forelimbs have led some authors to postulate that *Terrestrisuchus* was a bipedal cursor (Benton 1990, 117). Bipedalism is well known in certain living lizards and at first glance might seem to be a good candidate for approximating sphenosuchian movement. Snyder (1954) discussed the characteristics of modern bipedal lizards and emphasized a number of features associated with bipedalism, including the disparity in limb size, a short trunk relative to a long and robust tail, the relatively narrow intraacetabular width of the pelvis, increase in size of the sacral ribs with a great degree of fusion, and a pronounced preacetabular process on the ilium.

Several lizards in three different families adopt a bipedal posture when running. One of the best known is the basilisk. In these forms, the hind legs still assume a sprawling gait and describe a wide lateral arc in the recovery step. Bipedalism has been assumed for many fossil taxa that bear at least a superficial resemblance to lizards. For instance, the

Permo-Triassic diapsid *Lacertulus* was described by Carroll and Thompson (1982) as a biped on the basis of much-reduced forelimbs. However, some of these features, such as the intraacetabular width, are often difficult, if not impossible, to assess in fossil taxa, where individual specimens have typically been crushed and distorted. Moreover, unlike the modern basilisk, sphenosuchians have a much better defined and deeper acetabulum, and in all probability a more erect posture.

Indeed, Parrish (1987) described adaptations for erect posture not only for spheno-suchian crocodiles, but all archosaurs with crocodile-normal tarsi. Parrish noted the prominent anterior flange on the iliac blade as well as the deeper acetabulum. This flange is associated with the forward migration of the origin of the iliofemoralis muscle group, which increases their mechanical advantage as femoral protractors. More importantly, the proximal tarsals in sphenosuchians, the astragalus and calcaneum, are strongly suggestive of an erect posture. Modern crocodiles, of course, have reverted to a sprawling posture, and this is viewed as a secondary adaptation to an aquatic lifestyle.

We have tentatively restored *Terrestrisuchus* as being capable of both bipedal and quadrupedal gaits, but the evidence is not necessarily that strong. Certainly the specimen from Cromhall with the shorter hindlimbs is not a good candidate for having been a bipedal sprinter!

Whether bipedal or quadrupedal, sphenosuchians were highly active predators, and it is conceivable that the four-chambered heart that is characteristic of crown group croco-dilians may be a relic of the origins of the group among active ancestors. It is even possible that the four-chambered heart was a function of elevated metabolic rates and reflects some measure of "warm-bloodedness" in these early crocodiles. Perhaps a more likely ex-planation is that the four-chambered heart is associated with the development of an erect bipedal posture. With the elevation of the head significantly above the level of the heart, it is possible that the increased pressure necessary to pump blood to the brain might be too high for delicate tissues such as the capillaries of the lung to withstand. As a consequence, a double pumping system would be essential.

Sphenosuchians and potential sphenosuchian relatives are known from a variety of Tri-assic localities. *Pseudohesperosuchus* from the Norian Los Colorados Formation of Ar-gentina is similar in size to *Terrestrisuchus*, although it is only known from a single fairly complete, but poorly preserved skull (Bonaparte 1972). The best known of the Triassic sphenosuchians is *Saltoposuchus connectens* from the Stubensandstein (Norian) of Ger-many. It is actually similar to *Terrestrisuchus*, and it has even been suggested that the two are congeneric (Clark in Benton and Clark 1988). Sereno and Wild (1992) rejected this idea, citing some relatively minor differences between the two forms. However, the known specimens of *Saltoposuchus* are altogether larger than those represented in the fis-sure infills, and it may simply be that the *Terrestrisuchus* material originates entirely from immature individuals. The final Triassic sphenosuchian that ought to be mentioned here is *Hesperosuchus* from the Chinle Group. As noted in chapter 7, the most complete spec-imen originates from the famed Ghost Ranch Quarry. In overall appearance, it was very like *Saltoposuchus* and *Terrestrisuchus* and bore a double row of osteoderms running down the length of the back, with densely packed gastralia protecting the belly.

Gracilisuchus from the Chañares Formation of Argentina (chapter 5), although lacking some of the fundamental characteristics of crocodilians, such as an attenuated posterior process of the coracoid, elongate ulnare and radiale, and a quadrate contacting the prootic, also shows remarkable similarities to sphenosuchians, particularly in the orienta-tion of the squamosal. Another enigmatic form that bears a resemblance to the spheno-suchians is *Erpetosuchus* from the Lossiemouth Sandstone (chapter 8). Like *Gracil-isuchus*, it has a highly fenestrated skull with a large antorbital fenestra and a similar temporal region with an anteriorly inclined quadrate. And like *Gracilisuchus* it lacks elon-gated proximal carpals, the quadrate does not appear to have contacted the prootic, and there is no elongated posterior process on the coracoid. Whether either is related to the

Plate 10.2. During the Late Triassic, what is today southwest Britain was a low-lying coastal plain that gradually became inundated by an advancing sea. Outcrops of limestone became islands that were ultimately completely submerged during the Early Jurassic. In this environment a lone sphenosuchid crocodile avoids the surf, nimbly negotiating the rocks, as it searches for small sphenodontians seeking shelter. It is thought that this gracile predator may have usually walked on all fours, but rose up onto its hind legs when the need to move swiftly arose.

early radiation of the crocodiles must await further studies and new discoveries. Nor is it clear whether the sphenosuchians actually form a natural monophyletic group or simply a grade of organization.

Other pseudosuchians in the fissure deposits that are currently being described are also gracile animals that appear to be adapted to a highly cursorial existence. One of these is similar in overall proportions to *Terrestrisuchus*, although for an archosauromorph it has relatively short cervical vertebrae, and in this respect is aberrant. However, the shape of the maxilla and squamosal is almost indistinguishable from those of *Terrestriscuschus*. The crurotarsal ankle joint is typical of all pseudosuchians.

It is also worth reiterating here that several relatively small, long-legged, gracile archosaurs are known from the Triassic. As we have seen, the Chañares has a particularly diverse number of such forms. As well as *Gracilisuchus*, there are *Tropidosuchus* and the poorly known dinosauromorphs, *Lagerpeton*, *Marasuchus*, and *Lagosuchus*. This trend toward gracile bodies and at least semierect postures seems to be a recurring theme among Late Triassic archosaurs. Whether changes in posture coupled with more active lifestyles are closely linked to a trend toward homoeothermy has yet to be tested in any way, but the concept that the crocodilian four-chambered heart is a relic of an even earlier "warm-blooded" ancestry is an intriguing scenario.

Other archosaurs in the Complex A fissures include the prosauropod dinosaur *Thecodontosaurus* known from Tytherington and Pant-y-ffynon (Kermack 1984), drepanosaurs and possible pterosaurs (Fraser 1988a; Renesto and Fraser 2003) and the poorly known dinosauromorph *Agnostiphys* (Fraser et al. 2002).

Variable Distribution of Taxa within the Complex A Assemblages

Although the broad similarity between the Complex A assemblages was recorded early on, it was also recognized that there are definite differences among the individual localities, and that these are much greater than those observed in the Complex B assemblages (see below). Thus at Emborough and Batscombe, the gliding kuhneosaurs were the most common elements. Sphenodontians form the predominant component of most of the Cromhall and Tytherington assemblages, but whereas *Planocephalosaurus* is the number one tetrapod at Cromhall, at Tytherington, this honor goes to *Diphydontosaurus*. The Pant-y-ffynon assemblages are unique in that they are apparently dominated by archosaurs. At Ruthin the most common taxon that has not been recorded from any of the other fissure systems is a procolophonid characterized by an unusual dermal armor. To add further to this variability, it is now known that even within one quarry, neighboring fissures (sometimes separated by distances of 1 or 2 meters) can have quite different faunal profiles.

At Cromhall Quarry, Fraser and Walkden (1983) noted that *Planocephalosaurus* is almost always the predominant tetrapod in each fissure, often accounting for about 50 percent of all the elements recovered. But the second most abundant taxon is rarely the same: in one fissure it might be *Diphydontosaurus*, in another it might be a species of *Clevosaurus*, and in a third a small rauisuchian. Moreover, there are a couple of fissure systems at Cromhall where *Clevosaurus hudsoni* is the most ubiquitous form, and in these instances, *Planocephalosaurus* is not even present.

It is significant that within each fissure system there are frequently marked vertical lithological changes, with the various rock types comprising marls, crinoidal limestones, breccias, and sandstones. These changes clearly reflect slightly different depositional conditions, yet the faunal profile from each of the different sediment types within one fissure remains remarkably constant. Fraser and Walkden subsequently argued that the faunal differences seen from one fissure to the next were not the result of taphonomic biases, but rather reflected faunal differences through time. For the most part, these are not snapshots of terrestrial ecosystems at any one instant in Triassic time.

The crinoid sands probably required hundreds or thousands of years to weather out of the limestone and accumulate in the various depressions and crevices in the ancient land surface. As these scattered residues built up, so too did the bones of generations of decomposed, scavenged, and predated tetrapods. Violent storms periodically rumbled through the region, generating flash floods that scoured the land surface, transporting the vertebrate remains and associated sediments before dumping them into the ever-widening fissures and caves.

At this stage we should consider what role predator activity might have played in sorting the bones. Mellett (1974) believed that most, if not all, microvertebrate fossil accumulations are the direct result of predators. He based his work on a detailed comparison between disaggregated scat of modern carnivores and a variety of different microvertebrate fossil collections. The critical similarities he observed were breakage at identical loci on certain bones, together with uniform size ranges. Another thing to look for would be tooth crowns from carnivorous reptiles. Reptiles regularly replace their teeth, so that in the carnivorous forms, including crocodiles and dinosaurs, tooth crowns would be regularly shed while feeding on their prey.

With this mind, we can look at the Complex A assemblages, and a couple of features speak against predators being the only controlling factor in their vertebrate composition. The first is that modern predators are often highly selective with respect to size and species preference. Localities such as Cromhall, Tytherington, and Pant-y-ffynon exhibit a wide range of bone sizes and species. Moreover, although shed crowns of the carnivorous crocodilian *Terrestrisuchus* are frequently recovered, they are found together with more complete remains of the same animal. A predatory reptile might leave its calling card, but it certainly wouldn't leave its own remains in its midden! However, this is not conclusive evidence because cannibalism is fairly common in predatory reptiles. We shall revisit the question of predator activity when we look at the characteristics of the Complex B assemblages, and we shall see a rather different story.

It seems unlikely that the polishing and breakage that we see in the bones lodged within the Complex A deposits result from having passed through the digestive tract of carnivores. Instead, it is likely due to the rough ride they experienced as they were transported to their final resting place by floodwaters.

There are two different assemblages at Cromhall Quarry (Walkden and Fraser 1993) that bear mentioning. The first occurs in a fenestral limestone that resembles a lacustrine carbonate described by Tucker (1978) from south Wales. It is characterized by polygonal cracks, marl pellets, brecciation, and root systems—features that are all indicative of exposure and pedogenesis (soil formation) (e.g., Freytet and Plaziat 1982). In addition, the laminated texture of the fill is suggestive of an origin as an algal mat or a pond. This same sediment type, albeit dolomitized, also occurs at the top of one of the more typical fills at Cromhall. What is particularly interesting is that the Mesozoic cover sediments and the Carboniferous bedrock surface are heavily dolomitized, and this in turn indicates that the lacustrine phase immediately predates or codates the lowest sediments at the unconformity.

The composition of this faunal assemblage is also different from any of those we have discussed so far. First of all, the fauna is depauperate, with just four confirmed taxa. Second, the bones are in pristine condition and include articulated material. *Clevosaurus hudsoni* is the most common form, comprising over 60 percent of the fossil assemblage. Then we have a procolophonid and a small pseudosuchian, both of which seem to be unique to this single fissure at Cromhall. As yet, neither taxon has been fully described, but the procolophonid has two or three quadratojugal spikes and is perhaps closest to *Hypsognathus* from the Newark Supergroup (see chapter 10). The pseudosuchian shares some characters with sphenosuchian crocodiles. The fourth taxon is *Diphydontosaurus*.

The second is a slotlike fissure that is intimately associated with cover sediments, and both contain disarticulated fish and sphenodontian (*Clevosaurus* and *Diphydontosaurus*) remains. The fill is relatively ordered with repeated grading of the sediments, perhaps indicative of being open under a body of water when the sediments were arriving. Salt pseudomorphs demonstrate that this water bordered a saline area, such as a giant playa or a transgressing sea. Indeed, the rich fish fauna is highly suggestive of the latter.

It is not unreasonable to postulate that these two somewhat different Complex A assemblages, and their hint of some nearby marine influences, represent something of a transitional stage. The next phase of fissure infilling took place within the context of even greater marine influences.

The Sea Advances—The Complex B Assemblages

At the same time that the sea began to advance, there seems to have been a concomitant influx of synapsids into the area. These included morganucodontids and kuehneotheriids, forms that are traditionally considered as true mammals. Perhaps the closer proximity of full marine conditions had an ameliorating effect on the climate, and this in turn may have been responsible for the observed faunal changes. Nevertheless, sphenodontians are still the most abundant tetrapods in the majority of Complex B sediments (which contain mammals). Furthermore, a read through much of the earlier literature on the fissure deposits gives the impression that local tetrapod diversity levels dropped off dramatically in the Early Jurassic. Because early workers focused their attention almost exclusively on the synapsids, the more abundant nonsynapsid remains tended to be overlooked, and collectively, they were assigned to one or at best two different taxa. However, more detailed studies looking at all aspects of these bone assemblages have revealed significantly greater diversity levels (e.g., Fraser 1988c).

Even so, few Complex B assemblages have diversity levels that are comparable with the majority of the Complex A assemblages. One reason for this difference might be a dramatic decrease in the size of the exposed landmasses, because smaller islands support less animal and plant diversity. But an equally plausible explanation can be sought among the potential agents of bone accumulation. Although water transport into the fissure systems is undoubtedly one such agent, as borne out by the rounding and polishing of many of the elements, there is good evidence in the case of the Complex B assemblages that the activities of predators were an equally potent force. Walter Kühne (1956) postulated that the concentration of *Oligokyphus* bones at Windsor Hill Quarry were the result of a predator. He used the relative proportion of different elements, patterns of bone damage, and the occurrence of tooth marks on some bones to support his claims. Evans and Kermack (1994) went one step further and suggested that the profile of many other Complex B assemblages could be similarly attributed to the actions of predators. Evans then went on to make some interesting observations on the differences between a couple of the Complex B assemblages. She offered a convincing line of reasoning to support her hypothesis that selective predator action contributed to the makeup of all the Complex B assemblages.

In this respect, Evans showed that research on some Holocene fissure assemblages in the Caribbean is particularly instructive. First, a selective predator would certainly explain why assemblages are dominated by a few common species. The vertebrate assemblages in the modern Caribbean fissures are known to be the direct result of owl predation, and although they contain a broad spectrum of different forms, in all instances, one or two taxa dominate the assemblage. In one case, the anoline lizard *Anolis* is the principal representative, and in another the rodent *Geocapromys* dominates. In the latter example, which is on New Providence, *Geocapromys* is the only native mammal on the island. The presence or absence of rare components varies from fissure to fissure on the same island, and they would seem almost to be accidental components of the assemblages.

A small area around St. Bride's Major in South Wales probably remained above water until early Sinemurian times. (Robinson [1957] actually named this area St. Bride's Island.) Evans (1994) noted that many of the Complex B assemblages in this region of South Wales were remarkably similar. Typically 60 percent of the bones could be attributed to the small sphenodontian *Gephyrosaurus*, with the remaining 40 percent consisting of the two mammaliamorphs *Morganucodon* and *Kuehneotherium*, with *Kuehneotherium* being the more uncommon of the two. This uniform distribution is consistent with the theory that most of the bone accumulations mirror the activities of one particular kind of selective predator or similarities in the faunae available to such predator or predators. However, Evans (1994) did note one slight departure from this faunal profile in one of the numerous fissures in Pant Quarry. Pant 4, as it was called, was dominated by a new clevosaurid sphenodontian together with *Oligokyphus*, rather than *Gephyrosaurus* and *Morganucodon*. Although *Morganucodon*, *Kuehneotherium*, and *Gephyrosaurus* were recorded from Pant 4, they were comparatively rare. To account for the differences, Evans postulated that Pant 4 may represent the activities of a different predator species and that more of the island vertebrates were represented because the fissure had a larger catchment area.

So what was St. Bride's Island actually like to live on? It was a fairly small island that at its maximum extent was probably about 20 square kilometers, and the climate was likely tropical or subtropical. There are no significant macro plant fossils, and spores indicate occurrences of club mosses, bennettitaleans, cycads, and ferns, but with the dominant plants being conifers.

A single beetle has been described by Gardiner (1961). It serves to give a tantalizingly vague glimpse of what was almost certainly an incredibly diverse insect fauna. Judging from the incredible Russian and North American Triassic insect assemblages of the Carnian, which include some extant insect families, it is reasonable to assume that the islands were populated by many of the same orders and families, including dipterans.

The fissures seem to give us a peek at an environment that is not particularly well represented in the Triassic fossil record. That is not to say that it was necessarily a specialized area of the world; similar communities might have been just as widespread. However, these communities might have been in paleoenvironments that may not have been conducive to the preservation of the animals and plants that lived within them.

If one is prepared to spend hours on end breaking down hundreds of pounds of fissure rocks in acetic acid baths, and then sifting through the hundreds of thousands of bone fragments that result from the washed and dried residue, a whole new perspective is opened up on the Late Triassic world. Even so, you would still gain just a mere glimpse of a small part of this world. For example, apart from a few palynomorphs from one or two of the infills, there are no good indicators of the vegetation cover—no remnants of foliage, wood, or fruiting bodies. Likewise, the invertebrate fossils are incredibly scarce. Occasionally conchostracan valves show up in some of the marls, and the only insect is the beetle elytron (Gardiner 1961).

This merely serves to underscore the fact that no single fossil assemblage can tell us the whole story of the Triassic world, or even provide a single chapter. Every Triassic locality is important for piecing together this fascinating puzzle. As we move across the Atlantic, we shall get another view of the northern hemisphere in the Late Triassic. Although there are a few fossils in common with the United Kingdom, mostly we shall see a different fauna. But rather than reflecting real and stark differences in the once-living faunas of North America and the United Kingdom, the few common fossils may merely represent the luck of the draw in the different depositional environments so far discovered and documented in the two regions.

chapter eleven

Time to Go Our Separate Ways

The Newark Supergroup

Introduction

At the same time that a special set of circumstances was leading to the preservation of a particular terrestrial ecosystem in what is now Britain, not far away, a different set of conditions began to prevail that would ultimately lead to the preservation of something very different. Today, a series of Early Mesozoic sediments are known to run down the eastern side of North America, and these have been the subject of intensive research for over 150 years. Deposited in a series of rift basins, the so-called Newark Supergroup is a testament to the beginning of the end of the great supercontinent Pangaea. With extensive crustal thinning, a rift zone developed along the margins of the North American and African plates. The initial rifts that opened in the earth's crust are still readily apparent in the distribution of Triassic (and Early Jurassic) sediments from Nova Scotia down to northern South Carolina, and in some ways, these rifts can be viewed as the "stretch marks" of the birth of the Atlantic Ocean. Eventually one of these great rents in the earth's surface became the primary rift, and the proto–Atlantic Ocean came into being. Today freeze frames of similar rifting systems can be observed in various places in the world, the best known being the Great African Rift Valley. Lakes Albert, Tanganyika, and Turkana in Africa, as well as the Dead Sea, have developed in the rift valley depressions, and they provide us with a model for a series of lakes that developed in the Triassic rift basins of eastern North America. A few million years from now, the two sides of the African Rift Valley will spread completely apart and become separated by a seaway. In this sense, the Newark Supergroup can in turn serve as a model for the possible future development of present-day rift systems.

But aside from being a vivid reminder of a restless earth and its constantly moving continents, the Newark Supergroup has also been the focus of changing ideas. For the better part of the last century, the sediments were considered to be entirely Triassic in age. This view changed when Paul Olsen and Peter Galton (1977) argued that the upper portions of the sequence actually were Jurassic. More importantly, as we shall see in chapter 12, the Newark Supergroup has become central to many discussions regarding the tempo of global faunal change across the Triassic-Jurassic boundary.

The Newark Supergroup sediments do not form a laterally continuous sequence, but instead constitute fairly discrete units roughly located in 12 principal basins. From north to south, the names of these basins are: (1) Fundy, (2) Deerfield, (3) Hartford, (4) Pom-

Plate 11.1. Spawned by basalt flows, smoke from forest fires plume into the sky in the Newark rift valley. In the foreground an injured aetosaur is shadowed by a large rauisuchian, while in the background phytosaurs bask on a lake shore and a group of small theropod dinosaurs troop to the water to take a drink.

Figure 11.1. Triassic basins of the Newark Supergroup.

peraug, (5) Newark, (6) Gettysburg, (7) Culpeper, (8) Taylorsville, (9) Richmond, (10) Farmville, (11) Danville, and (12) Deep River (fig. 11.1). Not all the basins are readily accessible, as major portions are buried onshore. Furthermore, extensive areas of related sediments lie buried offshore. Even those sediments exposed at the surface are not always readily accessible: dense vegetation, concrete jungles, and swift-moving rivers all conspire against the paleontologist. Nevertheless, as we shall now see, tucked away in some of the most unexpected places are fossiliferous sequences to rival any Triassic locality.

The sediments of the Hartford and Newark Basins were historically the first strata of the Newark Supergroup from which faunal remains were described. The footprint faunas described by Hitchcock (1858) were known as early as 1845, and also by that time, W. C. Redfield (1845) had described several fossil fish. Later on, even the great rivals Edward Drinker Cope and Othniel Charles Marsh got in on the act, contributing to the study of these sediments and associated fossils. They, in turn, were followed by such notable vertebrate paleontologists as Charles Gilmore, Richard Lull, Friedrich von Huene, and Joseph T. Gregory. Thus from early on, the sediments of the Newark Supergroup attracted the attention of some of the most eminent names in paleontology, and they continue to be the focus of intensive research programs.

Initially known variously as the "New Red" or "keuper," the name "Newark Group" was first proposed by Redfield (1856), who regarded all the sedimentary rocks in these rift basins as Triassic. Indeed, this was the widely held opinion until the 1970s, when the seminal paper by Olsen and Galton (1977) began to change our thinking. On the basis of vertebrate assemblages, they argued that the upper parts of the sequence in the Newark Basin were Early Jurassic. Palynological evidence provided additional support for this idea, and radiometric dates derived from the Newark basalts also yielded a Jurassic age. With further work, it became clear that parts of the Connecticut Valley and sections in Nova Scotia were also Early Jurassic. Significantly, many of the "Triassic" trackways that Hitchcock originally described from the Connecticut Valley are now regarded as Early Jurassic (Hettangian) in age.

With general acceptance that Newark Supergroup sediments straddle the Triassic-Jurassic boundary, their fossils soon became the focus of debates on the mode of terrestrial faunal change across the boundary. The Newark Supergroup is still considered the key sequence worldwide for such matters and is likely to be pivotal in unraveling the complexities of terrestrial faunal change at the end of the Triassic. What is so important is that high-resolution dating methods can be applied to this group of sediments. I shall therefore look at the geological setting of this sequence in more detail than any of the other Triassic assemblages that are covered in this book.

The range of localities, sediment types and age is immense, and clearly it is impossible to discuss all aspects of the Newark Supergroup—it deserves a book of its own. Fortunately it has just that, and the reader is referred to LeTourneau and Olsen (2003a, 2003b) for details.

Basin Formation and Sedimentation

The beginning of the breakup of Pangaea in the Triassic—and the ultimate separation of North America from Europe and Africa—was marked by tearing of the earth's crust from Greenland to Mexico along the axis of the future Atlantic Ocean. Nine major basins and several minor basins are exposed from Nova Scotia to South Carolina, and many more are buried. The basins closely follow the line of the Appalachian orogen and are filled with thousands of meters of continental sediment and basalt lava flows. Together, they represent a period of roughly 45 million years. Diabase plutons and dikes (apparently coeval with the basalt flows) extensively intruded and metamorphosed preexisting strata. These are tremendously important in providing us with absolute dates and benchmarks with which to date the rest of the Newark Supergroup. All the extrusive rocks in the exposed basins are thought to be early to middle Hettangian in age.

Most Newark Supergroup basins are half-grabens with boundary faults that appear in most cases to be reactivated Paleozoic faults. Each major basin has a separate series of lithologically defined formations, and collectively, they probably range from the Anisian to at least Sinemurian or Pliensbachian times. The Triassic-Jurassic boundary is present in all the northern basins and is recognized on the basis of palynostratigraphy and vertebrate fossils.

Two different types of depositional pattern can be recognized in the basin sediments, and both patterns can be observed in the same basin at different times during its development (fig. 11.2). The first of these patterns of sediment deposition can be essentially termed a fluvial, or river-dominated, system. At such times, a basin is said to be an open system. That is, water coming into the basin by way of streams and rainfall can escape at some point. What characterizes times of open basin sedimentation are widespread ancient channel systems. At the same time, large- to small-scale lenticular bedding is typically present, and intervals of conglomerates may stretch across the basin. Where they can be determined, paleocurrent patterns often trend in one direction across the basin, which indicates that a through current was present. Evidence for large lakes is lacking, although the presence of paludal deposits indicates some periodic ponding of water.

The second pattern is normally described as closed basin deposition. In this case, there is no significant outlet through which water can escape. The sediments representing this type of system are characterized by a systematic increase in grain size toward all the boundaries of the basin. The paleocurrent patterns trend toward the center of the basin, indicating that water flowed from all sides of the surrounding land.

Each of the Newark Supergroup basins tends to show a similar gross threefold stratigraphy that consists of a basal fluvial interval, followed first by a deeper water lacustrine (lake) interval, and finally by a shallower water lacustrine or fluvial interval. Such a sequence is a common pattern in many nonmarine extensional basins (Lambiase and Rodgers 1988), and it has typically been explained in terms of a standard pattern of varying basin subsidence rates. Initially the subsidence rate is slow, resulting in a hydrologically open basin and sediments that are almost entirely deposited by rivers flowing across the basin. Then there is a change toward much more rapid subsidence of the basin floor. As a consequence, the outflow of water from the basin is greatly restricted, and a deep lake forms. For a time, sedimentation rates do not keep up with basin subsidence, but finally, as subsidence rates wane, the lake and basin fill up with sediment, and there is a return to a hydrologically open basin and fluvial sedimentation.

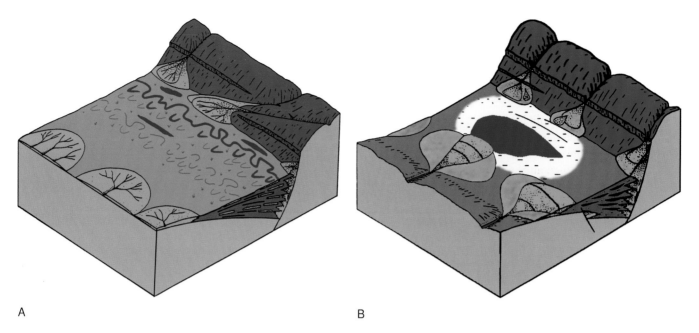

A   B

Figure 11.2. Diagram depicting the accumulation of sediments in a rift basin. (A) Deposition as a result of rivers
flowing through the basin.
(B) With the intermittent formation of a lake, dark, well-
laminated shales and mud-
stones typically form.

Because the Newark Supergroup basins are all part of the same rifting system, one
would expect to see some general controls over basin subsidence and therefore some
marked regional trends. However, as pointed out by Olsen and Schlische (Olsen 1988;
Schlische and Olsen 1988), the transitions between fluvial and lacustrine sedimentation
patterns apparently occurred at different times in the different basins. Consequently the
notion of a standard pattern of variation in subsidence rates that accounts for the basic
threefold basin stratigraphy cannot be readily applied to the Newark Basins. Instead, Olsen
and Schlische have sought to explain this pattern on the evolving geometry of the basin.

If the subsidence rate is slow and/or the sedimentation rate is high, then sediments will
initially fill the basin, leaving little or no marked depression in the land surface. In other
words, sediment accumulation rate equals subsidence rate. Consequently, excess water
and sediments will run off the surface and leave the basin through river channels. This
scenario explains the initial fluvial stage. As the basin continues to grow larger, however,
then the same volume of sediment added to the basin per unit time is spread over an increasingly larger depositional area. There comes a time when sediments only fill up the
basin to the level of its lowest outlet point. After that, as the basin continues to expand,
sediment deposition no longer keeps pace with expansion, and a lake will form between
the level of the depositional surface and the level of the lowest outlet point. At first this
may be a fairly shallow lake, but as the difference in elevation between the depositional
surface and the lowest outlet grows, the depth of the lake will continue to increase. If the
subsidence rate is fast or sedimentation rates are low (or both), then lacustrine deposition
will occur early in basin formation. Any excess water from rainfall will continue to flow
out from the lake, but some will also leave through evaporation. However, as the basin
continues to increase in size and the surface area of the lake also grows, evaporation rates
will increase. Eventually a point will be reached when there is no outflow and the rate of
water inflow matches the rate of water loss through evaporation. At this time, the lake
reaches its greatest depth. After that, as long as the basin continues to grow through subsidence, there is a trend toward decreasing lake depth, but an increased surface area with
concomitant increased evaporation of water from the lake surface. If, on the other hand,
basin subsidence slows or stops, then the basin will once again fill with sediment to its outlet, and fluvial sedimentation will resume.

We would expect that subsidence in all the Newark Supergroup basins began at the
same time, even though they varied in size. However, if the onset of lacustrine conditions

is tied to basin size and geometry, then we would expect to see the formation of lakes beginning at different times in the different basins. This is precisely the case.

As if this were not complex enough, superimposed on this basic pattern of sedimentation is another system of controls. Expressed in the sediment sequences within each basin there is a common theme of repetitive cycles that are a direct result of the continuous waxing and waning of ancient lake levels. These are known as Van Houten cycles (Olsen 1986). Each cycle can be split into three basic divisions. First, there is lake growth or transgression (division 1 sediments), followed by a period of deep-water lakes (division 2), and finally lowering of lake levels, ending with a low-stand sedimentary facies (division 3). These changes have been attributed to consistent patterns of climate fluctuation that directly affected both the rate of water inflow into the basin and evaporation rates. Thus, quite naturally, the high lake levels were attained at the wettest times in the climate cycle. These climate changes are thought to have been directly under the control of regular variations in the earth's orbit around the sun—a pattern of variability that persists even to this day. In just the same manner that a spinning top wobbles about its longitudinal axis, so the earth wobbles as it spins and moves in its orbit around the sun. This wobble, which occurs on a 20,000-year cycle, is widely considered to force the world's climate through phases of relatively high humidity alternating with drier ones. The cycle is termed the precession of the equinoxes. Moreover, as the northern hemisphere experiences more humid conditions, then the southern hemisphere sees more arid times, and vice versa.

The extent to which this cycle is expressed in the rock sequences is variable. In some of the southern basins it is weak, presumably because the extremes of humidity and aridity were not as pronounced in the more equatorial regions. By contrast, road cuts along Interstate 91 in Connecticut provide the motorist with a striking example of the uniformity and repetitive nature of the cycles. Another of my favorite places to get a feel for this cyclical pattern of sedimentation is the aptly named Cascade Creek, which flows alongside the Solite Quarry in southern Virginia. Here, small cascades are evenly separated by stretches of calmer water. The cascades occur at the contacts between division 1 and division 2. The less resistant laminated shales of the lake high stand are rapidly worn away by the stream's action, leaving a resistant shelf comprising division 1 and the base of division 3 from the previous cycle.

These 20,000-year cycles are not the only levels of cyclicity that can be seen in the sediment sequences. Apart from the precession of the equinoxes, other, higher-order cycles have been recognized. There are compound cycles of 40,000 years, as well as 100,000-year, 400,000-year and possibly even 2- and 6-million-year cycles. Each of these is each brought about by different components of the earth's orbital movements. Thus the 40,000-year cycle has been linked with the tilt in the earth's axis, and the 100,000- and 400,000-year cycles with the change in the ellipsoidal path of the earth around the sun.

Dating the Newark Supergroup Sediments

The regimented pattern of sedimentation in each basin permits us to accurately gauge the duration of each phase of sediment deposition, but by itself, it can neither provide a means of determining the age of a particular horizon relative to strata in another basin, nor, more importantly, can it provide us with an absolute age for the sediments. In the absence of these two time constraints, the Newark Supergroup would be just another Early Mesozoic sequence that hinted at altered rates of faunal turnover at the end of the Triassic that could never be properly tested. Fortunately, we have methods of constraining both the absolute and relative dates of sediments within the basins, and consequently, we have a frame of reference with which to compare other potential Triassic-Jurassic boundary sequences worldwide. Because this is a rift basin system, there were times when extensive lava flows erupted on the ancient land surface. The remnants of these lava flows comprise a series of basalts that occur in the Culpeper, Newark, Hartford/Deerfield, and Fundy Basins. From these basalts it is possible to obtain isotopic dates, and these all cluster around the 201-million-year mark. Interspersed with these lava flows are a number of sedimentary horizons. The ages of these sediments must also be approximately 201 million years.

Palynology

There have been numerous studies of the spores and pollen within the Newark Supergroup sediments (Cornet and Olsen 1985), and these have been used to assign relative ages to many of the sequences. For instance, a suite of palynomorphs, including various species of *Corollina*, is associated with the very end of the Triassic (latest Norian = Sevatian), and Gore (1988) used the presence of these palynomorphs in a section near Manassas, Virginia, to assign such an age to the deposits. Likewise, Cornet and Olsen (1990) noted that there are apparently 27 palynomorph species that have their last occurrences in the Carnian of the Newark Supergroup that also have their last occurrences in the Carnian of western Europe. The discovery of these palynomorphs in new exposures can therefore greatly assist in providing an upper age limit for the sediments.

Magnetostratigraphy

Any Boy Scout will tell you that a compass always points north, but that has not always been the case. It is now well known that over the course of earth's history, the magnetic pole has periodically flipped, reversing the direction of the magnetic pole. The duration of each reversal is in the order of hundreds of thousands to millions of years, but the alternation between magnetic states does not occur in any predictable pattern, and so the duration of each reversal is variable. By analyzing the hematite (iron-bearing mineral) in sediments of the Newark Supergroup, it is possible to determine whether the sediment was deposited during a period when the magnetic pole pointed north, as it does today (normal), or was reversed. The magnetic profile of any extensive section will show a sequence of magnetic flips.

We can depict this visually on a stratigraphic column by showing one magnetic pole, say north, in black, and the reverse pole in white. By itself, it may not be immediately apparent how this can be useful in dating individual sedimentary sequences, but by using the Van Houten cycles, it is possible to calibrate the duration of each phase of the magnetic pole. For example, if a continuous section of magnetic normal sediment exhibiting 20 Van Houten cycles is sandwiched between two blocks of magnetic-reversed sediments, then we can say that this particular normal sequence lasted for approximately 400,000 years. By analyzing extensive sediment sequences, we can discern a characteristic pattern of alternating normal and reverse poles that is akin to a fingerprint. This can then be matched against the fingerprints of other sections. We therefore have a powerful tool to cross-correlate different sections within a basin and from one basin to another. In certain sections, basalts provide us with absolute dates.

These dates can be used to precisely calibrate the age of many of the documented sections in all basins of the Newark Supergroup. Putting all these dating methods together provides us with a wonderful opportunity to assign dates to each horizon with a remarkably low margin of error.

The Fossils of the Newark Supergroup

The concrete jungle that is northern Virginia can be a surprisingly rich source of Triassic vertebrates. Every day, thousands of people land on the tarmac at Dulles Airport, unaware that buried beneath them are the remains of ancient vertebrates, possibly even dinosaurs! In the past, bones of phytosaurs were unearthed from this very spot (Weems 1979). In recent years, modifications to Interstate 66 have exposed great masses of red Triassic mudstones, and within them the occasional footprint of a small tetrapod. Unending subdivision and office construction projects have also exposed the same strata. Careful examination of the surfaces has yielded such treasures as small phytosaur tracks complete with skin impressions.

The Manassas Battlefield attracts thousands of Civil War buffs each year, but how many are aware that the site also documents a much earlier period in the history of life on earth? Staphylinid, or rove beetles, are today represented by over 30,000 living species, but one of the oldest known species in the world was discovered in the vicinity of the battleground (Gore 1988).

New York and its immediate environs conjure up glitzy images of Broadway, Times Square, and Fifth Avenue. Aside from the magnificent American Museum of Natural History, this is generally a city far removed from the natural world. Yet the sole specimen of the gliding reptile *Icarosaurus* was discovered in 1960 by Alfred Siefker and two companions in the Granton Quarry in west New York (Colbert 1970), just over the Hudson River from the American Museum of Natural History, where it is now housed.

All this goes to show that important fossils and significant fossiliferous sequences do not have to occur in exotic places. One of my favorite such examples is a site in southwest England. Most of the shoppers at a large hypermarket on the west side of Bristol park their cars as close to the entrance as possible so that they can join the crowds milling inside, completely unaware that the far side of the parking lot is one of the few accessible outcrops of the Rhaetic bone bed—a fossil treasure trove for elements of marine vertebrates from the Latest Triassic. So it is with the fossils of the Newark Supergroup; it is remarkable how many of them have originated in the most mundane places. Be prepared to expect the unexpected in the Newark Supergroup.

Trackways in the Newark Supergroup

Since Hitchcock first described dinosaur trackways from the Connecticut Valley, the Newark Supergroup has been renowned for the sheer abundance and variety of its tetrapod tracks. For a long time they have overshadowed the skeletal remains of their makers. Any single tetrapod has the potential to leave millions of tracks during its entire lifetime, but the same tetrapod will leave only one skeleton at the time of its death. It might therefore seem logical to expect that there would be a greater chance of finding tracks in the fossil record than skeletal remains. Of course, quite different depositional environments are required for the optimal preservation of tracks than for the optimal preservation of body fossils. The Newark Supergroup, preserving as it does a series of lacustrine systems that waxed and waned, documents numerous mudflat sequences situated around the margins of lakes. The lakes were undoubtedly the principal source of drinking water for many animals, and as the creatures made their daily trek across the mudflats, they left imprints in the soft sand and mud that were preserved forever. Seeing the delicate imprints of feet, complete with the clear impressions of scales, one could almost be forgiven for thinking that these were the prints of an animal that passed this way a few days ago rather than 210 million years earlier.

Some trackways tell intimate stories of everyday life in the Early Mesozoic. At Dinosaur State Park in Rocky Hill, Connecticut, large tridactyl tracks (*Eubrontes*) are the most common (Farlow and Galton 2003). The size of these tracks and their general morphology are consistent with the theropod dinosaur *Dilophosaurus*. *Dilophosaurus* is known from the Early Jurassic Kayenta Formation of Arizona but has not been described from the eastern United States. Some less distinct markings at Rocky Hill appear to comprise claw drags, lacking any accompanying pad impressions. Coombs (1980) interpreted these as faint traces left in submerged mud when a half-submerged theropod dinosaur swam through the shallow water, pushing the tips of its toes off the bottom to give it added momentum.

Another interesting example of tracks made underwater are those referred to *Apatopus*.

Baird (1957) suggested that the *Apatopus* ichnotaxon might have been generated by a phytosaur such as *Rutiodon*. Certainly the age ranges of the sediments that contain *Apatopus* are practically coincidental with those bearing phytosaurs. The single impressions, comprising four elongated grooves terminating in clear claw marks, are consistent with a swimming animal pushing one hind foot down into the mud in order to change direction. Unfortunately, complete articulated feet of phytosaurs are unknown, so presently it is not possible to compare the foot of the putative track maker with the track. Nevertheless, on the basis of disarticulated and somewhat incomplete material, Parrish (1986) suggested that the pedal morphology of phytosaurs is actually inconsistent with *Apatopus*. He favored the notion that a trilophosaur or a rhynchosaur was the maker of *Apatopus* footprints.

Another beautiful little vignette of daily life in the Early Mesozoic can be observed in another trackway in the Hartford Basin, although this time from Jurassic sediments. The type slab of the probable ornithischian, *Anomoepus curvatus*, documents two trackways running side by side. Over most of the slab, a dense pattern of raindrops is also superimposed on the tracks themselves. We can therefore deduce that the shower occurred after the animals had traveled across the wet mud. Moreover, a smooth band lacking raindrops cuts across the block, showing that this particular spot was covered by a small puddle at the time of the rain shower (Olsen 1989).

Olsen (1989) reports on the occurrence of a beautifully clear trackway of a small lizard-like creature from the Newark Basin Triassic sediments. Over 30 consecutive steps preserve the slightly meandering path of this small tetrapod, including a slight detour to avoid a tree stump that was directly in its path. The same trace also preserves a clear tail drag and some skin impressions.

It has been argued that the trackway assemblages reflect incredibly diverse tetrapod communities living on the eastern side of the North American continent during the Latest Triassic. There is certainly no denying the abundance of the tetrapod tracks, but monitoring diversity levels on the basis of these tracks is not a straightforward task (cf. Farlow and Pianka 2000). Although many different ichnotaxa have been erected to accommodate the different forms of the Newark Supergroup trackways, it must be remembered that many of them have essentially the same basic features. Farlow (2001) noted that all large theropod foot skeletons were alike, and so presumably were their prints. Although there are many different three-toed tracks referred to separate species of the ichnogenus *Grallator*, the distinctions may be made on little more than absolute size of the prints. Traditionally, the largest tridactyl tracks bearing claw impressions have been called *Eubrontes*, but there are really no discrete size classes for tridactyl tracks, and they grade into the size range typically referred to as *Grallator*.

The Culpeper Stone Quarry

Although some of the best-known trackway sites are situated in the northern basins, there are some substantial trackway localities further south. Not least of these is the Culpeper site in Virginia. The Culpeper Quarry, Culpeper County, is situated in the so-called Balls Bluff Siltstone and has yielded abundant dinosaur footprints at a variety of stratigraphic levels. The first discoveries at the site were made on an upper level of the quarry, where numerous poorly defined tracks were first described by Weems (1987). According to Weems, these represented the activities of at least six different kinds of animal around the margins of an ancient lake. However, the prints all lack pad impressions and show a large amount of individual variation, so that there are no real diagnostic characters that could not be attributed equally well to the effects of a variable substrate. However, in 1989, during normal quarrying operations, a deeper bedding plane was exposed that vividly displayed dinosaur activity along the margins of an ancient lake some 210 million years ago (Weems 1992a). This second horizon was eventually shown to contain over 2000 individual tracks belonging to bipedal theropod dinosaurs, and it provides us with an amazingly vivid insight into animal activity millions of years ago.

Some of the imprints are beautifully preserved and include pad impressions and claw marks. In places, trackways run parallel with each other, and it is difficult not to imagine all kinds of scenarios: was this pair of theropods part of a small group on the lookout for dinner beside the lake? Alternatively, were they solitary individuals passing by the lakeside at completely different times of the day? Weems (1992a) went so far as to suggest that some trails represented a male and female participating in some kind of courting ritual. Although this interpretation is impossible to substantiate, it does evoke a powerful image.

Newark Body Fossils

It will be convenient for the rest of the discussion to divide the basins into northern and southern examples. The principal northern basins are the Fundy in Nova Scotia, the Hartfield/Deerfield in Connecticut and Massachusetts, and the conjoined Newark, Gettysburg, and Culpeper Basins running from New Jersey through Pennsylvania and into Virginia (fig. 11.3).

The main southern basins are the Richmond Basin in Virginia, the Danville/Dan River Basin extending from southern Virginia into North Carolina, and the Deep River Basin of North Carolina. Of course, each basin has its own system of formation names, and in some instances, two different names have been applied to the same sequence merely because it crosses a state line.

FUNDY BASIN

The most northerly and probably the largest Newark Supergroup basin (it is not all exposed on land) is situated over the Bay of Fundy in Nova Scotia and New Brunswick, Canada. The Fundy Basin sediments are divided into five formations with just two—the basal Wolfville and the overlying Blomidon Formations—being Triassic. Extensive exposures of these rocks crop out along the eastern margin of the Bay of Fundy.

It is interesting that, unlike all other Newark Supergroup basins, the Fundy sediments are almost completely devoid of organic, carbon-rich shales (Olsen 1989). Indeed, the sediments frequently point to deposition under arid conditions and are more akin to those of the Argana Basin in Morocco than the rest of the Newark Supergroup. Nevertheless, there are exceptions, such as those containing an abundance of large unionid clams, pointing to at least some periods of fairly wet conditions.

The Wolfville Formation has a variety of skeletal elements representing a wide range of tetrapod groups. These include procolophonids, dicynodonts, rhynchosaurs, aetosaurs, rauisuchians, and traversodontids. It has also yielded a unique and peculiar animal. *Teraterpeton* has a long, thin, edentulous snout region. This beast is almost certainly an archosauriform because it has a clear antorbital fenestra, but it is unlike any other member of that group (Sues 2003). The procolophonids are particularly diverse and include the unusual *Scolaparia* (Sues and Baird 1998), which bears sculptured dermal scutes covering at least the neck region. Along with a particularly spinose cheek region, the dermal plates would have provided some protection against predatory attack. The teeth of *Scolaparia* are transversely broadened and chisel-shaped. Interestingly, they show a close resemblance to the teeth of *Tricuspisaurus* from the Welsh fissure locality of Ruthin. *Tricuspisaurus* may also have had a similar sculptured dermal armor. Other procolophonids from the Wolfville are based on fragmentary remains. These include *Haligonia*, with a particularly massive posterior tooth on both the maxilla and dentary, and *Acadiella*. The

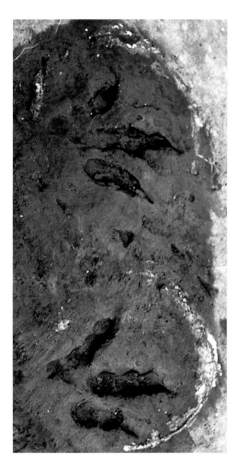

Figure 11.3. Two of the beautifully preserved theropod footprints at the Culpeper Stone Quarry.

age of the Wolfville is not well constrained, though Lucas considered at least part of it to be equivalent to the Otischalkian land vertebrate faunachron (LVF).

The Blomidon Formation (equivalent to the Revueltian LVF) has yielded *Hypsognathus*, the type of which originated from the Passaic Formation of the Newark Basin. *Hypsognathus* is one of the most derived procolophonids, and its cheek region bristled with spines that would have been an effective deterrent to attack.

HARTFORD BASIN

Although eight separate formations are recognizable in the Hartford Basin, only the lowest one, the New Haven Arkose, is regarded as Triassic. Red sandy mudstones of the New Haven Formation containing vertebrate fossils are part of a sequence of sedimentary deposits laid down by a meandering river, alternating with red overbank sediments. Intense bioturbation of the mudstone by roots and invertebrate burrows are indicative of a past when there was much biological activity.

Skeletal remains are relatively rare in the Latest Triassic (Norian to Rhaetian) strata of the Newark Supergroup series, and these red clastic deposits are assumed to be depauperate in vertebrate material. However, this notion is largely self-fulfilling because the rocks are consequently not systematically prospected for fossils. Recent fieldwork in the New Haven Formation of the Hartford Basin by Paul Olsen and colleagues has challenged this assumption. Descriptions of sphenodontian (Sues and Baird 1993), procolophonid (Sues et al. 2000), and archosaurian (Olsen et al. 2001b) material from the middle and upper New Haven Formation point to an increasingly diverse fauna. Admittedly much of the material is still incomplete, but it is sufficient to indicate a remarkable similarity to penecontemporaneous faunas from elsewhere in the northern hemisphere.

One of the archosaurs, *Erpetosuchus* species, is particularly significant. It is the first documentation of this enigmatic taxon aside from the holotype specimen from the Lossiemouth Sandstone (chapter 8), which is generally considered to be Carnian. On this basis alone, it is conceivable that the New Haven Formation is actually of Carnian age; however, that seems to be unlikely because it has been dated as Norian on three independent lines of evidence. First, the base of the New Haven Formation has produced a palynoflorule that has been dated as latest Carnian to early Norian (Cornet 1977), thus placing the middle and upper parts of the New Haven firmly within the Norian. Second, uranium-lead dating of sediments in the fossil-bearing horizon yielded a date of 211.9 ±2.1 million years (Wang et al. 1998), which on the basis of recent timescales once more places the New Haven well within the Norian stage. Last, a single natural mold of the dorsal dermal armor of a stagonolepidid archosaur is known from the middle New Haven. This characteristic specimen was referred by Marsh (1896) to a new species, *Stegomus arcuatus*, but Lucas et al. (1998) synonymized it with *Aetosaurus*. Furthermore, they considered *Aetosaurus* to be an index fossil for continental strata of early to middle Norian age (see also chapter 10).

The Erpetosuchidae was one of the families that Benton cited as a casualty in a putative end Carnian extinction event. As Olsen et al. (2001b) noted, the presence of *Erpetosuchus* in the New Haven Formation somewhat weakens claims that the most critical (or ecologically key) extinction event in the Late Triassic occurred at the Carnian-Norian boundary. For the time being, it is fair to say that there is strong evidence that the Erpetosuchidae extended well into the Norian. Indeed, it could be argued that *Erpetosuchus* is a Norian taxon, and that the Lossiemouth Sandstone is actually Norian in age (as was originally suggested by Benton and Walker [1985]). Acceptance of this argument would mean that the rhynchosaurs also extended into the Norian, which would further weaken the case for a major Carnian-Norian extinction—but more of that in the next chapter.

Newark Basin

The Newark Basin is the largest of all the completely exposed basins. Although it covers an area exceeding 2700 square miles, it also has a combined thickness of sediments of about 3¾ miles. Nine formations can be recognized in the Newark Basin, but only the lower three are Triassic. The lowermost Stockton Formation is a fluvial interval comprising red and brown sandstones, and is thought to be middle to late Carnian in age. Then comes the Lockatong Formation, characterized by gray and black fine-grained lacustrine sediments that include intervals of deep-water sedimentation. It has been divided into 11 separate members. The Lockatong is considered to be late Carnian (Adamanian LVF). Finally, in the Triassic portion of the sequence there is the Passaic Formation, which underlies the so-called Orange Mountain basalt. The Passaic consists mostly of red sequences representative of deposition within a relatively shallow lake, although there are still intervals of deep-water sedimentation. The Passaic has been divided into eight members, although some of these have not been given formal names.

One of the most intriguing aspects of the lake level changes in Newark Supergroup sediments is the observed changes in aquatic faunas over time. Subtle differences in fish faunas at different levels and at slightly different localities have been interpreted as colonization followed by extinction of fish populations as the lake levels first rose and then dried out. Once the lake levels began to rise again, slightly different races or species might colonize the lake before they too were ultimately lost as the lake dried out once more. Thus some Newark fish assemblages are characterized by groups of closely related, but not identical, fish species which are termed *flocks*.

The primitive ray-finned fish *Semionotus* has been recovered in large numbers from many deposits throughout the Newark Supergroup, in particular from the Jurassic sequences, and the best known come from the Newark Basin. Morphological variation in *Semionotus* is expressed in two principal ways: the nature of the dorsal ridge scales that run the length of the dorsal midline, and the overall body outline. Some species are deeper bodied than others, and the ridge scales may be small, spined, or globular. When lakes were shallow, the different species became extinct, before different flocks evolved after the lakes rose again. Changes in dominance of different fish genera are also seen from one cycle to the next. One beautiful example of this cyclic change has been described for an excavation in the Lockatong Formation at Kings Bluff, New Jersey. Here the so-called Yale Quarry, Weehawken, was opened up by Paul Olsen, Amy McCune, and colleagues, and yielded over 4000 fish and reptiles. Two particular cycles (5 and 6) produced different fish faunas. Whereas the older cycle 6 was dominated by *Semionotus*, together with the palaeonisciform *Synorichthys* (fig. 11.4), semionotids are virtually absent from cycle 5, and the palaeoniscid *Turseodus*, which is absent in 6, is the dominant taxon. The coelacanth *Diplurus* is found in both cycles, but is more abundant in 5. These changes could reflect changes in salinity (Olsen et al. 1989).

The recognition of these species flocks reflects another similarity between the modern East African rift valleys and the ancient Newark Supergroup rifts. The Newark had species flocks of semionotids, whereas the African Great Lakes have species flocks of cichlid fishes.

Another interesting feature of the Weehawken Yale Quarry is that diversity levels are inversely correlated with fish preservation. The best-preserved fish are found in the horizons with the lowest diversity. This phenomenon is also seen at the Solite Quarry (see below). The highest diversity of lake zones tends to be toward the margins, whereas the deeper waters tend toward lower diversity. This is especially true for deep lakes, which develop anoxic bottom waters and thereby lack benthic taxa. At the same time the absence of bioturbation and the anoxic bottom leads to ideal conditions for preservation of sinking carcasses.

The nearby Granton Quarry in New Bergen was a large quarry, also in the Lockatong

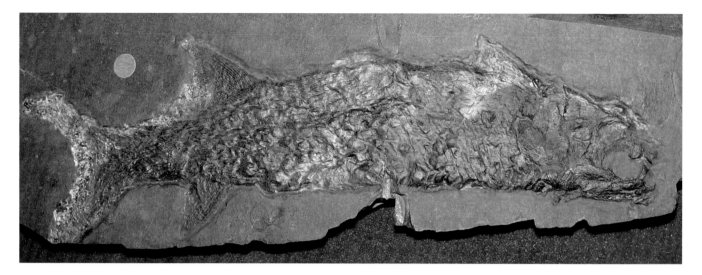

Figure 11.4. Complete fossil of the fish *Synorichthys*.

Formation, that was abandoned sometime in the middle of the twentieth century. Since then, it has been "reclaimed" to make room for commercial buildings. In the lengthy process of reclaiming this land, several interesting vertebrate fossils came to light, including the skull of a rutiodontine phytosaur (Colbert 1965), the gliding reptile *Icarosaurus* (Colbert 1970), the amphibious *Tanytrachelos* (Olsen 1979), and a small reptile that came to be informally known as the "deep-tailed swimmer." The Weehawken Quarry also yielded remains of this animal. It has since been described as *Hypuronector* (Colbert and Olsen 2001) and is another member of the obscure Drepanosurididae (see chapter 6). Colbert and Olsen (2001) compared the shape of the tail of *Hypuronector* to certain fishes as well as the deepened tail of newts and therefore considered it to be an adaptation to sculling in water (fig. 11.5). The enormously elongate chevron bones (hemal spines) give the tail a unique finlike shape. Almost identical isolated tail vertebrae have been found in the fissure fills at Cromhall Quarry (chapter 10), and it now seems that drepanosaurs had a widespread distribution in the Triassic.

Hypuronector had long, slender limbs, and Colbert and Olsen pointed out that these may have acted to compensate against the great depth of the tail, which would otherwise have got in the way when the animal was walking on land. In contrast, other authors (Renesto 2000) have argued that all drepanosaurids, including *Hypuronector*, were arboreal animals.

The great numbers of fish recovered from certain levels of the Newark Basin are impressive, but one of its most celebrated denizens is known from just a single specimen from the Granton Quarry. *Icarosaurus* (fig. 11.6) is similar to *Kuehneosaurus* (and *Kuehneosuchus*) known from extensive material from the British fissure deposits (see chapter 9). Indeed, I tend to consider them as three different species within the same genus. Certainly the presence of closely related forms such as the kuehneosaurs and sphenodontians on both sides of the Atlantic today does point to similar environments and the relative proximity of the two areas in Triassic times.

Figure 11.5. Restoration of the small drepanosaur *Hypuronector* (after Colbert and Olsen 2001).

CULPEPER BASIN

The Culpeper Basin is in many ways merely an extension of the Newark Basin. It is the southernmost one of the northern group and contains Jurassic sediments and basalt flows. It is divided into five formations, with the first three, the Reston Formation, Manassas Sandstone, and the Bull Run Formation, representing

the Triassic. The majority of these Triassic sediments represent fluvial and alluvial sedimentation, although the Balls Bluff Siltstone of the Bull Run Formation is notable for lacustrine sediments, including both shallow and deep-water deposition. The impressive dinosaur trackway at Culpeper Stone Quarry discussed above was exposed in this unit. Gore (1988) described a section in northern Virginia documenting the history of a saline lake during arid conditions. Around the edges of this lake, where freshwater streams entered, fish and a variety of arthropods lived.

Figure 11.6. Sole specimen of the gliding reptile *Icarosaurus*.

Deep River Basin

During Triassic times, the southern basins were situated at least in the subequatorial region, if not the equatorial zone. Consequently, we might expect this area to offer another test of the megamonsoon theory (see chapter 1). If Parrish (1993) is correct, there ought to be some evidence of seasonality in rainfall in these sediments and associated fossil assemblages.

The Deep River is the most southerly of the exposed Newark Supergroup rift basins and consists of three interconnecting sub-basins—from north to south the Durham, Sanford, and Wadesboro. Collectively, the rocks in these sub-basins have been divided into three distinct formations; however, all three units are not present in each of the sub-basins. This three-part division was recognized early on (Emmons 1856), but it was not until 1923 that the divisions were assigned formal names (Campbell and Kimball 1923). The basal part of the Pekin Formation has yielded the oldest palynomorph assemblages, with an age of early Carnian (Traverse 1986, 1987). The Cumnock Formation is restricted to the Sanford sub-basin, where it lies on top of the Pekin. It has been dated as middle-upper Carnian, particularly on the basis of its spores (Traverse 1986); however, this has been questioned, and a somewhat older age has been proposed by Hunt and Lucas (1990, 1994) on the basis of aetosaur remains (see below). The Cumnock is in turn overlain by at least 1000 meters of red beds, the Sanford Formation, that has not been dated by palynological means. Fish assemblages from the equivalent strata in the Durham sub-basin indicate a late Carnian age, and it probably extends into the Norian.

The Pekin Formation is mostly composed of fluvial sandstones, conglomerates, and siltstones that show evidence of deposition in a humid environment. This is particularly evident from the sandstone mineralogy and the nature of the pollen and spores. The question is, how seasonal was the humidity? Abundant plant fossils are suggestive of a verdant landscape. Shale and siltstone beds cannot typically be traced laterally over great distances, and they are consequently thought to be representative of small pond and lake deposits. Frequent, intense bioturbation and burrows also point to a fairly humid environment. The Boren Clay Pit near Gulf, North Carolina, has yielded a magnificent flora that includes ferns, horsetails, pteridophytes, cycads, bennettitaleans, and conifers. Cuticle, the protective waxy layer that covers the surfaces of land plants, is sometimes well preserved on the foliage. Where present, it permits examination of the stomatal morphology, which can be critical in distinguishing cycad foliage from that of bennettitaleans (Gensel 1986). Particularly abundant are leaves and stems of the horsetail *Neocalamites*, leaves of the bennettitaleans *Zamites*, *Otozamites*, and *Pterophyllum*, and foliage and stems of the cycads *Leptocyas* and *Pseudoctenis*. Ferns are quite diverse, with seven or eight different genera, including *Pekinopteris*. Within the palynomorph assemblage, fern spores are dominant. Cones of *Compsostrobus*, an early member of the Pinaceae, have also been documented from this locality (Delevoryas and Hope 1973). Together, the flora speaks in favor of a humid environment, and there is nothing in the sediments to contradict this.

Perhaps one of the more important plant fossils from the Pekin Formation is *Eoginkgoites*. From its name, one would be forgiven for assuming that this Triassic foliage genus was a ginkgo. Certainly it has at least the superficial appearance of the family, and so Bock (1952, 1969) classified it as such. However, its phylogenetic affinities remained equivocal until the mid-1970s, when Ash (1976) showed that on the basis of its leaf venation pattern (frequently anastomosing and bifurcating) and the nature of the stomata, that it should be referred to the Bennettitales. At least four species of *Eoginkgoites* are known from North America: aside from the Pekin Formation species, there are two from the Stockton Formation of Pennsylvania, and the fourth comes from the Chinle Formation of Arizona and Utah. Ash (1980) recognized the close resemblance between the flora from the Boren pit and the lower parts of the Chinle Formation and termed it the *Eoginkgoites* megafossil zone. Although vertebrates are rare at Boren, the nearby Pomona Quarry has produced phytosaur (?*Rutiodon* species), aetosaur (*Longosuchus* species) and dicynodont (?*Placerias* species) remains. Backfilled *Scoyenia* burrows are abundant at both Pomona and Boren. These are believed to represent feeding burrows, possibly the work of crayfish-like crustaceans (Olsen 1977). Frey et al. (1984) considered *Scoyenia*-like burrows to be a possible indicator of a warm and moist alluvial or lacustrine environment. The occurrence of phytosaurs provides additional support for a low-lying landscape dotted with ponds and lakes. A single tooth of an ornithischian dinosaur, *Pekinosaurus olseni*, was described by Hunt and Lucas (1994). Although the referral of the tooth to the Ornithischia has not been questioned, the age of the beds is still under some scrutiny. As mentioned the palynomorphs are indicative of a middle Carnian age, but Hunt and Lucas (1990, 1994) argue for a slightly earlier Carnian age, pointing to the presence of diagnostic elements (scutes) from the aetosaur *Longosuchus* (Otischalkian LVF). Elsewhere in the United States, *Longosuchus* would appear to be restricted to early late Carnian sediments (see chapter 8), and we are thus left with an unresolved age conflict.

The lower part of the Cumnock comprises mostly siltstones and fine-grained sands, but somewhere between 60 and 80 meters above the base are two significant coal seams. Above these seams, the units are predominantly black and gray shales. Together, the sediments point to a lacustrine environment surrounded by swamps and fed by streams and rivers flowing mostly from the north and northwest. Coal has been mined from the Sanford sub-basin since 1775 or even earlier. The first large-scale mine was begun in 1852 with the sinking of the shaft at the Egypt Mine. Clearly the coals were produced in a swamp, and it is interesting to note that palynomorph assemblages from the coals show a marked reduction in fern spores. An influx of gymnosperms with cypresslike adaptations for growth in submerged soils provides additional evidence for raised water levels. Such flooding may have destroyed large areas of fern habitat. Gymnosperms are far more abundant in these coals than they are in the Richmond Basin coals (see below), which may reflect deposition at different paleolatitudes or in slightly different environments. Above the coal beds, the black and gray shales contain numerous ostracod and conchostracan valves together with fish scales and coprolites.

Sediments of the Sanford Formation are similar to those of the Pekin Formation, consisting mostly of fluvial deposits. Toward the southeastern margin of the basin alluvial fan deposits interfinger with the fluvials.

In the Durham sub-basin are some interesting thin, red-and-green, platy-bedded siltstone and claystone units. They would appear to represent sediments deposited in perennial lakes that alternated with meandering streams. The presence of caliche, limestone, and chert in some of the lacustrine sequences led Wheeler and Textoris (1978) to consider a playa lake setting responding directly to alternating wet and dry seasons. However, Olsen et al. (1989) noted that there were no fossils in the carbonates, and cherts to indicate a lacustrine environment. Moreover, they showed that the caliche does not have the character of classic caliche that forms in soils subject to intense drying. Instead, Olsen et al. argued that the caliche in the Durham sub-basin formed under more humid conditions, in soils that were perennially moist. In this regard, one of the most intriguing fossils is a filmy fern (Axsmith et al. 2001). As its name implies, this is a delicate fern; today, filmy ferns are restricted to humid equatorial environments. Its discovery in the Deep River Basin is highly suggestive of humid conditions during the Carnian in this area. This is perhaps the single most compelling piece of evidence that rainfall occurred year round, at least in this part of the world. Although it does tend to speak against the megamonsoon climate regimen, the effects of elevation on rainfall should not be overlooked. It is still possible that the southern basins were at higher elevations than the northern basins. If so, then they would be expected to receive higher rainfall totals in the same way that mountain ranges do today.

One of the most productive fossil localities in the Durham sub-basin is the Triangle Brick Quarry near Durham. Sandstones and siltstones of the Pekin Formation dominate the exposed sections of the quarry, but there are also mudstone beds, and one of these has yielded an extensive flora and fauna. *Scoyenia* burrows are particulary abundant. The size of the faunal list is comparable to those of the Cow Branch Formation at the Solite Quarry and Lockatong Formation localities in the Newark Basin (e.g., Weehawken Quarry). Principal among this fauna are a variety of fishes, notably the palaeonisciforms *Turseodus* and *Cionichthys*, and the coelacanths *Osteopleurus* and something akin to *Pariostegus*. Invertebrates include the usual conchostracans (clam shrimps) and ostracods, with some reports of beetle fragments, too. There are also occasional remains of phytosaurs and aetosaurs, and most recently, exquisitely preserved traversodonts have been found. The flora, comprising horsetails and ferns as well as some conifers, also testifies to a damp environment of small ponds, lakes, and streams.

Recently the Durham sub-basin has produced a most remarkable agglomeration of tetrapod remains consisting of a large rauisuchian being caught in the middle of some kind of Southern-style all-you-can-eat buffet. Associated within the gut region of the rauisuchian are osteoderms of a small stagonolepidid, the snout and some limb elements of a traversodont cynodont (*Plinthogomphodon* Sues et al. 1999), toe bones of a large dicynodont, and a fragment of a possible temnospondyl amphibian. To cap it off, curled up underneath the hip region of the rauisuchian is the articulated partial skeleton of a sphenosuchian crocodile. The rauisuchian is closely related to *Postosuchus* from the Dockum Group (chapter 7), whereas the sphenosuchian is a new form called *Dromicosuchus* that is related to *Hesperosuchus* from the Petrified Forest Member of the Chinle Formation (chapter 7) and *Saltoposuchus* from the German Stubensandstein. Adding intrigue to this scene of Triassic carnage is a conspicuous gap in the dermal armor that should have protected the neck of *Dromicosuchus.* This gap closely matches the size and shape of isolated teeth of the rauisuchian. It would seem that, not content with its hors d'œuvre, the rauisuchian went after an entrée that gave it more than indigestion—the result being that the diner somehow came to die alongside its intended victim.

Danville/Dan River Basin, or East of Eden

Driving around the gently undulating tobacco country of the Piedmont of Virginia and North Carolina, it is hard to find much rock exposure anywhere. Old tobacco barns and cornfields dot the countryside, which in summer is threatened by complete obliteration by the interminable kudzu. However, scouring the drainage ditches of the back-country roads can be profitable in the search for tracks and fish remains, and Paul Olsen has shown that for the intrepid paleontologist, wading through some dubious creeks and working on the creek bed can be rewarding. Occasionally creeks and rivers have cut through Mesozoic strata to leave an inviting shale cliff to explore with a hammer. One such site along the Bannister River in Pittsylvania County, Virginia, has yielded dinosaur tracks at several different horizons, including one exceptionally delicate trackway assigned to the ichnotaxon *Bannisterobates boisseaui* (Fraser and Olsen 1996). The hind feet of the trackmaker were no more than 15 millimeters long. Although functionally tridactyl, there are the impressions of four digits on the pes, and there are even small manus impressions (fig. 11.7). Fraser and Olsen (1996) considered the trackmaker to be either a small, possibly juvenile, ornithischian dinosaur or a dinosauromorph akin to *Marasuchus* or *Agnostiphys.*

The Dan River/Danville Basin is a long, skinny basin straddling the North Carolina–Virginia border. Within the basin, three formations are generally recognized, but for a long time what these were actually called depended on which side of the state border you were standing on. In North Carolina, Thayer (1970) recognized a basal, mostly fluvial Pine Hall Formation separated from another fluvial unit, the Stoneville Formation, by the lacustrine Cow Branch Formation. In Virginia, Meyertons (1963) divided the sediments into the mostly lacustrine Leakesville Formation, the fluvial Dry Fork Formation, and the fluvial and lacustrine Cedar Forest Formation. These formations all intertongue with each other and were always recognized as being, at least in part, lateral equivalents. Of course in the Triassic, state and international lines were not recognized, and fortunately, we have recently begun to realize this! Consequently there has been a tendency toward creating a uniform nomenclature.

Whatever name is given the rocks exposed in the Virginia Solite Quarry, situated near the middle of the basin and smack bang on the Virginia–North Carolina State line, it is hard to escape the reality that this is one of the biggest prizes of the Newark Supergroup—the jewel in the crown. Today, situated some 5 miles east of the town of Eden, North Carolina, is a modest quarry. Driving down the dusty approach road, dodging the potholes, one detects a strong odor of yeast coming from the local yeast factory that in turn supplies the Miller Brewing plant on the outskirts of town. The infrequently used and somewhat overgrown railroad tracks lend a depressing air to the scene, and one would be forgiven for thinking that this was just another ordinary quarry operation trying to eke a profit out of this economically depressed part of rural North Carolina/southern Virginia. However, the Virginia Solite Company takes pride in the fact that it has developed a unique

Figure 11.7. Trace fossil *Bannisterobates boisseaui,* a miniature footprint of a possible dinosauromorph or early ornithischian dinosaur from southern Virginia.

procedure to recycle waste products, such as paint thinners and varnishes, in order to fuel its rotary kilns. The kilns are instrumental in producing lightweight aggregate that finds its way into cinder block and lightweight concretes. Virginia Solite brags that its product has been widely used east of the Mississippi in a variety of building projects, including the roofs of both the House and Senate wings of the U.S. Capitol building in Washington, D.C. But Virginia Solite has an even greater claim to fame. Alongside the long-abandoned and rusty machinery in a disused corner of the quarry lie some of the richest Triassic fossil deposits in the world.

The site was first discovered by Paul Olsen in 1974, when he stumbled across skeletons of the prolacertiform *Tanytrachelos* and the paleonisciform fish *Turseodus* in an abandoned part of the quarry. Meyertons (1963) had described the deposits as largely unfossiliferous, but Olsen showed that this was far from the case. On much closer scrutiny, some of the lacustrine black shales revealed tiny silvery specks on the surface, little more than 2 or 3 millimeters long. To the naked eye, these would not bear further attention, but bathed in alcohol and viewed under a microscope and fiber-optic ring light, these specks become the most perfect examples of Triassic insects known (Fraser et al. 1996). This must surely rank as one of the few occasions that alcohol brings out the best in a body!

Fossil Lake Danville extended for 40 miles or more away from Solite both to the northwest and to the southeast. As a result of the rifting, over time, land would have been forced upward on the northwestern margins of the lake basin, so that mountains formed that were perhaps 8000 to 10,000 feet high. Somewhat lower hills and mountains would have bordered the southeastern side of the lake. Although its water level was subject to periodic fluctuations, a lake persisted in the area for a few million years. Lake sediments consist of laminated black shales and siltstones that were deposited in and around the margins of the lake, and it is estimated that almost 500,000 years of sediment accumulation is exposed in the Solite Quarries combined. The remains of dead animals and plants sank to the bottom of the lake, and because mud on the lake bottom was apparently inhospitable to life, these dead organisms were incorporated into the sediments with the minimum of disturbance. The result is exquisitely preserved fossils, sometimes complete with soft-part anatomy, that provide us with a vivid window on life 220 million years ago.

The air is filled with a cacophony of noise from the chirping grasshoppers. Insects buzz over the glistening surface of the water, while water bugs struggle to free themselves from masses of plant debris floating at the surface. Deeper in the water column, moving shadows betray the presence of lurking fish waiting for a tasty morsel to sink from the surface. This picture is not unlike the quarry as it is today, where an abandoned portion of the quarry has filled in with water. It now provides a wonderful habitat for fish and aquatic insects in addition to a variety of flowers and songbirds. However, it could equally be a description of a scene from the Triassic.

As would be expected from a lake deposit, many fish have been recovered from the Solite sediments, and these include coelacanths, redfieldiids, and semionotids. However, by comparison with many other fossiliferous lake deposits, fish are not particularly abundant, and to date, the number of individuals recovered of an amphibious reptile, *Tanytrachelos*, exceed the total number of fishes. *Tanytrachelos* grew to a foot or so in length and, at the end of an exaggeratedly long neck, it possessed a small head armed with tiny, needle-sharp teeth (fig. 11.8). The long neck is one of many features that suggest close affinities with *Tanystropheus* and *Langobardisaurus* (Renesto 1994) from the Alpine Triassic (see chapter 6). On the basis of specimens preserving soft-part anatomy, it is clear that its skin was quite smooth and that the hind feet were probably webbed. The occurrences on some bedding planes of small footprints that closely match this pedal structure suggest that *Tanytrachelos* was in fact the track maker. These tracks sometimes exhibit clear impressions of webbing. *Tanytrachelos* has also been recovered from the Newark Basin, and isolated cervical vertebrae from the Chinle also hint at its presence there, but nowhere are they as abundant as at the Solite Quarry. It is readily apparent that the reptile was an aquatic insectivore swimming through the shallows and at the surface in search of water bugs and other small insects that were wafted out across the lake. But whereas evidence of the very species that likely composed an insectivore's diet is typically missing from most fossil assemblages, in the case of the Solite Quarry, it is present in wonderful detail.

In recent excavations, crews from the Virginia Museum of Natural History have recovered well over 3000 insects, and although the insect remains from the limited Triassic localities elsewhere in the world consist mostly of isolated wings, the Solite Quarry is notable for the preponderance of complete insects (Fraser and Grimaldi 1997, 1999, 2003; Grimaldi et al. 2004, 2006). Preserved as silvery carbonaceous films, for the most part the insects are small (many no more than 4 millimeters long), yet they often have exquisite

A

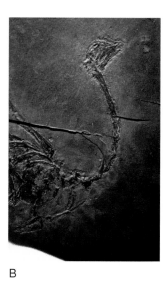

B

Figure 11.8. Little aquatic reptile *Tanytrachelos*, present in large numbers at the Virginia Solite Quarry. (A) Holotype specimen. (B) Second prepared specimen; its long neck is insignificant by comparison with that of its cousin *Tanystropheus* from the Middle Triassic of the Italian and Swiss Alps (chapter 4). (C) Sometimes soft-part tissue is preserved as silvery films around the bones.

C

A

B

C

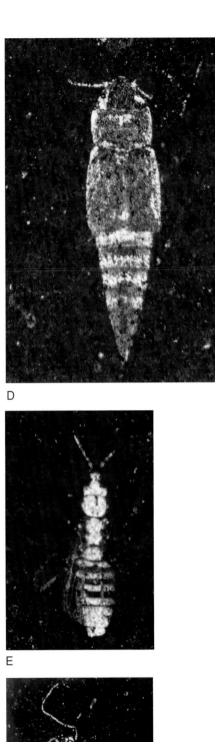

D

E

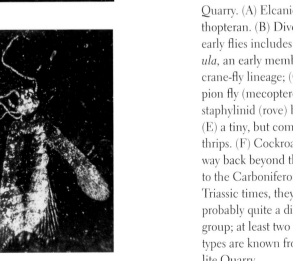

F

Figure 11.9. Triassic buzzing insects from the Solite Quarry. (A) Elcanid orthopteran. (B) Diversity of early flies includes *Architipula*, an early member of the crane-fly lineage; (C) a scorpion fly (mecopteroid); (D) a staphylinid (rove) beetle; and (E) a tiny, but complete thrips. (F) Cockroaches go way back beyond the Triassic to the Carboniferous. During Triassic times, they were probably quite a diverse group; at least two different types are known from the Solite Quarry.

microscopic details preserved, such as the microtrichiae (minute hairlike projections) covering the wings and body. Perhaps what is most remarkable about the insect assemblage is its undoubted modern flavor. It includes the oldest global records for six living families: belostomatids (predatory water bugs) (fig. 11.9A), naucorids (water bugs), psychodids (moth flies), anisopodids (wood gnats) (fig. 11.9B), tipulids (crane flies) (fig. 11.9C), and staphylinids (rove beetles) (fig. 11.9D). A thysanopteran (thrips) (fig. 11.9E) is the oldest definitive record of this small but important living order. The Coleoptera (beetles) and Diptera (true flies) (fig. 11.9F) are the oldest records of these orders from North America. To date, over 20 insect families have been identified, and as further excavations are carried out, this number continues to increase. The abundance of aquatic insects at the Solite locality contrasts sharply with most of the Madygen sites (chapter 6) where aquatic forms are relatively rare.

Apart from *Tanytrachelos*, the insects were also feasted on by another prolacertiform, but this one ventured into the aerial realm. To date, two specimens of a unique gliding reptile have been recovered from the site, but neither has so far been prepared. They occur in a silty matrix that also contains abundant coprolites and fishes, including most of the larger fish specimens. This matrix has proved exceptionally difficult to prepare, but the outlines of the tetrapods can be clearly discerned and many interesting features can be documented. Like the kuehneosaurs (see chapter 10), including the contemporaneous *Icarosaurus*, the Solite glider is characterized by greatly elongated "thoracic" ribs that in all probability supported a gliding membrane. However, unlike the kuehneosaurs, the Solite glider does not have particularly elongate transverse processes on the vertebrae, and this in turn suggests that its body must have been particularly narrow. It has a relatively long neck—certainly much longer than in the kuehneosaurs—that is comparable in proportions to that of *Tanytrachelos*.

At the Solite Quarry, the best-preserved and most abundant *Tanytrachelos* specimens originate in the most finely microlaminated beds. These are also the beds that produce the majority of complete insects (particularly the most terrestrial forms such as the staphylinid beetles). Contra Olsen (1989, etc.) the latest excavations at Solite revealed few articulated fish in these insect-rich horizons, and only the occasional isolated element, including a probable shark tooth. This is perhaps indicative of a nearshore depositional environment and is anomalous with the microlaminated sediments, which according to Olsen are indicative of deeper waters.

During periods of low water stands when the climate was drier, the Solite Quarry lay on the fringes of the lake. Here, among the mud cracks and ripple-marked surfaces, vertebrates left their tracks: *Tanytrachelos* scuttling across the open mud flats, small theropod and ornithischian dinosaurs coming down to the water to drink and hunt. Although the Solite Quarry sediments seem to be equivalent in age to the lake sediments of the Lockatong Formation, no specimens of *Hypuronector* have been recovered from Solite. Given the apparently similar depositional environment, this is a little curious.

This wonderfully rich tapestry of life that existed 220 million years ago was surely not a Garden of Eden isolated in time and space! Why, then, have we not found similar assemblages, particularly in similar types of sediments, elsewhere in eastern North America— for example, on the Lockatong? One answer may be that the special depositional environment required for detailed preservation occurs only on rare occasions. However, I suspect that the answer is a lot simpler—namely, that Triassic insect fossils are far more common than previously supposed, but we have yet to adopt the correct search image. After all, many of these fossils are hardly visible to the naked eye and are far less obvious, say, than a dinosaur bone. With this in mind, paleontologists have revisited outcrops of Triassic rocks in Virginia and North Carolina and have already identified another significant insect locality. The book on Triassic life is far from complete.

Because of the incredibly detailed preservation of insects and the ghosts of soft tissues on some of the vertebrates, one would be forgiven for expecting the plant fossils of the So-

Plate 11.2. With its powerful hindlimbs, a *Tanytrachelos* propels itself froglike along the leaf-strewn bottom of a Newark lake. Swimming over it is a *Turseodus*.

lite Quarry to have some of the most completely preserved examples of Triassic foliage anywhere in the world. Unfortunately, in some respects, paleobotanists might be disappointed with the Solite plants. As already mentioned, cuticle retains the impressions of the epidermal cells that it overlies and is typically resistant to decay. Because the epidermal cells of many plant groups typically display characteristic patterns, fossil cuticle can be useful for identifying plants and matching isolated plant parts. However, at some stage in their history, the sediments of the Solite Quarry were subjected to significant heating that has rendered the cuticle useless for taxonomic purposes. Despite the lack of cuticle, a diversity of conifers, ferns, cycadeoids, and ginkgophytes is still readily apparent.

Modern ginkgos are regarded by some as "living fossils." Supposedly the modern form has changed little from ancient members of the family. However, with its long, fanlike leaves, one look at *Sphenobaiera* from the Solite Quarry would dispel any notion that ginkgophytes have remained unchanged for millions of years. As in the case of the sphenodontians, this fact merely serves to emphasize that we know so little about earlier diversity levels of many groups, and so we simply try to hide our ignorance behind meaningless labels.

Despite the poor preservation, one or two new plant taxa have even been identified from the Solite Quarry. One of the most controversial of these is *Pannaulika triassica*. Identified on the basis of a single piece of foliage, the venation pattern is characteristic of flowering plants (Cornet 1993). Flowering plants are normally considered to have a Cretaceous origin, but the looping tertiary venation of *Pannaulika* is intriguing. We are still puzzled by this specimen, and until additional specimens are found, it is equally possible that *Pannaulika* is an unusual fern. Other foliage types preserved at the quarry include at least five genera of ferns, six conifers, and four cycadeoids. Fragments of foliage are common, and on occasion, there are dense mats of vegetation covering whole bedding planes, making it difficult to discern the individual shoots. Clearly, at times, the hills surrounding Lake Danville were cloaked with a lush and richly diverse vegetation.

RICHMOND BASIN

The fossils of the Richmond Basin have a long history of research, with even the illustrious Sir Charles Lyell passing his well-trained eye across the area. In the mid-1840s Lyell collected remains of the subholostean fish *Dictyopyge macrurus* (fig. 11.10) from the Black Heath Mine.

Sediments of the Richmond Basin range from the early to mid-Carnian, with the uppermost Otterdale Formation being of indeterminate age. It is apparently completely separate from the three other formations that are presently recognized in the basin. These formations are in turn named the Boscabel, Tuckahoe, and Turkey Branch Formations. The stratigraphic position of the Boscabel Formation is unclear. It consists of coarse, angular breccias and may in part be equivalent to parts of the other two formations. The Tuckahoe Formation comprises three members, including the middle Productive Coal Measures Member, which, as its name indicates, has produced significant amounts of coal. Coal was mined in the Richmond Basin from the early 1700s well into the 1930s (Goodwin et al. 1986). In all, over 8 million tons of coal have been commercially mined from the Richmond Basin, principally from five main areas (Robbins et al. 1988). The occurrence of coals is strongly indicative of deposition in a warm, humid environment. Throughout the Tuckahoe and Turkey Branch Formations, lacustrine sediments are prevalent—another indicator of continuously humid conditions. As Olsen et al. (1989) point out, aside from those of the Fundy Basin, these deposits represent the oldest in the Newark Supergroup and were probably deposited at a time when the North American plate was close to the paleoequator (Cornet and Olsen 1985). More recently, wells have been drilled exploring for oil and gas, but to date, no commercial operations have been developed.

All the coal mines of the Richmond Basin have yielded a magnificent array of plant fos-

sils. Early on, these were fully documented by Fontaine (1883) and Ward (1900). Both au-thors amassed large and important collections, but unfortunately these have since gone missing from the collections of the University of Virginia—a situation that its founder, Thomas Jefferson, would surely have found depressing! Fontaine's collection included material from the Winterpock, or Clover Hill, mining district. More recently, Cornet and Olsen (1990) reexamined the flora of the Richmond Basin. Although the sites are now overgrown, Cornet and Olsen were able to collect fossils from the Clover Hill area, which helps at least to lessen the loss of the historic collections. Typical plant fossils include cy-cadeoids (*Macrotaniopteris, Podozamites, Pterophyllum*), ferns (e.g., *Sphenopteris, Clathropteris, Cyathoforma, Acrostichites, Mertensides, Gleichenites*), ginkgophytes (*Sphe-nobaiera*), and horsetails (*Equisetites, Neocalamites*). This is only a fraction of the forms that have been identified, and it is worth repeating here a portion of the letter addressed to J. W. Powell (the director of the United States Geological Survey) that appears in the front of Fontaine's monograph: "The many difficulties attending the collection of fossils from these beds show that the plants here described form but a small fragment of what was evi-dently a very rich flora." Fontaine continues, "some of these areas were, at some time in their history, in the form of marshes, or had such a character as to permit the growth of abundant vegetation and the accumulation of considerable amounts of coal." Clearly there was no doubt in Fontaine's mind that he was sampling what was once a dense and luxuriant vegetation cover.

On the other hand, in studying the palynomorphs as well as the megaflora, Cornet and

Figure 11.10. Complete fossil of the fish *Dictyopyge*.

Olsen (1990) suggested that there was a distinct change in climatic conditions in this area during Carnian times. In this corner of Pangaea, the early Carnian was a time of high-diversity wetland vegetation dominated by pteridophytes and cycadophytes, but by the end of the Carnian, a low-diversity floodplain vegetation dominated by conifers was widespread. Cornet and Olsen argued that these changes were principally brought about by increased aridity, although they acknowledged that decreased temperatures brought about by uplift might also have been a contributing factor. As rainfall decreased, the upland floral communities expanded into the basins, and an intermediate ecological zone between perennial wetlands and seasonally dry floodplains created an opportunity for new taxa to appear. As a consequence, the greatest floral diversity levels occurred at the onset of these drier conditions in the middle Carnian, but by the end of the Carnian, many of these taxa had become extinct. By contrast, these changes in the Carnian flora of eastern North America did not take place elsewhere. As noted in earlier chapters, other parts of Pangaea experienced more humid conditions during the middle Carnian. This is certainly reflected in Europe, where floral diversity levels remained high throughout the Carnian, and here, some Late Triassic taxa even extended into the Jurassic. Cornet and Olsen postulated that the differences between eastern North America and Europe might be due in part to the ameliorating effects of the maritime climate that prevailed around the Tethys Sea.

Another remarkable site occurs close to the Powhite Parkway—one of the commuter's escape routes from downtown Richmond. Described as a "mudhole" by a colleague who will remain nameless, the Tomahawk locality may look unpromising, but grubbing around in the mud can sometimes yield a gold mine: this site has produced an incredible assemblage of mostly small vertebrate remains that has a significant bearing on many issues concerning Late Triassic tetrapod distribution.

There are partial skulls and abundant jaws of a small traversodont cynodont that was named *Boreogomphodon jeffersoni* by Sues and Olsen (1990). Sues et al. (1994) note that most of the jaws represent juvenile individuals, conceivably a sorting effect. *Boreogomphodon* is most like *Traversodon stahleckeri* from the Carnian Santa Maria Formation (Huene 1936) (Brazil; see chapter 9) and *Luangwa drysdalli* from the Anisian of Zambia (Kemp 1980). As noted in chapter 1, the traditional view regarding traversodont cynodont distribution was that they are almost exclusively found in Gondwana. But on the basis of the abundant *Boreogomphodon* remains, Sues and Olsen (1990) suggested that instead of representing geographic separation, the putative difference really reflected differences in stratigraphic age and that the apparent absence of traversodonts in Laurasia was really a reflection of poor stratigraphic sampling of the transition from Middle to Late Triassic.

Olsen et al. (2001a) have gone a step further and suggested that the traversodonts track the humid belt of the Triassic equatorial regions, and that as Pangaea drifted northward the distribution of the traversodonts also tended to migrate northward. If this was indeed the case, then why is there a complete absence of traversodonts in the Chinle? Olsen and Kent would argue that Chinle deposition was in a region that was positioned slightly north of the equatorial belt, and certainly to the north of the Richmond Basin. The discovery of another traversodont in the Deep River Basin, *Plinthogomphodon*, does lend support to the notion that the occurrence of traversodonts at the Tomahawk locality is not simply a function of the depositional environment. On the other hand it is difficult to argue that the St. Nicolas-de-Port (eastern France) and Gaume (Belgium) localities were situated in the equatorial belt during Triassic times, and they were every bit as far north as the region in which Chinle deposition took place. Yet unequivocal traversodont remains have been described from this area, although, admittedly, they are extremely rare (being known just from two teeth at Gaume; Hahn et al. 1988; Sigogneau-Russell and Hahn 1994).

Another cynodont, *Microconodon*, has also been found at the Tomahawk locality. Pre-

viously known only from an isolated dentary from the Cumnock Formation (Osborn 1886a, 1886b; Simpson 1926a, 1926b), Sues et al. (1994) found no reason to believe that the Tomahawk and Cumnock taxa were not conspecific.

Other remains from the "mudhole" include the remarkable teeth of *Uatchitodon kroeleri* (Sues 1991). These recurved, bladelike, serrate teeth are probably from an archosauriform. They possess distinctive grooves on both the labial and lingual sides that are quite reminiscent of the fangs of venomous snakes and of the poisonous lizard, the Gila monster. Administering poison to subdue your prey would seem to have a long history! Isolated *Uatchitodon* teeth have also been recognized at the *Placerias* Quarry in the Chinle.

Although fossils are not common in the Otterdale Formation, it is notable for petrified wood that has been compared with *Araucarioxylon* from the Chinle's Petrified Forest.

As work continues on the Newark Supergroup sediments, we are beginning to see that it is turning into one of the most productive areas for Triassic vertebrates anywhere in the world. This turnaround is in large part due to efforts of small field crews led by Paul Olsen and Hans-Dieter Sues.

Of course, rifting did not occur on just one side of what is now the Atlantic Ocean. Rift basins in North Africa and Europe also contain important fossiliferous sediments de-

Plate 11.3. In a lush tropical rain forest, a diverse variety of plants include the dipteridaceous fern *Clathropteris*, the cycads *Sphenozamites*, *Ctenophyllum*, and *Macrotaeniopteris*, the tree ferns *Cyathoforma* and *Todites*, and araucarian conifers. The crocodile-like pseudosuchian *Doswellia* wades through a murky pool under the watchful eye of a small *Boreogomphodon*.

posited at the same time as the Newark Supergroup. For example, the Argana Basin of Morocco is known to contain a wealth of vertebrate material, including spectacular specimens of phytosaurs and temnospondyls. However, by comparison with the Newark sediments, they have been poorly prospected until recently. I fully expect to see a complementary side to the rifting story coming out in the next decade.

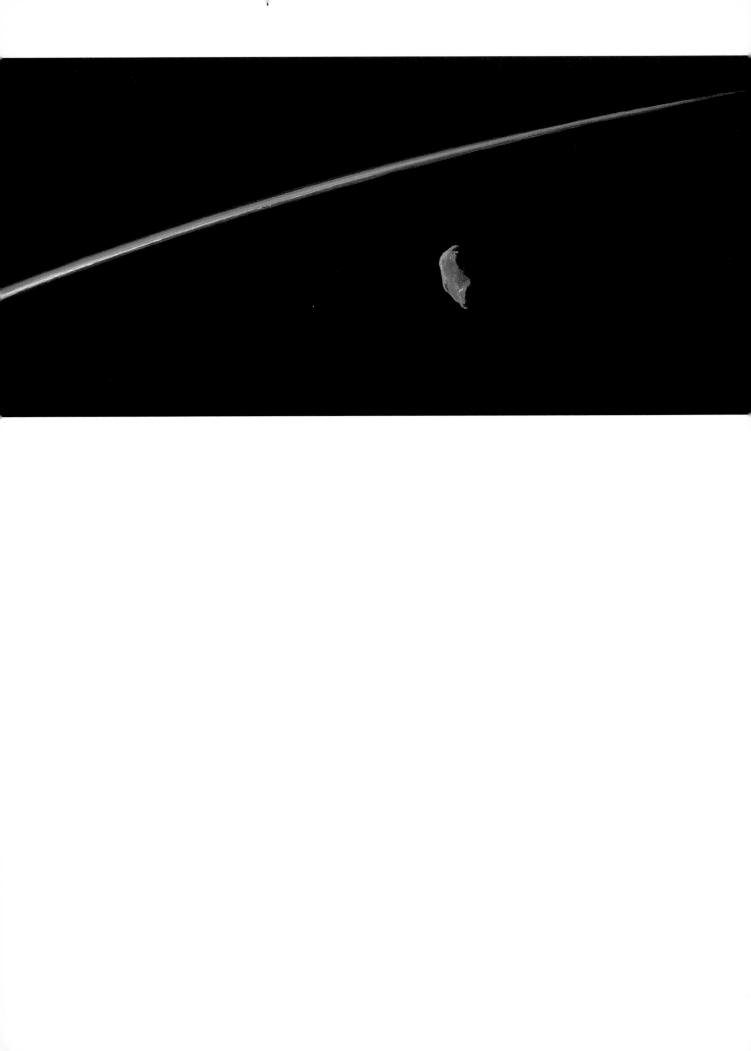

chapter twelve

A Catastrophic Finale?

The Mode and Tempo of Faunal Change

Out with the Old and in with the New

Although the end of the Triassic has not attracted the attention of the Cretaceous-Tertiary event or the end-Permian episode, it has long been argued to have witnessed one of the world's great mass extinctions. Raup and Sepkoski (1982) cited an end-Triassic event in support of their theory that the Phanerozoic was marked by a periodicity in mass extinctions, and that once every 26 million years or so, the earth's biota suffered a major perturbation. Hallam (1990) presented evidence for a significant end-Triassic event in the marine realm, a theme that Benton (e.g., 1983a, 1986, 1987, 1991) expanded on and compared with terrestrial faunal turnover over the same time interval. It is certainly apparent from earlier discussions that the close of the Triassic was marked by profound continental faunal changes with, for example, the appearance of many of today's major vertebrate and insect groups and the disappearance of many wonderfully bizarre forms that left no known descendants. The causes and timing of these changes have been the subject of considerable debate, and at the outset, it must be said that the arguments continue to rage (e.g., Weems 1992b; Padian 1994).

Accepting that around the Triassic-Jurassic boundary there was indeed a time of major extinctions, we can ask a number of questions. Were the extinctions gradual or catastrophic? Did they occur at the very close of the Triassic—in other words, right at the end of the Rhaetian—or, as some authors have suggested, did they reach their peak in the Carnian? What was the cause of the extinctions? Is there, for example, any evidence for major climate or sea level changes, or, if the data point to some kind of catastrophic event, is there evidence for a culprit—perhaps increased volcanism on a global scale or even an extraterrestrial impact? Interestingly, all these questions have been answered in the affirmative at one time or another.

Of course, to be able to confidently address these issues requires that we have a remarkably good and complete fossil record on a global scale. As we have seen, this is not the case. But as we have also seen, there are some extensive Late Triassic continental sequences in different parts of the world, and some of these include some extraordinary fossil preservation. For this reason, it is possible to assess regional data and see whether any patterns emerge.

As we saw in earlier chapters, the terrestrial inhabitants of early Pangaea were typified by forms such as dicynodonts and cynodont synapsids, basal archosaurs, procolophonids,

Plates 12.1–12.6. A giant extraterrestrial bolide is depicted seconds before impacting the earth. Shortly after impact, a growing vapor plume rises above the earth. It rises as a growing fireball that would have been visible from the moon.

and rhynchosaurs. And, as expected after such a catastrophic event as the Permian extinction, Early Triassic vertebrate assemblages are typified by relatively few taxa. But where these taxa do occur, they are frequently abundant. This situation continues well into Middle Triassic times, with the same basic players, the synapsids and archosaurs, dominating the landscape, albeit with gradually increasing diversity levels.

At some point during the latter part of the Triassic, there was a shift in emphasis among the terrestrial faunas. To reiterate, the earliest mammals, turtles, lissamphibians, lizards, and crocodiles are all documented from the Late Triassic. This was also the time of the first dinosaurs and pterosaurs. By earliest Jurassic times, a major change had taken place, and a whole new list of vertebrate groups is in place. To date, almost all investigations concerning the precise timing of the onset of change, how rapidly the change occurred, and what might have triggered the terrestrial faunal turnovers have centered on the vertebrate fossil record.

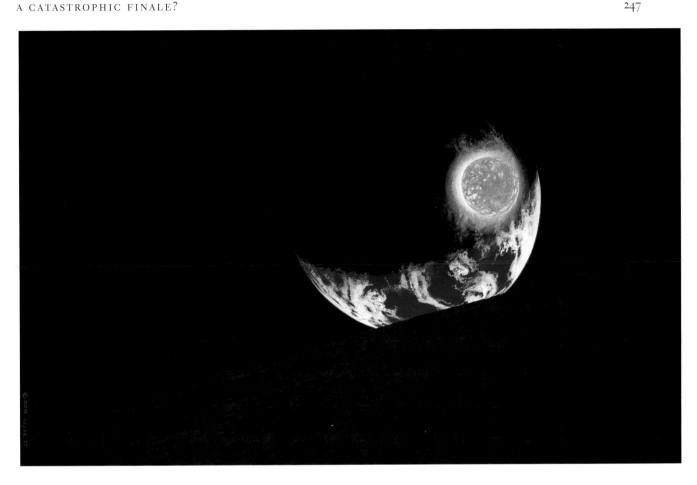

Land Vertebrates

Traditionally, the changeover from essentially archaic forms to those of modern aspect was seen as competitive replacement (Charig 1984). It was argued that the sprawling posture of basal therapsids, rhynchosaurs, and basal archosaurs was no match for the erect posture of the first dinosaurs and early crocodiles (Charig 1984). If this were the case, then one might expect to see a fairly gradual replacement of these groups, with significant temporal overlap between the archaic and modern forms. However, at least one author has suggested that higher-order taxa do not compete, and thus opportunistic replacement might be considered a more likely mechanism. In this scenario, the "neotetrapods" would only radiate once the "paleotetrapods" had become extinct and left vacant niches (Benton 1987). One might expect that this scenario would leave a definite signature in the fossil record, with little overlap between the archaic forms and their ultimate replacements. Therefore, it should be a relatively straightforward task to distinguish between the two mechanisms. As always, things are never simple: one of the difficulties is the lack of any consensus of opinion regarding precise age correlations for Late Triassic terrestrial faunal assemblages worldwide. Although the opportunistic replacement model has gained more support in recent years, another variable has been added to the mix. Some authors have postulated that at least one mass extinction occurred toward the end of the period, but the exact timing of any such event is the cause of additional disagreement. One school of thought argues for a major mass extinction at the end of the Carnian (220 million years ago), with a somewhat lesser event right at the end of the Triassic (205 million years ago) (Benton 1991), whereas the other favors only a single event at the close of the period (Olsen et al. 1987).

We know from preceding chapters that most of the richly fossiliferous Triassic sequences only document relatively short intervals of Triassic time. Consequently, one of the biggest obstacles to resolving these issues is the absence of any single continuous fossiliferous sequence documenting the critical interval, from the Ladinian (235 to 230 million years ago) or Carnian (230 to 220 million years ago) through the Sinemurian (201 to 195 million years ago) or Pliensbachian (195 to 188 million years ago). Of all the Triassic-Jurassic sequences known worldwide, the Newark Supergroup (chapter 11) might be considered to offer the best opportunity for testing hypotheses of Early Mesozoic terrestrial faunal change, at least on a regional basis. Collectively, the basins cover a time span from the Carnian to the Pliensbachian, contain more than 6000 meters of fluviatile and lacustrine sediments, and can be cross-correlated with each other. As discussed in the preceding chapter, ages based on palynologic and magnetostratigraphic data, tied in with radiometric dates, provide a well-constrained time frame for deposition of the sediments (Kent et al. 1996).

In light of what we saw in chapter 11, it is curious that Newark Supergroup sediments have something of a reputation for being unfossiliferous, particularly for vertebrate skeletal remains in the Norian sections. And because the Newark Supergroup has been widely studied, this putative paucity of fossils is cited as a true reflection of the once-living Triassic communities. Consequently, it has been suggested that this poor Norian record is real evidence of a sharp drop in diversity at the end of the Carnian (Benton 1991). At the same time, it must be remembered that footprint assemblages are relatively common, indicating widespread tetrapod populations. Moreover, at times of low diversity levels, such as in the aftermath of the end-Permian extinction, vertebrate body fossils are still common, although limited in the numbers of different taxa. Thus it would be dangerous to consider the relative paucity of skeletal material in the Norian of the Newark as a direct indicator of low diversity levels. Because of these continued uncertainties, no one has been able to build a really convincing case to support one theory over another. The discussion founders on whether the fossil record represents a real reflection of events or not. Although the competitive replacement model may have fallen into disfavor, there is as yet no resolution on the timing of a possible mass extinction event.

As work continues on Newark Supergroup sediments, as well as elsewhere, new data constantly force us to reevaluate the situation and reconsider various hypotheses on faunal turnover. The discovery of the erpetosuchid skull in fluvial facies of the Hartford Basin (see chapter 11) certainly raises a large red flag. What is of real interest is that this particular fossil was initially mistaken for a paleosol carbonate nodule. Furthermore, fluvial facies are only infrequently prospected for fossils. Taken together, this strongly suggests that we do not know the Newark Supergroup as well as we think we do, and that the apparent paucity of fossils in parts of the Newark Supergroup might in reality be merely a function of adopting an inappropriate search image.

The Forgotten Invertebrates

But a more glaring gap in all these analyses on terrestrial faunal turnover is the dearth of discussions concerning the invertebrate record. After all, although today large vertebrates might be the most obvious component to human eyes, it is unquestionably the invertebrates, and in particular the insects, that dominate terrestrial animal communities. Undoubtedly insects were just as critical to continental ecosystems during the Mesozoic as they are today, and to rely totally on vertebrates for our analyses of terrestrial faunal turnover at the Triassic-Jurassic boundary would be unwise. Yet for a long time, we were not in a very good position to engage in any meaningful discussion about terrestrial invertebrates.

At first it might seem natural that insects would be poorly represented as fossils. After all, a judicious swat with the hand, and a delicate-winged mosquito, complete with its piercing mouthparts, becomes nothing more than a nasty, if somewhat satisfying, smear. How well could we expect insects to fare against the rigors of fossilization? But despite their delicate nature, insects generally do have an extensive fossil record. So why have Triassic terrestrial arthropods largely been ignored? It turns out that an exceptionally large part of the insect fossil record as a whole is restricted to lagerstätten, and consequently through the Phanerozoic, the insect record is discontinuous. Until relatively recently, this was particularly true for the Triassic, and by and large, there was a huge void in Triassic insect collections. For a long time, the only areas in the world for which there were any detailed descriptions of Triassic insects were a few localities in Kyrgyzstan and Kazakhstan (chapter 6) as well as one or two in New South Wales and Queensland, Australia.

But of course, we now have the Grès à Voltzia sediments (chapter 4) and the Solite Quarry (chapter 11), both of which have phenomenal insect preservation: the Grès à Voltzia provides a vivid window on an estuarine environment from the Ladinian-Carnian, and the Solite locality preserves a marginal lacustrine ecosystem from a slightly younger period. Ongoing field studies in the Molteno Formation are also yielding a great abundance of insects (chapter 9), with claims for as many as 175 different species (mostly beetles, bugs, and cockroaches) (Cairncross et al. 1995). If we add to this deposits in northern Italy and Switzerland (chapter 6) that have also hinted at much greater insect riches, then suddenly the prospects for more inclusive analyses of terrestrial faunal turnover become much brighter (Bechly 1997).

Collectively, these sites contain the earliest records for many higher taxa that prosper today, including true flies (Diptera), caddis flies (Trichoptera), thrips (Thysanoptera), and water bugs (Hemiptera). What is particularly striking is the close resemblance to the living members of their respective orders and families. A thrips from Solite has the same long, narrow body and short wings with the distinctive fringe of long hairs around their margins as does its modern-day cousins.

It is therefore possible that the Late Triassic insect record has an even more modern flavor than the vertebrate record. That does not mean to say that all Triassic insects adhered to the modern-day blueprints. There were also a plethora of archaic forms, and some, such as the Titanoptera, that were unique to the Triassic (chapter 6). Overall, the terrestrial arthropods seem to mirror the vertebrates in representing a mixture of the ancient and modern.

Perhaps a more significant feature of these terrestrial invertebrate records is the great diversity of some of the modern-day insect orders. For example, six (including three extant) families of dipterans have been recorded from Solite (Fraser et al. 1996), implying that the order actually originated much earlier in the Triassic. This in turn suggests that there was some extensive temporal overlap between the modern and archaic faunas. The occurrences of Late Triassic insects are still too spotty to provide unequivocal evidence regarding the nature of terrestrial faunal turnover and any putative mass extinction events, but they do strongly suggest that the shift from ancient to more modern forms was more protracted than the vertebrate studies alone would lead us to believe. This, however, poses another question: were faunal turnovers among terrestrial invertebrates synchronous with those among the vertebrates? Furthermore, were any terrestrial faunal changes linked with those that have been documented in the marine fossil record?

As well as providing a much-needed window on terrestrial arthropods in the Late Triassic, the Virginian quarry also underscores the erpetosuchid find and the danger of assuming that a relative paucity of fossils in certain Newark Supergroup sediments reflects low diversity levels in the once-living communities. The rocks in the quarry were long considered to be largely barren of fossils, despite repeated searching. We have since learned that the critical factor in finding fossils is the adoption of an appropriate search image. One-millimeter silvery impressions on black shale in a dusty quarry are readily overlooked. The angle of incident light is critical if we are even to distinguish anything, and some kind of optical magnifier is essential. It is premature to say that parts of the Newark Supergroup are scarce in fossils; the new evidence points to the exact opposite conclusion.

Faunal Turnover in the Late Triassic Seas

There is no question that there is almost always a bias in the fossil record toward shallow marine organisms, and the Triassic is no different in this respect. It is fair to say that a lot more is known about life in Triassic seas during this same time interval. Debates concerning Late Triassic extinctions have also focused on the magnitude of an end-Carnian event as opposed to an end-Norian or Rhaetian event. There are many reasons for the lack of consensus. First, it would seem that different groups of organisms were perhaps affected at different times. Hallam (1981) provides strong evidence that bivalves experienced a major peak in extinctions right at the Triassic-Jurassic boundary. On the other hand, Johnson and Simms (1989) argue that crinoids suffered major losses mostly during the early and middle Carnian, and the bryozoans (Schäfer and Fois 1987) and echinoids (Smith 1990) were likewise decimated during the Carnian. Stanley (1988) has documented an apparent major change in reef organisms around the middle to late Carnian. Before this time, reefs were dominated by calcareous algae, bryozoans, and calcisponges. By contrast, in the Norian, scleractinian corals are found to be the main components of the reefs.

If the extinctions were spread throughout the Carnian and Norian, then it seems that we should perhaps be looking for more than one agent of faunal turnover. However, many argue that the Carnian extinctions are not real. Sepkoski and Raup (1986), for example, say that the Carnian extinction peak is really just an artifact of gaps in the fossil record. They state that a handful of exceptionally rich early and middle Carnian fossiliferous beds coupled with subsequent late Carnian and Norian marine horizons being somewhat fossil poor could together give the false impression of an extinction. Others have argued that even if a Carnian extinction event is a real phenomenon, it was in all likelihood not global in its extent, but merely regional (Hallam 1990). Some authors have even suggested that many groups were in gradual decline throughout the Late Triassic (Tanner et al. 2004).

A Topsy-Turvy World

The patterns of Late Triassic terrestrial faunal turnover were undoubtedly complex. Faunal changes had clearly started many millions of years before the end of the Triassic (205 million years ago). What might have been the driving forces behind these faunal changes? The opposing competitive and opportunistic replacement models are surely too simplistic to account for the patterns that are beginning to emerge. Why did the predecessors of our modern terrestrial vertebrates survive while many other groups perished? Just as we struggle with the apparent random pattern of extinctions at the end of the Cretaceous, we may never fully understand the selectivity at the close of the Triassic. Even so, there are some clues to factors that might have played at least a partial role in this critical faunal change.

As we have seen, Triassic geography and climate were very different from today and must have had a profound impact on the Triassic world. The unification of all the continents into the single supercontinent Pangaea, the distinct possibility of megamonsoon climate conditions (see chapter 1), and the absence of polar icecaps naturally had far-reaching implications. I have discussed how this may have affected vegetation cover and how this in turn would have controlled faunal distributions. Some analyses of floral communities indicate that equatorial regions had more xeric forms by comparison with today and were subject to seasonal rainfall (Parrish 1993, 1998). In contrast, other studies suggest that low latitudes did not experience seasonal rainfall and were humid all year round. Part of the reason for these conflicting data and viewpoints may involve errors in basic assumptions concerning the relative positions of the continents during this critical time frame. Pangaea continued to drift northward and rotate into Jurassic times, so that some records of apparent faunal changes may simply reflect shifting climate and vegetation belts.

A Cataclysm?

On top of the gradual shift in faunal emphasis, there still remains strong evidence in the vertebrate fossil record for at least one abrupt interval of extinction. There is wide, although not unanimous, agreement that the sudden demise of the procolophonids, aetosaurs, and phytosaurs, among other groups, points to some kind of catastrophic event exercising a particularly dramatic effect (Fraser and Sues 1994).

A collision of a large extraterrestrial body with the earth has been put forward as a possible explanation for major environmental change right at the Triassic-Jurassic boundary (Olsen et al. 1987); the Manicouagan impact crater in Quebec has been cited as a candidate for the smoking gun. However, when it was shown that this particular impact was more likely to have occurred during the early part of the Norian (approximately 214 million years ago) (Hodych and Dunning 1992), its possible role in any extinction event fell into disfavor. Its reported age is compatible with neither an end-Triassic event nor the proposed end-Carnian event (220 million years ago). Even so, recovery of shocked quartz in marine sequences at the Triassic-Jurassic boundary in Italy still points to a possible impact at this time (Bice et al. 1992), and Walkden et al. (2002) documented an impact ejecta layer in southwestern Britain. Just because we have not yet identified a suitable impact structure does not rule out a terminal Triassic event.

Spray et al. (1998) postulated a multiple impact event at the end of the Triassic. They cited five impact structures, including the Manicouagan, which may have been produced within hours of each other. They noted that the Saint Martin impact site in western Canada, Manicouagan in eastern Canada, and Rochechouart in France plot at virtually the same paleolatitude, and they hypothesized that this apparent crater chain resulted from a series of projectiles colliding in rapid succession with the earth. These three impact craters range from 25 kilometers (Rochechouart) to 100 kilometers (Manicouagan) across. The other two impact sites are Oblon in the Ukraine (15-kilometer diameter) and Red Wing in North Dakota (9-kilometer diameter).

Clearly such a bombardment from outer space would have had far-reaching effects upon life on land. Nevertheless, Kent (1998) questioned whether Manicouagan and Rochechouart could have formed at the same time, noting that the melt rocks of the two sites had opposite magnetic fields. At Manicouagan the melt rocks are of normal polarity, but they are reversed at Rochechouart. Because it takes a few thousand years for a geomagnetic reversal to take place, this perhaps tends to indicate that at least these two impact structures did not occur within hours of each other. Spray et al. (1998) countered this criticism, invoking the size difference between the two impact sites and the resultant difference in the length of time it would take for the melt layers of the two impacts to cool below their Curie points (the temperatures at which the magnetic minerals acquire their magnetic fields). Although at the smaller Rochechouart site the melt rocks may have passed through their Curie points 100 years or so after impact, Spray and colleagues suggested that the melt rocks of the larger Manicouagan impact would have taken much longer—on the order of 10,000 years after impact—before they passed through their Curie points and acquired their magnetic fields.

Olsen et al. (2002) have since added further fuel to the impact theory through extensive studies of numerous sites in the Newark Supergroup. They marshaled three lines of evidence that they felt strongly supported the impact scenario. First, they noted a spike in fern spores right at the Triassic-Jurassic boundary in sediments in Pennsylvania. Ferns are notable for quickly moving in after a disaster has befallen an area, and a similar fern spike characterizes the Cretaceous-Tertiary boundary in North America. Second, on the basis of an extensive database of tetrapod footprints, they observed that in the tens of thousands of years immediately before the boundary, nondinosaurian footprints markedly declined in relative abundance, while at the same time dinosaurs went from constituting 20 percent to 50 percent of the track makers. Moreover, they noted that the size of the meateating dinosaurs dramatically increased over a similar time frame. Olsen et al. (2002) estimated from the size of the tracks that theropod dinosaurs may have doubled their mass. However, Olsen and his colleagues found no noticeable change in the type of footprints or their general size across the Carnian-Norian boundary. Thus this extensive set of footprint data does not support a Carnian extinction event in any way, but it is suggestive of a sudden faunal change at the end of the Norian.

By themselves, neither of the two lines of reasoning presented so far points to an impact. It is the third line of evidence that Olsen et al. (2002) outlined in support of a catastrophic event that they believe specifically points to an impact. Not only did they document a fern spike in certain sediments, but they also recorded elevated iridium levels. Unfortunately, although the extraordinary high iridium anomaly at the Cretaceous-Tertiary boundary is convincing evidence of an extraterrestrial event, the iridium levels that Olsen et al. reported were no more than a third the figure for the Cretaceous-Tertiary spikes. This discrepancy tends to leave the door open to other possible agents of doom because it is quite feasible that the iridium at the Triassic-Jurassic boundary might have been the result of volcanic eruptions.

The incipient breakup of Pangaea might provide some clues to the possible role of volcanism in the faunal and floral changes. It has recently been shown that the extensive volcanism associated with the initial separation of North America from Africa and Eurasia was of brief duration, and it may be causally associated with a mass extinction marking the very close of the Triassic (Hames et al. 2000). Indeed, there is considerable support for the concept that the end-Triassic extinctions can be attributed to the eruption of flood basalts—the so-called Central Atlantic magmatic province (CAMP) (McHone and Puffer 2003). There can be no questioning the power of a volcanic eruption. The devastating effects of a single eruption of Vesuvius on the ancient city of Pompeii are well documented, as is the more recent eruption of Mount St. Helens.

What, then, would be the result of repeated and widespread eruptions that produced basalt flows up to 3000 feet thick extending over 2.5 million square kilometers such as the Siberian or Deccan Traps? In particular, it has been suggested that the resultant release of significant amounts of carbon dioxide gas into the atmosphere could have led to catastrophic greenhouse warming (e.g., McElwain et al. 1999; Olsen 1999). The extent of CAMP, covering eastern North America, West Africa, and down into South America, rivals the Siberian and Deccan Traps (Olsen and McHone 2003). On the other hand, in a study of pedogenic calcite in Late Triassic paleosols, Tanner et al. (2001) found no evidence of vastly elevated atmospheric carbon dioxide levels at that time and felt that it was unlikely that volcanic outgassing of carbon dioxide was responsible for the extinctions. Nevertheless, they did not rule out some other volcanic effect, such as the release of atmospheric aerosols or tectonically driven sea-level changes, as the causative agent.

It is worth bearing in mind that the extraterrestrial impacts and greatly increased volcanic activity are not necessarily mutually exclusive hypotheses in the search for any smoking gun. Perhaps massive impacts of extraterrestrial bodies triggered great flows of

flood basalts, and perhaps the world's great mass extinctions all have a common cause. The end-Cretaceous extinction seems to be contemporaneous with the Chixalub Crater and the Deccan Traps. And although there is no great impact crater associated with the end-Permian event, such sites could easily have been obliterated by the Siberian Traps themselves.

It is a distinct probability that broad climate change in the Carnian (and even Ladinian) may have been responsible for some gradual terrestrial faunal change. This climate change may, for instance, have been a driving force in the appearance of many modern insect groups. At the same time, catastrophic short-term events seem to have been contributing factors in shaping the terrestrial faunas of the Early Jurassic.

At this stage, we are still a long way from being able to say that the impact of extraterrestrial bodies with the earth brought about both the rise and demise of the dinosaurs. After all, at the Cretaceous-Tertiary boundary crocodiles survived relatively unscathed. Yet if a similar extraterrestrial impact accounted for both the end-Cretaceous and end-Triassic extinctions, why did the phytosaurs, the ecological vicars of the Cretaceous crocodiles, not survive the end-Triassic? Moreover, in both the end-Triassic and end-

Plate 12.7. Immense forest fires would have raged as a result of lightning strikes on tinder-dry forests. Much bigger fires may have ravaged the earth as a result of abnormal volcanic activity or bolide impacts.

Plate 12.8. Super basalt flow.

Cretaceous extinctions, faunal changes were already well underway before the catastrophe hit. Consequently, the magnitude of any impact's effects is always going to be difficult to gauge.

Whatever the cause of the end-Triassic faunal changes, it was time to move on to a new era in terrestrial life. This would be a world dominated by dinosaurs, but also one in which the seeds for modern-day terrestrial ecosystems were being incubated.

appendix 1. global biostratigraphy and age correlation of triassic fossil assemblages

One of the first difficulties encountered when reading about the Triassic (or any other time period, for that matter) is the bewildering nomenclature used to describe the ages and sequences of the various strata and fossil assemblages. Although some names have a global application, others are regional, and still others pertain only to a very localized area.

Putting the different sediment sequences and fossil zones together in a coherent fashion is a complex task. The first difficulty is that terms assigned to rock units are typically separate from those assigned to actual intervals of time. Although there might be a broad correspondence between rock type (lithostratigraphic) and geologic-time units, this is not always the case, and terms for rock units cannot be used interchangeably. Why is it possible to encounter the exact same lithological unit in different regions that is not necessarily of equivalent age in all areas? The reason for the discrepancy takes a bit of explaining. Consider, for instance, a large inland lake that expands over time. Let's imagine the lake grows in a northerly direction. The sediments brought in by river systems and deposited around the shoreline will be of one characteristic type, whereas those forming in the much deeper waters at the center of the lake will have a different character. But as the lake expands with time and part of the shoreline also shifts northward, the rock types associated with these two depositional environments will also migrate outward across the region. Consequently, the shoreline facies (rock type) will be younger the further north they are encountered. Similarly, the age of the deep lake facies will be dependent on its geographical position.

The Triassic and Jurassic rocks of the U.S. southwest afford us real examples of such spatial and temporal distribution of lithologic units. Any visitor to Canyonlands National Park will be very much aware of the extensive and massive cliff-forming Wingate Sandstone that is responsible for much of the natural beauty of the area. The Wingate Sandstone is an eolian (deposited by wind) unit laid down in a desert environment, and it can be seen stretching continuously for miles in a number of places. It is the lowermost of four formations that together constitute the Glen Canyon Group (the other three being the Moenave Formation, the Kayenta Formation, and the Navajo Sandstone). It is almost impossible for us to comprehend how wind alone could be solely responsible for transporting and depositing such an enormous thickness of sediment. The Wingate Sandstone is not known for its vertebrate fossils, unlike the Moenave, Kayenta, and Navajo Sandstone Formations. Overlying the Chinle Formation (see chapter 7), the entire Glen Canyon

Group was traditionally regarded as Triassic in age (e.g., Galton 1971; Colbert 1981). For the Wingate Sandstone, two separate members had originally been recognized, the lower Rock Point Member and the overlying Lukachukai Member. These two units are separated by an unconformity. This means that there was a period of erosion or a time of nondeposition marking the boundary between the two different rock types that we see today. Recognizing that an unconformity existed, Peterson and Piperingos (1979) placed the Rock Point Member in the Chinle Formation, thereby restricting the Wingate Sandstone to just the Lukachukai Member. As additional work was carried out on the vertebrate fossils, the age of the Glen Canyon Group began to be questioned, so that Olsen and Galton (1977) assigned it an Early Jurassic age. Most recent authors have now largely accepted a Jurassic age. Although the four units of the Glen Canyon are lithologically distinct, there are also significant differences within each across their geographic range. Even more significant is the fact that they reportedly intertongue at their mutual contacts (Harshbarger et al. 1957). In this case, parts of the Wingate Sandstone are equivalent in age to parts of the Moenave Formation. Some authors consider the Wingate to be partially Triassic and partially Jurassic.

If correlation of rock units in the great wide-open exposures of the western United States presents us with challenges, what hope have we when it comes to such well-vegetated and urbanized areas as the eastern United States or western Europe? As before, with additional research will come changes in our understanding of the temporal relationships among strata. The Danville/Dan River basin of the eastern United States is a good case in point and offers another glimpse of the stratigraphic problems that we shall encounter. This particular sedimentary basin extends from North Carolina across the state line to Virginia. Within North Carolina, three distinct formations were recognized. They were believed to be time-stratigraphic units ranging (in order of oldest to youngest) from the largely fluvial Pine Hall Formation, through the lacustrine Cow Branch Formation, to the fluvial Stoneville Formation. Three formations were also recognized in Virginia, but rather than integrate them with the scheme used in North Carolina, a completely different nomenclature was used (Meyertons 1963). Here, then, was an example of a very localized system that in effect regarded the state line as some kind of natural break! The Virginia sediments were differentiated largely on grain size, with the Leakesville Formation comprised principally of claystones, siltsones, and sandstones; the Dry Fork Formation consisting mostly of graywackes and arkoses; and with shales and conglomerates predominant in the Cedar Forest Formation (see Appendix 2). Although Meyertons (1963) considered the Cedar Forest Formation to be the youngest, he regarded the Leakesville and Dry Fork Formations as time equivalents. The stage was therefore set for a great deal of confusion. Fortunately, Thayer (1970) undertook a closer study of the geology and rescued us from the earlier follies. He noted that the rocks of the basin were actually not time-stratigraphic units but intertonguing lithologic facies. The result was the creation of a single acceptable scheme that pertains to both North Carolina and Virginia (Lutrell 1989). Three formations can still be recognized in parts of the basin, notably the southwest and northeast portions, and here the lacustrine Cow Branch Formation, comprised predominantly of shales, serves to divide the Pine Hall below from the Stoneville above. However, in the central part of the basin, the lacustrine units pinch out, and consequently, where the Cow Branch is absent, it is impossible to recognize but one formation. Here it is termed the Dry Fork Formation, and it is essentially equivalent to the sandstone facies of the Pine Hall and Stoneville Formations.

In the absence of absolute ages based on radiometric dating, one of the most important ways to cross-correlate sediments is through biostratigraphy. Because of a relative abundance of fossils, marine sequences are the easiest to work with. Relative dating of terrestrial sequences is more problematic, and the choice of index fossils to help correlate sediments from one area to another is critical. In addition to having a wide geographic range and a limited distribution in time, a good index fossil is easy to identify and relatively common. Ideally, a number of different index fossils can be used to corroborate the age of a given unit. If we examine one particular sedimentary sequence in a given area, we may see a very specific change in faunas through time. For example, the replacement of phytosaur A by phytosaur B may be accompanied by the replacement of aetosaur A by aetosaur B. Alternatively, this change might manifest itself in a more profound way; for example, there may be a transition from an assemblage dominated by mammal-like reptiles to one dominated by archosaurs. In both instances, an initial reaction might be to suggest that this shift is directly age related, perhaps associated with some sort of shift in climate type. The next step is to try and test this idea by examining other regions elsewhere in the world that contain assemblages from the same time interval. If we document similar faunal changes, then a possible inference would be that some significant global change occurred that directly affected terrestrial faunas at this time. However, to complicate the issue, we are faced with the dilemma created by the fact that during the Triassic, the continents slowly moved northward. In essence, this has the effect of continually moving our stable point of reference, and if we are not careful, it is all too easy to become completely disoriented. It is conceivable that biotic provinces remain stable through time, but shift in latitude as the continents move. The faunal changes described above could simply mean that the original biotic province—represented by phytosaur A and aetosaur A—just moved northward and out of our frame of reference. For that reason, if at all possible, faunal distribution needs to be tied to the distribution of the continents.

When it comes to correlating sediments on a global scale, there will inevitably be the problem of being unable to equate terrestrial deposits with marine sequences and vice versa. The terrestrial equivalents of the Middle Triassic marine Muschelkalk are the most difficult to determine, and potential index fossils must not only be relatively widespread and of short temporal duration, but include at least one or two offering a possible tie-in with marine sequences. This situation is further compounded by the presence of relatively few biostratigraphically useful plants or tetrapods in the Lower Triassic Buntsandstein.

So, bearing in mind that the correlation of different rock sequences is anything but a straightforward task (even under the ostensibly optimum situation such as the American southwest), we can now present a general overview of the division of Triassic sediments worldwide.

The word *Triassic* is derived from the Greek *trias*, meaning triad, and is named for the three-part division of the sediments in the Germanic basin: the Buntsandstein (terrestrial), Muschelkalk (predominantly marine), and the Keuper (mostly terrestrial). These three rock-type subdivisions are equivalent to the Lower, Middle, and Upper Triassic series. These latter terms are time-stratigraphic units (i.e., rock units with boundaries based on geologic time), and in turn they correspond to the Early, Middle, and Late Triassic epochs (geologic-time units). Further subdivision of these three basic series/epochs into a number of stages/ages is typically undertaken. The standard stage-level division of the Triassic is based on marine sequences, but over the years, a number of different systems have been used. These largely reflect the geographic location of the researchers or the rocks in question. Although the stage names for the Middle and Upper Triassic have remained relatively stable, the same cannot be said for the Lower Triassic. For instance, many authors chose to recognize only a single stage, the Scythian, for all Lower Triassic rocks (fig. A1.1). On the other hand, Lower Triassic sediments have been divided between as many as four stages. Such discrepancies typically result from variation of rock types from continent to continent or even localized usage from basin to basin. At present, however, many workers split the Lower Triassic into two stages, the Olenikian and the Induan, and this is the scheme (the standard global chronostratigraphic scale) followed in this book.

As with any problem, it is always a good idea to start on a small scale: examine the local situation before attempting to describe the big picture. In fact, a number of provincial biochronologies have been developed by a variety of authors. Bonaparte (1966) proposed

Figure A1.1. Division of the Triassic into stages with the sequence of some key events.

Period	Series	Stage	Time (mya)	Events
JURASSIC				
TRIASSIC	Upper	Rhaetian	205	← Extensive volcanism. Final disappearance of procolophonids, aetosaurs, phytosaurs, etc. ← Second Triassic mass extinction?
		Norian	210	← Manicouagan Impact structure
		Carnian	220	← First Triassic mass extinction? Decrease in diversity levels of some terrestrial tetrapod groups? ← Insect assemblages including the earliest members of many modern-day orders and families.
	Middle	Ladinian	230	
		Anisian	235	
	Lower	Scythian	240	Depauperate terrestrial tetrapod assemblages comprising synapsids and basal archosaurs.
PERMIAN			248	Greatest mass extinction of the Phanerozoic era.

a biostratigraphic framework for the Triassic of Argentina; Lucas (1993) set out a scheme for China; and with a variety of coworkers, Lucas also proposed various tetrapods as index fossils for the Triassic of North America (e.g., Lucas and Hunt 1993; Huber et al. 1993).

Of course there have been many attempts to look at the big picture. Romer (1966), for example, recognized three basic faunal assemblages in the terrestrial Triassic. His A-type assemblages of the Lower Triassic are composed of nonmammalian therapsids, and these are perhaps best typified by the tetrapods of the Karoo Basin of South Africa (see chapter 3). The Middle Triassic B-type assemblages are dominated by traversodont cynodonts and are usually associated with rhynchosaurs. Finally the more recent C-type assemblages of the Upper Triassic are dominated by archosaurs, particularly dinosaurs.

Many authors have adopted, but also modified, Romer's scheme since its publication. For instance, Cooper (1982) proposed a more complex six-part division largely on the basis of the distribution of various dicynodonts, although very few workers have subsequently followed Cooper's proposal.

A number of provincial and global biochronologies have been published, many of them in the last 30 years or so. Each divides the Triassic into different zones and introduces new terms to confuse the uninitiated. Lucas (1998) has published the most comprehensive global scheme to date. He splits the Triassic into eight so-called land vertebrate faunachrons (LVF). Each LVF is defined by the first appearance of one

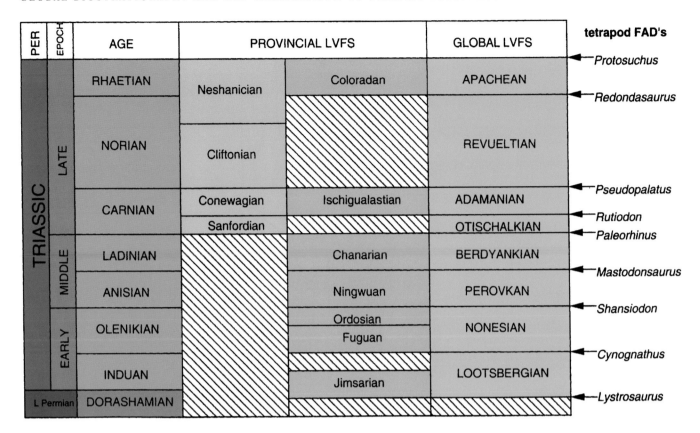

PER	EPOCH	AGE	PROVINCIAL LVFS		GLOBAL LVFS	tetrapod FAD's
TRIASSIC	LATE	RHAETIAN	Neshanician	Coloradan	APACHEAN	Protosuchus
		NORIAN	Cliftonian		REVUELTIAN	Redondasaurus
		CARNIAN	Conewagian	Ischigualastian	ADAMANIAN	Pseudopalatus
			Sanfordian		OTISCHALKIAN	Rutiodon / Paleorhinus
	MIDDLE	LADINIAN		Chanarian	BERDYANKIAN	Mastodonsaurus
		ANISIAN		Ningwuan	PEROVKAN	Shansiodon
	EARLY	OLENIKIAN		Ordosian	NONESIAN	
				Fuguan		Cynognathus
		INDUAN		Jimsarian	LOOTSBERGIAN	
L Permian		DORASHAMIAN				Lystrosaurus

Figure A1.2. Zonation of the Triassic by means of terrestrial vertebrates (after Lucas 1998).

characteristic tetrapod, but a whole range of different taxa is used to recognize each time zone. These time zones are shown in figure A1.2, and they are referred to periodically throughout this book. Without going into all the details here, it is worth looking at one or two of these LVFs.

The earliest of Lucas's LVFs, the Lootsbergian (named for the Lootsberg Pass in the Karoo Basin), is defined by a very distinctive genus of dicynodont, *Lystrosaurus* (see chapter 3). As for all the LVFs proposed by Lucas, the Lootsbergian is also defined by a number of additional characteristic terrestrial vertebrates (fig. A1.3). These include a second dicynodont, *Myosaurus*, the cynodont *Thrinaxodon*, the procolophonid *Procolophon*, the archosaur *Proterosuchus*, and the archosauromorph *Prolacerta*.

After the Lootsbergian in Lucas's scheme comes the Nonesian LVF, which is principally identified by genera such as *Cynognathus*, *Diademodon*, and *Kannemeyeria* (fig. A1.4). Lucas correlates the Nonesian with the Olenikian stage of the standard global chronostratigraphic scale. This is based on the occurrence of the temnospondyl amphibian *Paratosuchus* in marine strata that are well accepted as Olenikian (Lozovsky and Shiskin 1974). Lucas also makes a case for the correlation of his Revueltian LVF with the Norian on the basis of the occurrence of the stagonolepid *Aetosaurus* in marine sequences of the Italian middle Norian.

There is a lot of merit to this scheme, although undoubtedly there must be further refinement as new discoveries are made or as ranges and distributions of certain fossils are altered. Significantly, it is an inherently fluid system. However, because each LVF is defined by the first appearance of a particular taxon, the discovery of an older record of that taxon does not automatically alter the utility of the system; it simply pushes the onset of the LVF to an earlier time. Some of the LVFs cover relatively long periods of time, and one future goal will be to subdivide them further. Currently, various assemblages known to be of slightly different ages are lumped together in the same LVF. For example, the Tomahawk locality of Virginia is included with the slightly younger tetrapod strata of the

Figure A1.3. Some of the more common vertebrates from the Induan (earliest Early Triassic). Sometimes referred to as Lootsbergian fossils (Lucas 1998). (A) Skull in dorsal and ventral view of the temnospondyl *Wetlugasaurus*. (B) Skull of the procolophonid *Procolophon* in lateral and dorsal views. (C) Side view of the skull of *Proterosuchus* (sometimes referred to as *Chasmatosaurus*). (D) Skull of temnospondyl *Tupilakosaurus* seen from above. (E) Skull of the dicynodont *Lystrosaurus* in dorsal view. (F) Skull of *Thrinaxodon* (after Lucas 1998).

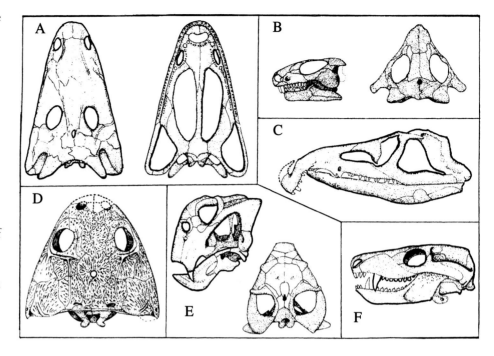

Pekin and Cumnock Formations of North Carolina (see chapter 11) within the Otischalkian LVF.

It seems likely that it will be possible to use vertebrate assemblages in a stratigraphic context on a regional scale, but whether they can ultimately be used to form a high-resolution global biostratigraphic framework is less clear.

One particular aspect that has come under some scrutiny is the use of one or two fossils that on phylogenetic considerations could be older than current records would indicate. For example, the use of the phytosaur *Rutiodon* as a marker for the Adamanian (the later part of the Carnian) has been questioned on the grounds that its putative sister taxon extends further back in time and therefore, from a cladistic standpoint (see Appendix 4), so should *Rutiodon*. For the present, we can only go by our observations, and the numer-

Figure A1.4. Representative terrestrial vertebrates of the Olenikian (late Early Triassic). Referred to as Nonesian fossils by Lucas (1998). Skulls of the temnospondyls *Parotosuchus* (A) and *Eryosuchus* (B) in dorsal and ventral views; skulls of the cynodonts *Cynognathus* (C), *Diademodon* (D), and *Trirachodon* (E); and skull of the dicynodont *Kannemeyeria cristarhynchus* in lateral and dorsal views (F).

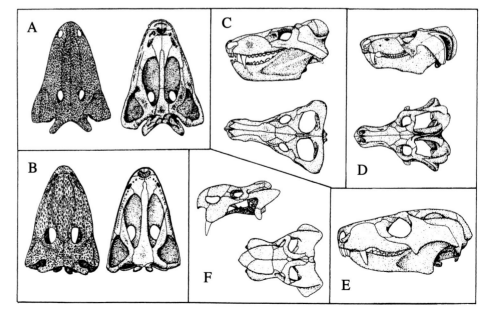

Ammonoid zones	Palynological zones	Tetrapod zones	Chinle pollen zones	Newark pollen zones	Phytosaur biochrons	Aetosaur biochrons
Crickmayi	(Rhaetian)	Norian L2		"Upper Balls Bluff-Upper Passaic" Palynofloral zone		
Amoenum	Sevatian					
Cordilleranus		Norian L1				
Columbianus	Upper Norian (Alaunian)	Norian M2				*Redondasuchus* Biochron
Rutherfordi			III			
Magnus	Lower Norian	Norian M1		"Lower Passaic-Heidlersburg" Palynofloral zone		
Dawsoni	(Lacian)	Norian E			*Pseudopalatus* Biochron	*Typothorax* Biochron
Kerri						
Macrolobatus	Tuvalian	Carnian L2	II	"New Oxford-Lockalong" Palynofloral zone	*Rutiodon* Biochron	*Stagonolepis* Biochron
Welleri						
Dilleri		Carnian L1	I	"Chatham-Richmond-Taylorsville" Palynofloral zone	*Paleorhinus* Biochron	*Longosuchus* Biochron
Nansoni	Julian	Carnian M				
Obesum	Cordevolian	Carnian E				
Sutherlandi	Langobardian	Ladinian				
Maclearni						
Maginae						

Figure A1.5. Zonation of the Triassic by means of palynomorphs.

ous characteristic fossils of *Rutiodon* that have been recovered all point to its utility as an index fossil for the Adamanian. Another more critical criticism focuses on whether many of the key index fossils actually have global, or even anything more than regional, distributions. Rayfield et al. (2003) consider the documentation of index taxa in many localities to be questionable because it is often based on scrappy material. Moreover, they suggest that many taxa are not as temporally restricted as Lucas and colleagues indicate. Ultimately, it may be that the LVFs as currently defined have somewhat more restricted use, but for our purposes, they do at least provide some useful reference points.

At the risk of adding the proverbial last straw to the camel's back, I shall introduce one final scheme of nomenclature into Triassic stratigraphy. The ease with which pollen and spores are dispersed through the air, and their great abundance in many sediments make them an ideal candidate for biostratigraphic correlations. Unfortunately, palynologists use a completely different set of terms to break down the Triassic on the basis of pollen and spore assemblages (fig. A1.5). Again, these are referred to from time to time throughout the book.

This list is still not exhaustive; there are other markers, such as ammonoid zones, which are of paramount importance when dealing with marine sediments. Nevertheless, for the terrestrial Triassic, the principal stratigraphic frameworks have been mentioned, and although these names can be confusing, they will help us locate in time various assemblages of fossils. With these terms at hand, we can begin to examine the physical aspect of the Triassic world—a world very different to the one we know today.

appendix 2. bare bones sedimentology

Upon the death of an animal, scavengers can soon break down a carcass and, within a very short period of time, the action of microorganisms can completely destroy all traces of the body. Thus, for any organism living in the world today, the chances of becoming a fossil are not very good. However, burial by sediment will greatly increase the chances of the organism being fossilized because it will significantly reduce the processes of decay. But for the land-dwelling animals and plants that are the subject of this book, the chances of entering the fossil record are particularly slim. Land is a region of weathering and erosion rather than sedimentation, and more specialized conditions are required for the preservation of terrestrial organisms. Sediment deposition on land is limited to the areas of rivers and lakes, blowing sand, or, more rarely, volcanic ashes.

Careful study of a fossil can give us an insight into the habits of an extinct animal. But if we fail to study the sediment that encases the fossil, we would be guilty of ignoring some of the critical clues concerning the organism's environment and lifestyle. Although fossils form the focus of this book, the picture would be far from complete without some mention of the Triassic sediments. The following is intended to help the reader unfamiliar with sedimentological terms.

The term *sediment* strictly refers to any solid material that has settled down from a state of suspension in a liquid. Thus R. D. Oldham (1879) stated that sediment was "anything settling down from suspension in water. Sedimentary rocks are rocks made up of sediment, loosely applied to all stratified rocks." Today the term *sediment* is usually applied to material originating from weathering of rocks and transported by, suspended in, or deposited by air, water, or ice. In addition, material that is accumulated by other natural agents—in particular chemical precipitation or secretion by organisms—forming layers on the earth's surface are also considered sediments.

Sediments can be divided into four categories on the basis of their origin. First, there are the *terrigenous clastics*. A *clast* is a particle or grain of rock, and *terrigenous* means they are produced from the earth. Literally, then, we are talking about particles of rock produced from the earth. Terrigenous clastic sediments can be further subdivided on the basis of the size of the clasts or the constituent fragments. Those with the coarsest grains are the conglomerates, next come the sandstones, and then finally the mudstones, which are composed of the finest grains of rock. These finest rock grains are silt particles (4 to 62 μm) and clay particles (<4 μm).

Second are the *biochemical-biogenic-organic* sediments. These are sediments laid down directly as a result of the action of organisms. The best known are limestones and coals. Limestones result from the accumulation of calcium carbonate "skeletons" of a variety of animals, including corals, brachiopods, mollusks, and bryozoans. Coals form from the accumulation and compression of decaying organisms, mostly plants.

Third are the *chemical precipitates*. Evaporites and ironstones are the principal types. *Evaporites* are composed of minerals produced from a saline solution through the extensive evaporation of the solvent. A good example is where salt is precipitated along the margins of a salt lake or a restricted area of seawater. In terrestrial environments, sediments predominated by carbonates indicate deposition in shallow lakes situated in wide-open, often arid plains where streams and rivers do not reach. On the other hand, lacustrine sediments rich in clastics indicate more confined basins with significant input of streams and rivers into the lake.

Finally there are the *volcaniclastic* (or *pyroclastic*) sediments. These refer to the particles of debris that fly out from active volcanoes. This is where the distinction between igneous and sedimentary rocks can sometimes become blurred. Judging from those illustrations of Triassic scenes where volcanoes are disgorging vast quantities of smoke and ash into the atmosphere, we would expect to find volcanoclastic rocks common in Triassic sequences. There are some well-known vertebrate-bearing volcanoclastic sediments such as the Chañares (chapter 5), but they are probably no more abundant in the Triassic than any other period in the Phanerozoic. Tuffs are a specialized form of volcaniclastic rock formed of compacted beds of volcanic ash. This should not be confused with tufa, which is a chemical precipitate of calcium carbonate that typically forms as a crust around springs.

The physical and chemical nature of the sediment can tell us a great deal about the depositional environment. The size of the clasts, or particles, making up the sediment are important. The bigger the clast, the greater the amount of energy needed to transport it. So we can say that massive conglomerates required much more energy to be deposited than a fine-grained claystone. The degree of wear on individual clasts and the extent to which they are sorted are also important characteristics. Sediments that are made up of well-rounded grains and are well sorted (all sediment grains of similar size) are said to be mature. Eolian (wind-blown) and beach sands are examples of mature sediments. Sediments that are poorly sorted (a range of grain sizes) and consist of angular grains are said to be immature, and we would expect to see these in sandstones washed down by rivers.

The chemical properties of sediments can be important in providing information about the geology and climate of the source area. For children playing on a beach in summer, the chemical composition of sand is not something requiring any attention. But to the geologist, sands can be very different, and they take on a variety of different forms and names. For instance, *graywackes* are dark coarse-grained sandstones that are poorly sorted. They are typically formed in an environment where erosion, transportation, deposition, and burial are rapid. *Arkoses* are also coarse-grained sandstones that are usually derived from the rapid breakdown of granites, and they are consequently rich in feldspar. They are commonly pink to red in color and tend to be indicative of a semiarid depositional environment. *Arenites* are relatively "clean" sandstones—that is, they have a relatively simple mineralogical composition and have little in the way of matrix material. Quartz arenites characterize eolian (wind-accumulated) sandstones.

Chemical and physical changes normally occur in sediments after their initial deposition. Such alteration to the sediment is known as *diagenesis*. Those changes that take place before consolidation of the sediment into hard rock are referred to as early diagenetic changes. One example of diagenesis is the alteration of limestone into dolomite through the replacement of the calcium by magnesium. When that takes place soon after deposition, the sediment is normally fine-grained and is often associated with structures associated with supratidal conditions, such as evaporites and desiccation cracks. On the other hand, late dolomitization often obliterates the original structure of the limestone so that any fossils become difficult to discern. Under semiarid conditions, early dolomitization occurs on intertidal flats.

The color of a sediment may provide us with some hints about the ancient environment. Black sediments are characteristic of high organic content. Terrestrial sediments are sometimes reddened through hematite (iron oxide) pigmentation. The presence of red

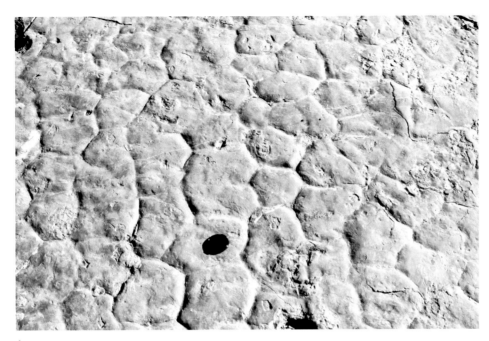

A

B

Figure A2.1. (A) Mud cracks preserved around the margin of a Triassic lake that existed in the Culpeper area of Virginia. (B) Modern-day mud cracks in mud around the edge of a small pond.

rocks is suggestive of deposition in an oxidizing environment. However, as discussed in chapter 1, not all red beds are terrestrial, and we should use color with some caution.

Sedimentary units are divided up into *beds*. These are layers of varying thickness and character, and they are produced by changes in the pattern of sedimentation. A *bedding plane* is the planar surface that visibly separates each bed. Bedding planes may represent long or short breaks in sedimentation. Beds that exhibit little in the way of internal structure are said to be massive. They are characteristic of a sudden dumping of large amounts of sediment. Other beds exhibit laminations and indicate more gradual accumulation of sediment

Sediment surfaces can exhibit a variety of physical features such as ripple marks (fig. A2.1) or desiccation features (mud cracks). Ripple marks can be formed by waves or currents, and they indicate that the sediment was laid down in relatively shallow water. Often ripple marks and mud cracks occur in successive beds and are indicative of fluctuating water depths, as would be seen along a lake shoreline. Such sediments may also display rain spots consisting of small, shallow, circular depressions with slight rims. They are produced through raindrops hitting soft, fine-grained sediments.

Synaeresis cracks are short narrow gashes that are spindle-shaped or triradiate—a bit like bird feet impressions. Like mud cracks, they result from dewatering of the sediment, but they occur underwater and typically result from changes in salinity.

Tool marks and grooves might be produced as a result of objects being dragged across soft and unconsolidated sediment. In streams and rivers, pieces of wood or empty shells may have bounced across the soft sediment of the channel floor.

Of course, a soft sediment might be home to any number of creatures; burrows and trackways of crawling animals are sometimes very common. A sediment that has a preponderance of trace fossils is said to be *bioturbated*.

Although a careful analysis of the sediments containing fossils can tell us a great deal, there are limitations. For instance, mudstones can be deposited in practically any environment. A few examples are floodplains, lakes, deltas, the deep-sea floor and a low-energy coastal shelf. And we must always keep in mind that the depositional environment may have been very different from the environment in which the organism actually lived.

It is rare for any organism to be preserved intact at the site of its death. All soft tissue (e.g., muscle, skin, internal organs) not eaten by scavengers is susceptible to bacterial decay, even when covered by sediment. Agents of sediment transport—for example, a river in flood—will also act to break up and take the resistant remains away from the place where the animal died. As a consequence, the remains of an organism can end up in an environment very different from the one in which it actually lived. Animals from very different communities can end up becoming fossilized together in the same deposit. If we are to accurately put together the lifestyle of an extinct animal, we must fully understand the manner of burial and the origin of the remains. The study of this branch of paleo-ecology is called *taphonomy*.

appendix 3. the anatomy of a tetrapod

For living vertebrates, we have a whole range of options at our disposal to use as characters in phylogenetic analyses. These include hard-part and soft-part anatomy, features of behavior, and a range of molecular data. By contrast, when it comes to fossils, for the most part, we only have hard-part anatomy to go on. Nevertheless, because of the complexities of the vertebrate skeleton, a plethora of characters can usually be identified, and subtle variations in the shape and positional relationships of the different bones are critical. Some knowledge of the relationship of the different elements of the skeleton will therefore be helpful to the reader. There is no basic pattern of organization that is common to all vertebrates. After all, the presence or absence of different elements is part of what constitutes the different grades of vertebrate organization. As a basic example, fish do not have limbs; only the tetrapods (four legs) have limbs. Mammals have only a single element forming the lower jaw, the dentary. In reptiles and amphibians, the dentary is still the principal tooth-bearing bone in the lower jaw, but there are a variable number of additional elements, with amphibians typically having a greater number than reptiles. On the other hand, amphibians and reptiles have only one bone in the middle ear region, the stapes, whereas mammals have three: the stapes, incus, and malleus. The incus and malleus have actually been derived from elements that are found in the lower jaw of reptiles. In other words, they are homologous ossification centers, but coopted for very different functions. The defining feature of a vertebrate is the presence of a backbone (composed of individual vertebrae) and a well-organized head that is encased in the skull bones. The skull frequently becomes the focus of attention for phylogenetic analyses, perhaps because of its complexity. Furthermore, many parts of the skull are actually quite durable.

Jaws

The tooth-bearing bones are particularly important because teeth are the most readily fossilized part of any vertebrate. Moreover, although in the different vertebrate groups there are a number of tooth-bearing elements, there are almost always three principal ones—in the lower jaw the dentary, and in the upper jaw the premaxilla and the maxilla. Thus in the bony fish *Semionotis*, the temnospondyl amphibian *Metoposaurus*, the dinosaur *Coelophysis*, and the mammal *Morganucodon*, the paired dentaries meet in the midline (the symphysis) at the front of the skull and bear prominent teeth along much of their margins. Likewise, in the upper jaw, the front of the skull comprises the paired premaxillae that are situated below and in front of the openings for the external nares (the nostrils), and behind them the maxillae. A notable exception to the rule is found in the lungfish, where the principal tooth-bearing elements are those of the palate, and the marginal elements are lost.

Eye and Cheek Region

In side view, the eye socket (orbit) is typically surrounded by four bones (fig. A3.1B). In front is the prefrontal; behind is the postorbital. Between these two bones below the orbit is the jugal, and above is the postfrontal. In some vertebrates, a fifth bone, the lachrymal ("of tears"), lies between the jugal and prefrontal along the anterior border of the orbit. The tear ducts may penetrate through this bone. In many reptiles, the prefrontal and postfrontal are small bones, and the frontals squeeze between them to enter into the dorsal margin of the orbit. Behind the postorbital the major bone of the cheek is normally the squamosal. In some vertebrates, a separate ossification, the intertemporal, may separate the postorbital and squamosal. Below the squamosal is the quadrate. Typically the cranium articulates with the lower jaw through the quadrate bone. The articular is the element of the mandible involved in this critical articulation. However, in mammals, in which these bones have been greatly modified and are now to be found in the inner ear, the articulation is directly between the squamosal and the dentary.

Skull Roofing Elements

When viewed from above, along the midline of the skull there are typically a series of three paired sets of bones (fig. A3.1A). Starting at the front of the skull are the nasals. These extend behind and separate the nares. Behind the nasals are the frontals and the parietals. In reptiles and mammals, these are essentially the only bones contributing to the skull roof. However, in fishes and amphibians there are usually additional prominent elements. Principal among these are paired postparietals (naturally lying behind the parietals), and lying to the sides are the supratemporals and tabulars.

Palate

Turning the skull upside down and looking down onto the cranium, we see the bones of the palate. There are four pairs of bones that together form much of the palate in all vertebrates. Starting at the front of the snout are the vomers. They meet each other along the midline, and their lateral margins form the borders to the internal nares (choanae). Behind the vomers are the palatines, which extend back from the internal nares and along the sides of the palate. In turn, stretching back behind the palatines are the ectopterygoids. But frequently the most prominent elements are the pterygoids. They typically extend back from the vomers and separate both palatines and ectopterygoids. Usually along the midline the paired pterygoids separate from each other to leave a long, narrow gap, the interpterygoid vacuity. In many fishes and amphibians, as well as some reptiles, the palatal bones may bear teeth, often very numerous and, particularly on the lateral elements, sometimes as large as those on the marginal tooth-bearing elements.

Braincase

The braincase of vertebrates is usually very well ossified and composed of a number of different bones. Contributing to its floor are the basisphenoid anteriorly and the basioccipital posteriorly. The basisphenoid is sheathed ventrally by the parasphenoid, a thin bone that sends forward a slender rodlike process into the interpterygoid vacuity. The walls are formed by two bones in the otic (ear) region: the prootic anteriorly and the

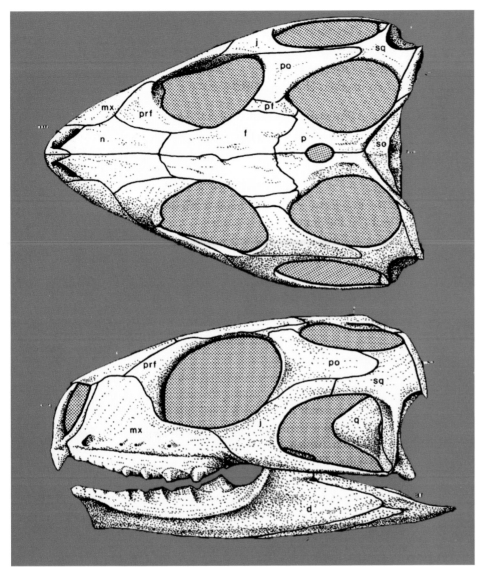

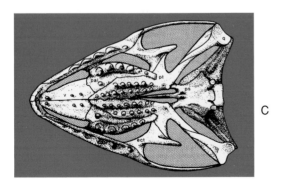

Figure A3.1. Skull of the sphenodontian *Clevosaurus* in (A) dorsal, (B) lateral, and (C) ventral views (after Fraser 1988b).

opisthotic posteriorly. These two ossifications often fuse together, and within them are to be found the canals of the inner ear. The posterior (occipital) portion of the vertebrate braincase comprises four bones that together surround the foramen magnum—the opening through which the spinal cord passes. Above the foramen magnum is the single supraoccipital, to either side are the paired exoccipitals, and below is the basioccipital that, as we have seen, forms part of the floor to the braincase.

Vertebral Column

The vertebral column provides the main supporting structure to the vertebrate body while giving it some degree of mobility (fig A3.2). Vertebrae of tetrapods, such as reptiles and mammals, largely consist of two principal components: the disk-shaped centrum and the neural arch that sits on top and is often fused to the centrum. The dorsal nerve cord passes down the body above the series of articulating centra, but passes through the neural arches. The arch is extended dorsally as a neural spine. The spines support the dorsal muscles. The successive neural arches articulate with each other through a complex system of overlapping processes called zygapophyses. The anterior zygapophyses (prezygapophyses) typically terminate in a disk-shaped surface facing upward and inward, while the posterior zygapophyses (postzygapophyses) have a corresponding articulating surface directed downward and outward. Wedged between the centra may be small crescentic ossifications called intercentra. In the tails of

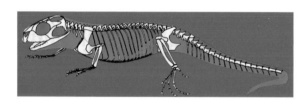

Figure A3.2. Restoration of the skeleton of the sphenodontian *Clevosaurus hudsoni* (after Fraser 1988b).

many tetrapods, the intercentra bear extensive Y-shaped projections and are known as hemal arches or chevron bones. The stem of the Y supports the major muscle blocks of the tail; the main blood vessels pass between the bifurcating arms of the Y. Although in reptiles and mammals the centrum is the principal ossification center below the nerve cord, in many amphibians and fishes, the intercentrum was a much more important ossification center, while the centrum (pleurocentrum) was much smaller and usually paired. Number of vertebrae vary from one group of vertebrates to another, and there is also variability in the degree to which they are modified into sections, such as the neck, thoracic, lumbar, sacral (hip), and caudal (tail) regions. The first two vertebrae—the atlas and axis—typically tend to be substantially modified in tetrapods to accommodate their key role in supporting the head.

In tetrapods, ribs may attach to the vertebrae by two separate articulations. The main head, the capitulum, attaches to the intercentrum in early vertebrates, but in those forms with a reduced intercentrum, the capitulum tends to articulate with the leading edge of the centrum. The second head of the rib, the tuberculum, generally articulates with the neural arch.

Girdles

Normally vertebrates possess two pairs of appendages—fins in fishes and legs in tetrapods. In both cases they are supported on the body by a discrete series of bones called the girdle elements. The shoulder girdle, or pectoral girdle, is prominent in most vertebrates, including bony fishes. The principal element of the pectoral girdle of fishes is the paired cleithrum, which is vertically orientated. On each side below the cleithrum is a smaller clavicle (collarbone). Additional ossifications may be present above the cleithrum, and they include bones variously known as the supracleithrum, postcleithrum, and scapulocoracoid.

Tetrapods may also have cleithra and interclavicles, but they are often rudimentary or even absent. On the other hand, they may also have an interclavicle—a single primitively diamond-shaped bone that separates the clavicles on the ventral surface. In birds and some of their theropod ancestors, the clavicles and interclavicle typically fuse to form the furcula, or wishbone. But in tetrapods, the most important bones of the pectoral girdle are the coracoid and scapula. These latter two elements lie on each side above the interclavicle (when present). The scapula is the more dorsal of the two bones. Where the scapula and coracoid meet is an excavated depression, the glenoid, for the reception of the humerus (the bone of the upper arm).

The pelvic girdle (or hip) is very small in fishes and generally consists of a single triangular plate embedded in the muscle block. The two sides of the girdle are usually in ventral contact (pelvic symphysis), but they do not directly articulate with the vertebral column. By contrast, in tetrapods, there are three separate bones, but they are normally firmly articulated with each other and with the vertebral column through the sacral ribs. The most dorsal element is the ilium. Below it are the pubis anteriorly and the ischium posteriorly. Where the three bones meet there is normally a very well-defined concavity, the acetabulum, which is the socket that receives the proximal head of the femur (the first leg bone). Below the level of the acetabulum, the pubis and ischium deflect inward to meet their counterparts in a midline symphysis. The two ilia articulate with the sacral ribs on either side.

Limbs

A characteristic feature of tetrapods are the limbs. Each limb typically has a single proximal element (propodial) that articulates with two more distal elements (epipodials), and these in turn articulate with a series of small bones in the wrist (carpus) or ankle (tarsus) (fig. A3.2). Finally, the digits (fingers and toes) radiate from the carpals and tarsals.

In the forelimb, the proximal element is the humerus, and the two epipodials are termed the radius and ulna. The ulna lies lateral to the radius, and typically the radius is the more robust of the two. The ulna is notched at the proximal head to receive the humerus at the elbow. Projecting above this notch is the olecranon process. To this spur of bone (our "funny bone"), the tendon of the triceps attaches, and this muscle is the one primarily responsible for extending the forearm.

The carpus (wrist) comprises a series of small elements that, at least primitively, can be divided into three rows. First are three proximal elements, the radiale, intermedium, and ulnare, which articulate directly with the radius and ulna. Then there is a row of distal carpals with one for each digit (finger). Finally, wedged between these two rows are a series of central carpals, or centralia. Although four were present primitively, this number is frequently reduced. For each finger there is a proximal element, the metacarpal, and then a variable number of phalanges. Usually the last phalanx, the ungual, is modified into a claw or hoof.

In the hindlimb, the proximal element is the femur (thighbone). It may bear a series of ridges or tuberosities (trochanters) that serve for the attachment of muscles. The distal end comprises a double articular surface for the head of the tibia, which is the major weight-bearing bone of the shin. Lateral to the tibia is the fibula, the second epipodial of the hindlimb. Frequently the fibula is a slender, even splintlike, bone.

The arrangement of bones in the tarsus (ankle) mirrors that of the carpus in many respects. Primitively, there was a series of three proximal elements, the tibiale, intermedium, and fibulare, followed by a middle series of centralia, and last the distal tarsals. In reptiles and mammals, however, there are only two proximal tarsals, namely the astragalus and calcaneum. Furthermore, as in the carpus, loss of certain other tarsal elements is not uncommon. Each digit (toe) has a proximal metatarsal and then a variable number of phalanges.

Spare Bones

Although the principal elements have been described above, certain groups of vertebrates have additional bony elements. For instance, many reptiles have "belly ribs" or gastralia, a herringbone pattern of splintlike elements that form a basket to support and protect the intestines. Bony armor also comes in all shapes and sizes, from the scutes and spikes of certain archosaurian reptiles to the shell and carapace of turtles. Then there are strange new bones, such as the postcloacal bones found in tanystropheid reptiles (chapters 6 and 11).

appendix 4. names, grades, and clades

As T. S. Eliot said, "The naming of cats is a difficult matter, it isn't just one of your holiday games." So it is also with biological nomenclature. Two branches of scientific study are key components of the discussions in this book. One is taxonomy, and the other is systematics. Taxonomy is the practice of assigning names to organisms and placing them together with like groups of organisms, whereas systematics is concerned with the study of diversity between and within groups of organisms. It is worthwhile to briefly consider the different approaches to these two branches of biology.

Taxonomy

A sophisticated, complex communication system is one of the hallmarks of *Homo sapiens*, and the power of speech can never be underestimated. At the same time, poor communication is often worse than no communication at all. "Say what you mean and mean what you say" is wonderful advice, but there is no universal everyday language, and people from different countries still have difficulty understanding others, no matter how clearly, slowly, and loudly they may speak. Of course, each language has common names for the different animals and plants that also populate the earth. Naturally, these animals and plants are not particularly obliging and do not follow the artificial boundaries of the world's countries and languages. However, once Europeans began the Great Age of Discovery and started to explore the world and travel great distances from their homeland, they discovered completely new and exotic faunas and floras. Naturally, the explorers sought to document and catalog the new things they saw and to share this information with the people back home. Inevitably, as new information was disseminated in the scientific literature, it was not long before confusion started to spread, because each language often had different names for exactly the same organism. The scientists of the 1600s recognized the necessity of a single scientific language to identify different animals and plants. Caspar Bauhin and John Ray introduced the concept of genus and species, and in the mid-1700s, the Swede Carolus Linnaeus introduced the more ordered hierarchical divisions of kingdom, class, order, family, genus, and species. In this system, each group is nested within a series of larger and larger groups, and this taxonomic system is still in widespread use today.

Systematics

Systematics is a discipline of biology concerned with the identification of groups of organisms on the basis of their evolutionary relationships. Two different methods are widely

Figure A4.1. (A) Cladogram showing the interrelationship of three generalized taxa, A, B, and C, in which B and C are sister taxa and more closely related to each other than they are to A. (B) Similar three-taxon statement depicting the relationship of three Lepidosauromorph taxa.

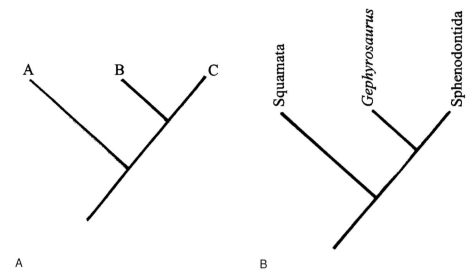

used today by paleontologists in systematic research. The first, termed *evolutionary systematics* or *gradistics*, is based on morphological similarity between organisms and adheres closely to the Linnaean taxonomic system. Membership within a group is based on degree of similarity to other members of the group. All members of a group must have a common ancestor that is also deemed to be a member of that group. However, not all descendants of that ancestor are necessarily included in the group. There is a degree of subjectivity in this system. Thus when descendants are "sufficiently" far removed from the ancestor and they have reached a stage of organization in which they are not really that similar in appearance, they are assigned to a different group. Only those organisms of a similar grade of development are grouped together, and organisms of a "higher" or more "advanced" grade of development are excluded. Thus although we accept that mammals and birds have each evolved from some separate reptilian ancestor, their level of organization is so distinct from the animals we call reptiles that under the gradistic method, mammals and birds are placed in completely distinct classes. This practice can be regarded as the classical approach to systematics.

Within the last half century, a different approach has become prevalent. In the second half of the twentieth century, the German entomologist Willi Hennig developed a more objective technique. In this system, known as *phylogenetic systematics* or *cladistics*, one of the fundamental tenets is the recognition that most species originate as the result of splitting events, and that for any three taxa, two must always share a more recent common ancestor than the third. In other words, there is always a simple dichotomous branching pattern of relationships between related groups. This branching pattern can be shown graphically in what is termed a *cladogram* (fig. A4.1). The two groups that share a splitting event with each other are called sister groups. So in the example in figure A4.1A, B and C are sister groups to each other, and A is the sister group to B + C together. The more specific example in figure A4.1B shows that *Gephyrosaurus* is the sister taxon of the Sphenodontida, whereas the Squamata forms the sister group to *Gephyrosaurus* and Sphenodontida combined.

For any given cladogram, there are essentially three different ways that the various taxa can be combined together. Groups that are composed of a single ancestor and all descendants of that ancestor are referred to as *monophyletic* (single branch). In the example of figure A4.2A, the groups composed of C and D, [(C + D) + B], and {[(C + D) + B] + A} are all monophyletic. Groups that are composed of a single ancestor but not all of its descendants are termed *paraphyletic* (almost a branch) groups. These are the grades of organization that are fundamental to gradistic systematics. Thus in figure A4.2A, the group of

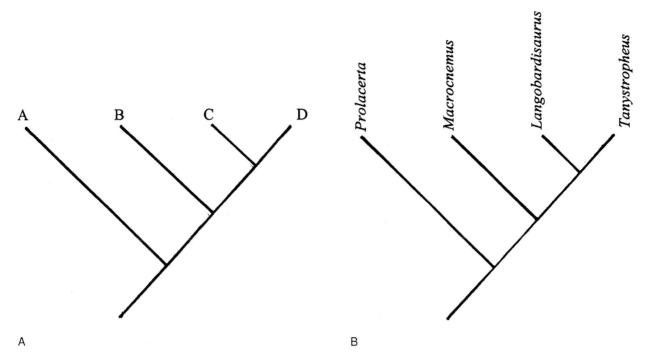

A

B

Figure A4.2. (A) Cladogram
depicting four successively
more closely related taxa, A,
B, C, and D. (B) Specific example of the four taxon statement showing the relationship of four protorosaurs.

[A + (B + C)] is paraphyletic because it excludes D. The third type of group is known as
polyphyletic (multiple branches). Combining A and D together forms a polyphyletic
group because A and D do not have a common ancestor that is not also the common ancestor of B and C.

Phylogenetic Analyses

In order to determine the relationships of different groups of animals, we must first search
for physical features of the body (characters) and then describe them in a manner that can
be readily understood by the scientific community. Then we must look at how each of
these characters is distributed among various groups. Those characters that are widespread and found in all the groups under study are considered primitive. Because these
characters are common to all the organisms being examined, they do not help to resolve
the cladogram. On the contrary, when a particular character is only found in a few members of the group under study, then it is likely that it evolved in a relatively recent common ancestor and that it is a derived ("advanced") character. Shared derived characters
(termed *synapomorphies*) can therefore help to resolve a cladogram. However, we also
have to remember that two or more groups of organisms can independently acquire very
similar features simply because they share a similar lifestyle. Such convergences can
sometimes act as red herrings and lead us to the wrong interpretation.

Ghost Lineages

Sometimes cladograms can lead us to make hypotheses about the temporal range of certain groups of fossils that are not necessarily apparent from the fossil record. For the example in figure A4.1, let's say that we know that the oldest known fossils of taxon B are 100
million years, and those of taxon C only 50 million years old. Because our phylogenetic
analysis indicates that they are sister groups, it means that B and C split from each other at
a particular moment in geological time. It must therefore follow that there should be examples of taxon C extending at least as far back as those of taxon B—in other words, 100
million years. When our studies indicate that examples of a particular taxon should be in

certain aged sediments and they are not, we refer to them as ghost lineages. Of course, if in the future we find a distinct member of taxon C in sediments that are 120 million years old, then the situation reverses and taxon B has a ghost lineage between 100 and 120 million years.

A good example of a ghost lineage that appears in this book is that of the lizards. Phylogenetic analyses strongly favor a sister-group relationship between lizards and sphenodontians. The oldest fossil sphenodontians are over 220 million years old, but currently, the oldest known unequivocal fossil of a lizard is only about 170 million years old. We are therefore forced to assume that lizards also go back at least 220 million years, but we have yet failed to identify them in the fossil record. The exciting and fun part of paleontology is that with continued field expeditions, new sites are regularly discovered, and renewed excavations at well-established localities often yield unexpected new specimens. As a result, we continue to test and refine our phylogenies. It is particularly satisfying when new discoveries support earlier hypotheses, and a challenge when we are forced to rethink our position.

glossary

Acrodont   Type of dentition in which the teeth are fused to the summit of the jaw ramus.

Allochthonous   Applied to rocks and fossils that have not been formed or deposited in situ.

Annelid   A phylum of worms characterized by a definite head and a segmented body. They are found in marine, freshwater, and terrestrial environments.

Apomorphic   Applied to a character that has been derived as a novelty from a preexisting character state.

Arenaceous rocks   A group of detrital sedimentary rocks, typically sandstones, in which the particle sizes range between $\frac{1}{16}$ and 2.0 millimeters.

Argillaceous rocks   A group of fine-grained detrital sedimentary rocks comprising clays, shales mudstones, siltstones, and marls. Two grades are normally recognized: the silt grade, with particles between $\frac{1}{16}$ and $\frac{1}{256}$ millimeters, and clay grade, with particles less than $\frac{1}{256}$ millimeters.

Arkose   Arenaceous rock that contains significant quantities of feldspar in addition to quartz. They are normally thought to be indicators of erosion under arid conditions, and of rapid burial.

Authigenic   Generated in place. With respect to minerals, they have formed on the spot that they have been found.

Autochthonous   A fossil that has been buried where the organism once lived. When applied to rocks, it means that they have been moved very little from the location in which they were originally formed.

Benthic   A term that refers to the bottom of a body of standing water, the benthic terrain includes all of the bottom ground from the shoreline to the greatest depths. The benthos is the sum total of the animals and plants living on the bottom of a sea or lake.

Bioturbation   The disturbance of sediments through the action of biological agents.

Bisporangiate   Bearing two types of sporangia (microsporangia and megasporangia) in the same fructification.

Book lung   A type of respiratory organ that occurs in spiders. They are located in the abdomen and are composed of many fine leaflike surfaces.

Byssate   Referring to bivalve mollusks that are attached to the substrate by many strands of strong, horny threads (byssus).

Caliche   A calcareous material typically found in soils of arid or semiarid regions that accumulates in layers near to the surface. It forms when soil solutions rise to the surface, evaporate, and leave their salt contents on the surface.

Carbonate rock   A sedimentary rock formed by the precipitation (either by organic or inorganic means) from aqueous solution of carbonates of calcium, magnesium, or iron (e.g., limestones and dolomites).

Carbonate platform   A flat or gently sloping underwater surface comprising carbonate rocks (limestones and dolomites) that extend from the shore out to sea.

Carbon ratio   The ratio of carbon-12, the most common isotope of carbon, to either carbon-13 or carbon-14.

Carpals   Small bones at the base of the hand. (Not to be confused with the botanical term *carpel*, which is the female reproductive organ of flowering plants.)

Chaetae   The bristles on appendages of annelid worms. Sometimes they occur in multiple bundles, as in polychaetes, or more irregularly on each segment, as in oligochaetes.

Chelicera   One of the first pair of appendages of an arachnid. In many arachnids, the chelicerae are developed into a pincer- or clawlike structure (chelate).

Cingulum   Literally meaning *girdle,* cingulum is the term applied to a raised crest that runs around the edge of certain mammal teeth.

Clastics   Referring to rocks built up of fragments of preexisting rock.

Claystone   A fine-grained sedimentary rock that has the texture and composition of a shale, but lacks the fine lamination fissility. It forms a plastic mass when wet.

Conglomerate   A coarse-grained clastic sedimentary rock. Fragments range widely in size and are larger than 2.0 millimeters. Conglomerates consist of granules, pebbles, cobbles, or boulders set in a fine-grained cementing matrix.

Coracoid   A bone in the shoulder (pectoral) girdle. Positioned on the ventral side, it attaches to the scapula blade.

Corm   A short, bulbous, subterranean stem.

Coronoid   One of the bones in the lower jaw of many tetrapods. Some tetrapods have more than one coronoid. They are situated behind the tooth row and contribute to the coronoid process.

Crurotarsan   An archosaurian reptile such as a crocodile that has an ankle where the articulation comprises a ball-and-socket joint between the two principal ankle bones.

Deccan Traps   An enormous area of basaltic lava flow in southeast India. They extend across an area over 2.5 million square kilometers. The word is derived from the Swedish word *trappa,* meaning staircase, and refers to the steplike appearance of these massive lava flows. They are approximately 65 million years old and are possibly associated with the end-Cretaceous extinction event.

Diagenesis   The chemical and physical changes that sediments undergo after their accumulation and deposition, but before consolidation occurs. The changes take place while the sediments are close to the earth's surface.

Dicot   An abbreviation of the term *dicotyledon,* one of the two main lineages of flowering plants: the other is the monocotyledons. A cotyledon is a leaf that develops from the embryo plant. In dicots there are two such leaves, whereas in monocots there is a single leaf.

Dike   A body of rock that cuts across the structure of adjacent rock.

Durophagous   Eating tough, coarse material.

Edentulous   Lacking teeth.

Elytra   Modified pair of fore wings in a beetle that form a protective shield over the more typical flight wings. The singular is elytron. The colors of elytra give beetles their characteristic patterns.

Endochondral bone   The type of bone structure that occurs in the center of a bone.

Eolian   Pertaining to the agency of the wind. Eolian deposits are those that have been transported and arranged through the action of the wind.

Epicontinental   Pertaining to the continental shelf.

Epidermis   Outermost layer of cells on a multicellular organism. In animals, it is the skin. It forms a protective layer on the surface of a plant that is covered in aerial parts by a noncellular waxy cuticle.

Epipelagic   *Pelagic* refers to free-swimming organisms that are independent from the shore or the sea bottom. *Epipelagic* refers to the pelagic environment down to a depth of 100 fathoms—that is, in the upper portion of the pelagic realm.

Epipodial   The paired elements of the limbs that lie beyond the first joint. In the forearm, these are the radius and ulna, and in the hindlimb, these are the tibia and fibula.

Errant polychaete   A group of polychaete worms that have many body segments. The head and posterior end differ markedly from the rest. The mouth typically has a number of paired jaws.

Estivation   Dormancy during periods of heat and drought. It is analogous to hibernation in cool environments.

Evaporite   A nonclastic sedimentary rock that is produced as a solvent (water) evaporates, thereby concentrating its minerals. Two examples are rock salt and gypsum.

Facies   The combined lithological and paleontological characteristics of a unit of sedimentary rock. The origin and the depositional environment of the formation can be inferred from this.

Facies change   A lateral or vertical change in the lithological and paleontological characteristics of contemporaneous deposits. This reflects a change in the nature of the depositional environment.

Feldspar   Most widespread silicate mineral group, constituting 60% of the earth's crust.

Fossorial   Pertaining to animals adapted to digging and burrowing.

Fructification   Structure in plants that bears spores or seeds.

Gondwana   The supercontinent of the southern hemisphere that included South America, Africa, Madagascar, India, Australia, and Antarctica.

Graben   An elongate downthrown block that is bounded by faults on its long sides. A half-graben is bounded by faults on only one of its long sides.

Graywacke   Term applied to a dark, very hard, coarse-grained sandstone. Typically it consists of poorly sorted and very angular grains of quartz and feldspar.

Halophyte   Plant tolerant of very salty soil. Halophytes are found around seashores, tidal river estuaries, salt marshes, and alkali desert flats.

Heterocercal   In fish, a tail in which the vertebral column turns upward and extends into the dorsal lobe of the tail. In such instances, the dorsal lobe is often larger than the ventral lobe. Such a tail is found in Chondrichthyans and also many fossil bony fish.

Heterodont   Possessing a dentition comprising teeth of more than one shape (e.g., incisors, canines, and molars).

Homodont   Possessing a dentition in which the teeth are all of a similar shape.

Homoplasy   The appearance of a similar structure in quite distinct lineages. It includes convergent and parallel evolution.

Homologous   Corresponding in origin, general shape, or position.

Hydrophyte   A plant that grows in water, either emergent, submerged, or floating.

Hyoid   Bones that support the floor of the mouth, one of which supports the tongue.

Hyomandibular   A dorsal portion of the hyoid arch that contributes to the jaw attachment in many fish. It is homologous with the stapes (ear bone) of tetrapods.

Infaunal   Pertaining to aquatic animals living within rather than upon the bottom sediment. Epifauna refers to the organisms living on the surface of the seafloor.

Lacustrine   Formed in, produced by, or generally pertaining to a lake. Lacustrine sediments are those deposited at the bottom of a lake.

Lagerstätte   A fossil deposit of exceptional preservation. It is broadly divided into two main categories: concentration lagerstätten and conservation lagerstätten. The former is applied to deposits that have exceptional abundance of fossils. In the latter, quantity doesn't matter, but quality does. In some conservation lagerstätten, fossils are actually very sparse, but those that are present are spectacular. Solnhofen is a good example of a conservation lagerstätte.

Laurasia   The protocontinent of the northern hemisphere. It comprised North America, Greenland, Europe, and Asia (excluding India). Its southern counterpart was Gondwana.

Marl   A calcareous mudstone.

Medusa   The free-swimming body form in Cnidarians—a jellyfish. It resembles an umbrella or a bell floating with the convex side uppermost.

Membrane bone   Membrane or dermal bone refers to bone that develops directly and is not preformed in cartilage. Largely restricted in tetrapods to various bones of the skull and pectoral girdle.

Microsporophyll   Modified leaf that bears microsporangia. In flowering plants, it is the stamen.

Microsporangia   Structure that produces microspores.

Monocots   Abbreviated term for monocotyledon, a flowering plant that produces a single seed leaf from the embryo.

Monophyletic   Of a single origin. A group that has originated from and includes a single stem taxon, and all its descendants. This stem taxon may be known or hypothesized.

Mudstones   Similar to shales but lack the bedding plane fissility. They are nonplastic.

Neomorph   A completely new structure with no known homologue.

Notarium   Rigid bony structure resulting from the fusion of dorsal vertebrae. It acts to provide support to the forelimb. It is a characteristic feature of pterosaurs, where the neural spines, transverse processes, and centra of nearly the whole dorsal series may fuse. It provides a rigid anchor for the wing.

Olecranon   The bony prominence on the ulna at the elbow joint.

Operculum   Literally meaning a lid or cover, in gastropod snails it refers to the horny plate on the foot that seals the animal inside the shell when the foot is withdrawn. In fish, it is a hard, bony cover that covers the outside of the gill chamber.

Opisthotic   One of the bones of the skull that contributes to the wall of the braincase.

Orogen   A region of the earth that has been subjected to folding; a belt of deformed rocks. Orogeny is a period of mountain building, and orogenesis is the process of mountain building. Orogenesis leads to the formation of deformed belts that constitute ranges of mountains.

Outgroup   A taxon that is outside the monophyletic group under consideration. It serves to act as a stable reference point in phylogenetic considerations. Comparing the distribution of characters between the outgroup and members of the monophyletic taxon helps to determine which characters are derived and which are plesiomorphic, or primitive.

Ovulate organs   Structures in plants that produce unfertilized seeds (ovules).

Paleosol   An ancient buried soil horizon.

Paludal   Pertaining to a marsh.

Palynomorph   A microscopic organic body that has resistant walls. They include acid insoluble microfossils such as pollen, spores, cysts, and dinoflagellates.

Panthalassa   The protoocean that surrounded Pangaea.

Paraphyletic   A term used to describe a group of organisms that is thought to have arisen from a single stem taxon. However, the group in question does not include all the species believed to have evolved from that stem taxon.

Parapodium   In a polychaete worm, the paired fleshy appendages that project from the sides of the body segments.

Parenchyma   In plants, tissue composed of thin walls interspersed with a system of intercellular spaces.

Parietal   One of typically paired skull bones that lie toward the back of the skull and roof the brain-case.

Pedipalp   In arachnids, the second pair of head appendages.

Pedogenesis   The processes whereby soils are formed.

Phytophage   A plant-eating animal.

Pinna   In plants, the primary division of a compound leaf or fern frond, a leaflet. In mammals, it is the extension of the outer ear that is supported by cartilage.

Playa   A flat, completely enclosed desert lake basin composed of evenly stratified bands of fine clay, silt, and sand. After rain, water accumulates as a shallow lake, but it quickly evaporates, leaving deposits of soluble salts.

Plesiomorphic   A term used to describe a character state that is shared by a number of different groups of organisms, but still derived from a common ancestor. It describes the original preexisting state of the character. The opposite term is *apomorphic*.

Pleurodont   Type of dentition in which the teeth are set in a groove along the margin of the jaw, with the outer edge of the groove higher than the inner edge. It typically occurs in lizards.

Pluton   A body of rock that has formed beneath the earth's surface through the consolidation of magma. It is also used to include bodies of rock that have formed beneath the surface of the earth by mineral replacement of older rocks.

Prootic   A bone that is positioned in front of the opisthotic and contributes to the side of the brain-case.

Quadrate   Bone in the skull of tetrapods that forms the principal articulation with the mandible.

Red beds   Sedimentary strata deposited in a continental environment. The presence of ferric oxide (hematite) is responsible for their predominantly red coloration. Typically at least 60% of a given succession must be red for the term to be applicable.

Sabkha   A smooth and flat, typically saline plain. After a heavy rain, it may be occupied by a marsh or a temporary shallow lake. A sabkha may be coastal or continental. It refers to a salt flat that is only occasionally inundated by water.

Shale   A rock with a well-marked bedding plane fissility, and that does not become plastic when wet.

Siberian Traps   Continental flood basalt approximately 250 million years old that some authors consider to be associated with the great Permian mass extinction. They extend over an area almost 2.5 million square kilometers.

Siltstone   Similar to a mudstone, but with slightly coarser grains.

Sister groups   The paired groups that result from the splitting of a single parent (ancestral) group.

Sterile foliage   Foliage that lacks any reproductive structures such as spores or ovules.

Stoma   A pore in plant epidermis. Stomata present in large numbers on the leaves permit gas exchange.

Taphonomy   The branch of paleontology that studies the manner of burial and the formation of fossils.

Test   A protective shell that covers the cells of certain protozoans, and the bodies of invertebrates such as echinoids.

Tethys   A large sea that separated the supercontinents of Laurasia and Gondwana in the eastern hemisphere.

Thecodont   A dentition in which the teeth are implanted in deep sockets.

Trochanter   In vertebrates, a raised ridge on a long bone, especially the femur to which muscles attach. In insects, it is the second joint on the leg.

Unconformity   A surface of erosion or nondeposition that separates younger strata from older rock units.

Xeric   Pertaining to dry conditions.

Xeromorphic   Plants with structures associated with adaptations to dry conditions.

Xerophyte   Plant tolerant of prolonged drought.

references

Affer, D., and G. Teruzzi. 1999. Thylacocephalan crustaceans from the Besano Formation, Middle Triassic, N. Italy. In *Third International Symposium on Lithographic Limestones*, ed. S. Renesto, 20(Suppl.):5–8. Rivista del museo civico di scienze naturali.

Alcober, O. 2000. Redescription of the skull of *Saurosuchus galilei* (Archosauria: Rauisuchidae). *J. Vertebr. Paleontol.* 20:302–316.

Alessandrello, A., P. Arduini, G. Pinna, and G. Teruzzi. 1989. New observations on the Thylacocephala (Arthropoda, Crustacea). In *The early evolution of Metazoa and the significance of problematic taxa*, ed. A. M. Simonetta and S. Conway-Morris, 245–251. Cambridge: Cambridge University Press.

Anderson, J. M., and H. M. Anderson. 1993a. Terrestrial flora and fauna of the Gondwana Triassic. Part 1, Occurrences. In *The nonmarine Triassic*, ed. S. G. Lucas and M. Morales, 3–12. New Mexico Museum of Natural History and Science Bulletin 3.

———. 1993b. Terrestrial flora and fauna of the Gondwana Triassic. Part 2, Co-evolution. In *The nonmarine Triassic*, ed. S. G. Lucas and M. Morales, 13–25. New Mexico Museum of Natural History and Science Bulletin 3.

Anderson, J. M., and A. R. I. Cruickshank. 1978. The biostratigraphy of the Permian and the Triassic. Part 5, A review of the classification and distribution of Permo-Triassic tetrapods. *Palaeontol. Afr.* 21:15–44.

Anderson, J., H. Anderson, P. Fatti, and H. Sichel. 1996. The Triassic explosion (?): A statistical model for extrapolating biodiversity based on the terrestrial Molteno Formation. *Paleobiology* 22:318–328.

Arucci, A. 1990. Un nuevo Proterochampsidae (Reptiliana-Archosauriformes) de la fauna local de los Chañares (Triasico Medio), La Rioja, Argentina. *Ameghiniana* 27:365–378.

Ash, S. R. 1972. Late Triassic plants from the Chinle Formation in northeastern Arizona. *Palaeontology* 15:598–618.

———. 1976. The systematic position of *Eoginkgoites. Am. J. Bot.* 63:1327–1331.

———. 1980. Upper Triassic floral zones of North America. In *Biostratigraphy of fossil plants*, ed. D. Dilcher and T. N. Taylor, 153–170. Stroudsburg, Pa.: Dourden, Hutchinson, and Ross.

———. 1986a. Fossil plants and the Triassic-Jurassic boundary. In *The beginning of the Age of Dinosaurs*, ed. K. Padian, 21–30. New York: Cambridge University Press.

———. 1986b. Early Mesozoic land flora of the northern hemisphere. In *Land plants: Notes for a short course*, ed. T. W. Broadhead, 143–161. University of Tennessee Department of Geological Science.

———. 2001. New cycadophytes from the Upper Triassic Chinle Formation of the southwestern United States. *Paleobios* 21:15–28.

Ash, S. R., and G. T. Creber. 1992. Paleoclimatic interpretation of the wood structures of the trees in the Chinle Formation (Upper Triassic), Petrified Forest National Park, Arizona, USA. *Palaeogeogr. Palaeoclimatol. Palaeoecol.* 96:299–317.

Ax, P. 1984. *Das phylogenetische system*. Stuttgart: Gustav Fischer Verlag.

Axsmith, B. J., M. Krings, and T. N. Taylor. 2001. A filmy fern from the Upper Triassic of North Carolina (USA). *Am. J. Bot.* 88:1558–1567.

Baird, D. 1957. Triassic reptile footprint faunules from Milford, New Jersey. *Bull. Mus. Compar. Zool.* 117:449–520.

Bakker, R. T. 1986. *The dinosaur heresies: New theories unlocking the mystery of the dinosaurs and their extinction.* New York: William Morrow.

Baksi, A. K., and E. Farrar. 1991. $^{40}Ar/^{39}Ar$ dating of the Siberian traps, USSR: Evaluation of the ages of the two major extinction events relative to episodes of flood-basalt volcanism in the USSR and the Deccan Traps, India. *Geology* 19:461–464.

Barghusen, H. R., and J. A. Hopson. 1970. Dentary-squamosal joint and the origin of mammals. *Science* 168:573–575.

Barry, T. H. 1968. Sound conduction in the fossil anomodont, *Lystrosaurus. Ann. S. Afr. Mus.* 50:275–281.

Bechly, G. 1997. New fossil odonates from the Upper Triassic of Italy, with a redescription of *Italophlebia gervasuttii* Whalley, and a reclassification of Triassic dragonflies (Insecta: Odonata). *Riv. Mus. Civ. Sc. Nat. "E. Caffi"* 1997, 31–70.

Becker, L. R. J. Poreda, A. G. Hunt, T. E. Bunch, and M. Rampino. 2001. Impact event at the Permian-Triassic boundary: Evidence from extraterrestrial noble gases in fullerenes. *Science* 291:1530–1533.

Bennett, S. C. 1996. The phylogenetic position of the Pterosauria within the Archosauromorpha. *Zool. J. Linnean Soc.* 118:261–308.

Benton, M. J. 1977. *The Elgin reptiles.* Aberdeen: Aberdeen People's Press.

————. 1983a. Dinosaur success in the Triassic: A non-competitive ecological model. *Q. Rev. Biol.* 58:29–55.

————. 1983b. The Triassic reptile *Hyperodapedon* from Elgin: Functional morphology and relationships. *Phil. Trans. R. Soc. Lond. B* 302:605–720.

————. 1986. The Late Triassic tetrapod extinction events. In *The beginning of the Age of Dinosaurs: Faunal change across the Triassic-Jurassic boundary*, ed. K. Padian, 303–320. New York: Cambridge University Press.

————. 1987. Progress and competition in macroevolution. *Biol. Rev.* 62:305–338.

————. 1990. *Vertebrate palaeontology.* London: Unwin-Hyman.

————. 1991. What really happened in the Late Triassic? *Hist. Biol.* 5:263–278.

————. 1999. *Schleromochlus taylori* and the origin of dinosaurs and pterosaurs. *Phil. Trans. R. Soc. Lond. B* 354:1423–1446.

Benton, M. J., and J. M. Clark. 1988. Archosaur phylogeny and the relationships of the Crocodylia. In *The phylogeny and classification of the tetrapods*, Vol. 1, *Amphibians, reptiles, birds*, ed. M. J. Benton, 295–338. Systematics Association Special Volume 35A. Oxford: Clarendon Press.

Benton, M. J., and D. J. Gower. 1997. Richard Owen's giant Triassic frogs: Archosaurs from the Middle Triassic of England. *J. Vertebr. Paleontol.* 17:60–73.

Benton, M. J., and P. S. Spencer. 1995. *Fossil reptiles of Great Britain.* London: Chapman and Hall.

Benton, M. J., and A. D. Walker. 1985. Palaeoecology, taphonomy and dating of Permo-Triassic reptiles from Elgin, north-east Scotland. *Palaeontolgy* 28:207–234.

Benton, M. J., G. Warrington, A. J. Newell, and P. S. Spencer. 1994. A review of the British Middle Triassic tetrapod assemblages. In Fraser and Sues, *In the shadow of the dinosaurs*, 131–160.

Berman, D. S., and R. Reisz. 1992. *Dolabrosaurus aquitalis*, a small lepidosauromorph reptile from the Upper Triassic Chinle Formation of North central New Mexico. *J. Paleontol.* 66:1001–1009.

Bernasconi, S. M. 1994. Geological and microbial controls on dolomite formation in anoxic environments: A case study from the Middle Triassic (Ticino, Switzerland). Contributions to Sedimentology 19.

Bice, D., Newton, C. R., McCauley, S., Reiners, P. W., and McRoberts, C. A. 1992. Shocked quartz at the Triassic-Jurassic boundary in Italy. *Science* 255:443–446.

Bock, W. 1952. Triassic reptilian tracks and trends of locomotive evolution. *J. Paleontol.* 26:339–355.

————. 1969. The American Triassic flora and global distribution. North Wales, Pa.: Geological Research Series 2–3.

Bonaparte, J. F. 1966. Chronological survey of the tetrapod-bearing Triassic of Argentina. *Breviora* 251:1–13.

————. 1972. Los tetrápodos del sector superior de la Formacion Los Colorados, La Rioja, Argentina (Triásico Superior). *Parte Opera Lilloana* 22:1–183.

————. 1996. *Dinosaurios de América del sur.* Buenos Aires: Museo Argentino de Ciencas Naturales "B. Rivadavia."

Boureau, E. 1964. *Traité de Paléobotanique*, Vol. 3, *Sphenophyta, Noeggerathiophyta.* Paris: Masson et Cie.

Braun, J., and W. E. Reif. 1985. A survey of aquatic locomotion in fishes and tetrapods. *Neues Jahrbuch Geol. Paläontol. Abhandlungen* 169:307–332.

Briggs, D. E. G. 1999. Decay and mineralization in soft-tissue fossilization. *Third International Symposium on Lithographic Limestones*, ed. S. Renesto, 20(Suppl.):53–55. Rivista del museo civico di scienze naturali.

Brongniart, A. 1828. Prodrome d'une histoire des végétaux fossils. *Dict. Sci. Nat.* 57:16–212.

Broili, F., and J. Schröder. 1934. Beobachungen an Wirbeltieren der Karooformation. V. Über *Chasmatosaurus vanhoepeni* Haughton. *Sitz. Bayer. Akad. Wiss. München* 1934:225–264.

Broom, R. 1913. On the South African pseudosuchian *Euparkeria* and allied genera. *Proc. Zool. Soc. Lond.* 1913:619–633.

Brown, R. W. 1956. Palm-like plants from the Dolores Formation (Triassic) in southwestern Colorado. U.S. Geological Survey Professional Paper 274:205–209.

Buffetaut, E. 1993. Phytosaurs in time and space. In *Evolution, ecology and biogeography of the Triassic reptiles*, ed. J. M. Mazin and G. Pinna, 39–44. Paleontologia Lombarda, n.s., 2.

Bürgin, T., O. Rieppel, P. M. Sander, and K. Tschanz. 1989. The fossils of Monte San Giorgio. *Sci. Am.* 260:74–81.

Cairncross, B. J., M. Anderson, and H. M. Anderson. 1995. Palaeoecology of the Triassic Molteno Formation, Karoo Basin, South Africa—Sedimentological and palaeontological evidence. *S. Afr. J. Geol.* 98:452–478.

Calzavara, M., G. Muscio, and R. Wild. 1981. *Megalancosaurus preonensis* n. g., n. sp., a new reptile from the Norian of Friuli, Italy. *Gortania* 2:49–64.

Camp, C. L., and S. P. Welles. 1956. Triassic dicynodont reptiles. Part 1. *Mem. Univ. Calif.* 13:255–341.

Campbell, M. R., and K. W. Kimball. 1923. The Deep River coal field of North Carolina. North Carolina Geological and Economic Survey Bulletin 33:1–95.

Campbell, I. H., G. K. Czamanski, V. A. Fedorenko, R. I. Hill, and V. Stepanov. 1992. Synchronism of the Siberian traps and the Permian-Triassic boundary. *Science* 258:1760–1763.

Carroll, R. L. 1988. *Vertebrate paleontology and evolution.* New York: W. H. Freeman.

Carroll, R. L., and P. Thompson. 1982. A bipedal lizardlike reptile from the Karoo. *J. Paleontol.* 56:1–10.

Casamiquela, R. M. 1967. Un nuevo dinosaurio ornitisquio triásico (*Pisanosaurus mertii*) de la Formación Ischigualasto, Argentina. *Ameghiniiana* 5:47–64.

Charig, A. J. 1984. Competition between therapsids and archosaurs during the Triassic period: A review and synthesis of current theories. *Symp. Zool. Soc. Lond.* 52:597–628.

Chatterjee, S. 1974. A rhynchosaur from the Upper Triassic Maleri Formation of India. *Phil. Trans. R. Soc. Lond. B* 267:209–261.

———. 1980. *Malerisaurus*, a new eosuchian reptile from the Late Triassic of India. *Phil. Trans. R. Soc. Lond. B* 291:163–200.

———. 1985. *Postosuchus*, a new thecodontian reptiles from the Triassic of Texas and the origin of tyrannosaurs. *Phil. Trans. R. Soc. Lond. B* 309:395–460.

———. 1986. *Malerisaurus langstoni*, a new diapsid reptile from the Triassic of Texas. *J. Vertebr. Paleontol.* 6:297–312.

———. 1991. Cranial anatomy and relationships of a new Triassic bird from Texas. *Phil. Trans. R. Soc. Lond. B* 332:277–342.

———. 1993. *Shuvasaurus*, a new theropod. *Natl. Geographic Res. Explor.* 9:274–285.

Clark, J. M. H.-D. Sues, and D. S. Berman. 2001. A new specimen of *Hesperosuchus* from the Upper Triassic of New Mexico and the interrelationships of basal crocodylomorph archosaurs. *J. Vertebr. Paleontol.* 20:683–704.

Cluver, M. A. 1971. The cranial anatomy of the dicynodont genus, *Lystrosaurus. Ann. S. Afr. Mus.* 56:35–54.

Colbert, E. H. 1965. A phytosaur from North Bergen, New Jersey. *Am. Mus. Novitates* 2230:1–25.

———. 1970. The gliding Triassic reptile *Icarosaurus. Bull. Am. Mus. Nat. Hist.* 143:85–142.

———. 1981. A primitive ornithischian dinosaur from the Kayenta Formation of Arizona. *Bull. Mus. N. Ariz.* 53:1–61.

Colbert, E. H., and P. E. Olsen. 2001. A new and unusual aquatic reptile from the Lockatong Formation of New Jersey (Late Triassic, Newark Supergroup). *Novitates* 3334:1–24.

Conway Morris, S. 1985. The Middle Cambrian metazoan *Wixwaxia corrugata* (Matthew) from the Burgess Shale and *Ogygopsis* Shake, British Columbia, Canada. *Phil. Trans. R. Soc. Lond. B* 307:507–582.

Coombs, W. P., Jr. 1980. Swimming ability of carnivorous dinosaurs. *Science* 207:1198–1200.

Cooper, M. R. 1982. A mid-Permian to earliest Jurassic tetrapod biostratigraphy and its significance. *Arnoldia Zimbabwe* 9:77–104.

Cope, E. D. 1889. On a new genus of Triassic Dinosauria. *Am. Nat.* 23:626.

———. 1887a. The dinosaurian genus *Coelurus. Am. Nat.* 21:367–369.

———. 1887b. A contribution to the history of the Vertebrata of the Trias of North America. *Proc. Am. Philosophical Soc.* 24:209–228.

Cornet, B. 1977. The palynostratigraphy and age of the Newark Supergroup. Ph.D. thesis, Pennsylvania State University.

————. 1989. The reproductive morphology and biology of *Sanmiguelia lewisii*, and its bearing on angiosperm evolution in the Late Triassic. *Evol. Trends Plants* 3:25–51.

————. 1993. Dicot-like leaf and flowers from the Late Triassic tropical Newark Supergroup rift zone, USA. *Modern Geol.* 19:81–99.

Cornet, B., and P. E. Olsen. 1985. A summary of the biostratigraphy of the Newark Supergroup of eastern North America with comments on Early Mesozoic provinciality. In *Simposio sobre floras del Triassico tardío, su fitogeografía y paleoecología*, ed. R. Weber, 67–81. Memoria III Congreso Latinamericano de Paleontologia México. Instituto de Geología, Universidad Nacional Autonoma de México.

Cornet, B., and P. E. Olsen. 1990. *Early to Middle Carnian (Triassic) flora and fauna of the Richmond and Taylorsville basins, Virginia and Maryland, USA*. Virginia Museum of Natural History Guidebook 1. Martinsville: Virginia Museum of Natural History.

Corsin, P. and M. Waterlot. 1979. Paleobiogeography of the Dipteridaceae and the Matoniaceae of the Mesozoic. In *Fourth International Gondwana Symposium*, Vol. 1, ed. B. Laskas and C. S. Rao, 51–70. New Delhi: Hindustan Publishing.

Cousminer, H. L., and W. Manspeizer. 1976. Triassic pollen date Moroccan High Atlas and the incipient rifting of Pangea as middle Carnian. *Science* 191:943–945.

Crane, P. R. 1988. Major clades and relationships in the "higher" gymnosperms. In *Origin and evolution of gymnosperms*, ed. C. B. Beck, 218–272. New York: Columbia University Press.

————. 1993. Time for the angiosperms. *Nature* 366:631–632.

Crompton, A. W. 1974. The dentitions and relationships of the southern African Triassic mammals, *Erythrotherium parringtoni* and *Megazostrodon rudnerae*. *Bull. Br. Mus. (Nat. Hist.)* 24:397–437.

Crompton, A. W., and N. Hotton III. 1967. Functional morphology of the masticatory apparatus of two dicynodonts (Reptilia, Therapsida). *Postilla* 109:1–51.

Crompton, A. W., and F. A. Jenkins Jr. 1979. Origin of mammals. In *Mesozoic Mammals: The first two-thirds of mammalian history*, ed. J. A. Lillegraven, Z. Kielan-Jaworowska, and W. A. Clemens, 59–72. Berkeley: University of California Press.

Crompton, A. W., and Z. Luo. 1993. Relationships of the Liassic mammals *Sinoconodon*, *Morganucodon oehleri*, and *Dinnetherium*. In. *Mammal phylogeny*, ed. F. S. Szalay, M. J. Novacek, and M. C. McKenna, 30–44. New York: Springer-Verlag.

Cruickshank, A. R. I. 1972. The proterosuchian thecodonts. In *Studies in vertebrate evolution*, ed. K. A. Joysey and T. S. Kemp, 89–119. Edinburgh: Oliver and Boyd.

Crush, P. J. 1984. A late Triassic sphenosuchid crocodilian from Wales. *Palaeontology* 34:131–157.

deBeer, G. S. 1937. *The development of the vertebrate skull*. Oxford: Clarendon Press.

deBraga, M., and O. Rieppel. 1997. Reptile phylogeny and the interrelationships of turtles. *Zool. J. Linnean Soc.* 120:281–354.

Delevoryas, T., and R. C. Hope. 1973. Fertile coniferophyte remains from the Late Triassic Deep River basin, North Carolina. *Am. J. Bot.* 60:810–818.

————. 1975. *Voltzia andrewsii*, n. sp.: An Upper Triassic seed cone from North Carolina, USA. *Rev. Palaeobot. Palynol.* 20:67–74.

————. 1987. Further observations on the Late Triassic conifers *Compsostrobus neotericus* and *Voltzia andrewsii*. *Rev. Palaeobot. Palynol.* 51:59–64.

Demathieu, G. 1989. The appearance of the first dinosaur tracks in the French Middle Triassic and their probable significance. In *Dinosaur tracks and traces*, ed. D. D. Gillette and M. G. Lockley, 201–207. New York: Cambridge University Press.

Demathieu, G., and H. Haubold. 1978. Du problem de l'origine des dinosauriens d'après données de l'ichnologie du Trias. *Geobios* 11:409–412.

Demko, T. M., R. F. Dubiel, and J. T. Parrish. 1998. Plant taphonomy in incised valleys: Implications for interpreting paleoclimate from fossils plants. *Geology* 26:1119–1122.

Dilkes, D. W. 1998. The Early Triassic rhynchosaur *Mesosuchus browni* and the interrelationships of basal archosauromorph reptiles. *Phil. Trans. R. Soc. Lond. B* 353:501–541.

Dobruskina, I. A. 1980. The stratigraphical position of the plantbearing sediments of the Eurasian Triassic. *Trans. GIN AN SSSR N* 346:164.

————. 1994. *Triassic floras of Eurasia*. Österreichische Akademie der Wissenschaften Schriftenreihe der Erdwissenschaftlichen Kommissionen Band 10. Vienna: Springer-Verlag.

————. 1995. Keuper (Triassic) flora from middle Asia (Madygen, Southern Fergana). New Mexico Museum of Natural History and Science Bulletin 5:1–49, xlvi pl.

Doyle, J. A., and M. J. Donoghue. 1986. Seed plant phylogeny and the origin of angiosperms: An experimental cladistic approach. *Bot. Rev.* 52:321–431.

————. 1993. Phylogenies and angiosperm diversification. *Paleobiology* 19:141–167.

Dubiel, R. F., J. T. Parrish, J. M. Parrish, and S. C. Good. 1991. The Pangaean megamonsoon—Evidence from the Upper Triassic Chinle Formation, Colorado Plateau. *Palaios* 6:347–370.

Ellenberger, P. 1970. Les niveaux paléontologiques de première apparition des mammifères primordiaux en Afrique du sud et leur ichnologie. In *Proceedings Papers, 2nd Gondwana Symposium, CSIR Pretoria*, 343–370.

————. 1972. Contributions à la classification des pistes de vertebras du Trias: Les types du Stormberg d'Afrique du sud (I). *Palaeovertebrata Mem. Extraordinaire* 1972:1–117.

————. 1974. Contributions à la classification des pistes de vertebras du Trias: Les types du Stormberg d'Afrique du sud (II). *Palaeovertebrata Mem. Extraordinaire* 1974:1–141.

Emmons, E. 1856. *Geological report of the midland counties of North Carolina*. New York: Putnam.

Estes, R., and O. A. Reig. 1973. The early fossil record of frogs: A review of the evidence. In *Evolutionary biology of the Anurans: Contemporary research on major problems*, ed. J. L. Vial, 11–63. Columbia: University of Missouri Press.

Evans, S. E., and K. A. Kermack. 1994. Assemblages of small tetrapods from the Early Jurassic of Britain. In Fraser and Sues, *In the shadow of the dinosaurs*, 271–283.

Farlow, J. O. 1976. Observations on a captive tuatara (*Sphenodon punctatus*). *J. Herpetol.* 9:353–355.

————. 2001. *Acrocanthosaurus* and the maker of comanchean large-theropod footprints. In *Mesozoic vertebrate life*, ed. D. H. Tanke and K. Carpenter, 408–427. Bloomington: Indiana University Press.

Farlow, J. O., and P. M. Galton. 2003. Dinosaur trackways of Dinosaur State Park, Rocky Hill, Connecticut. In LeTourneau and Olsen, *The great rift valleys of Pangea in eastern North America: Sedimentology, stratigraphy and paleontology*, 2:248–263.

Farlow, J. O., and E. R. Pianka. 2000. Body form and trackway pattern in Australian desert monitors (Squamata: Varanidae): Comparing zoological and ichnological diversity. *Palaios* 15:235–247.

Flynn, J. J., J. M. Parrish, B. Rakotosamimanana, W. F. Simpson, R. L. Whatley, and A. R. Wyss. 1999. A Triassic fauna from Madagascar, including early dinosaurs. *Science* 286:763–765.

Fontaine, W. M. 1883. Contributions to the knowledge of the older Mesozoic flora of Virginia. U.S. Geological Survey Monograph 6:1–144.

Fraser, N. C. 1985. Vertebrate faunas from Mesozoic fissure deposits of southwest Britain. *Modern Geol.* 9:273–300.

————. 1986. New Triassic sphenodontids from southwest England and a review of their classification. *Palaeontology* 29:165–186.

————. 1988a. Rare tetrapod remains from the Late Triassic fissure infillings of Cromhall Quarry, Gloucestershire. *Palaeontology* 31:567–576.

————. 1988b. The osteology and relationships of *Clevosaurus* (Reptilia: Sphenodontida). *Phil. Trans. R. Soc. Lond. B.* 321:125–178.

————. 1988c. Latest Triassic terrestrial vertebrates and their biostratigraphy. *Modern Geol.* 13:125–140.

Fraser, N. C., and M. J. Benton. 1989. The Triassic reptiles *Brachyrhinodon* and *Polysphenodon* and the relationships of the sphenodontids. *Zool. J. Linnean Soc.* 96:413–445.

Fraser, N. C., and D. A. Grimaldi. 1997. Who else lived in the Late Triassic? The world of the early dinosaurs as illustrated by a fossil lagerstaette in Virginia. In *Dinofest International: Proceedings of a symposium held at Arizona State University*, ed. D. I. Wolberg, E. Stump, and G. D. Rosenberg, 191–198. Philadelphia: Academy of Natural Sciences.

————. 1999. A significant late Triassic lagerstaette from Virginia, USA. In *Third International Symposium on Lithographic Limestones*, ed. S. Renesto, 20(Suppl.):79–83. Rivista del museo civico di scienze naturali.

————. 2003. Late Triassic continental faunal change: New perspectives on Triassic insect diversity as revealed by a locality in the Danville basin, Virginia, Newark Supergroup. In LeTourneau and Olsen, *The great rift valleys of Pangea in eastern North America: Sedimentology, stratigraphy and paleontology*, 2:192–205.

Fraser, N. C., and P. E. Olsen. 1996. A new dinosauromorph ichnogenus from the Triassic of Virginia. *Jeffersoniana* 7:1–17.

Fraser, N. C., and H.-D. Sues, eds. 1994. *In the shadow of the dinosaurs*. New York: Cambridge University Press.

Fraser, N. C., and G. M. Walkden. 1983. The ecology of a Late Triassic reptile assemblage from Gloucestershire, England. *Palaeogeogr. Palaeoclimatol. Palaeoecol.* 42:341–365.

————. 1984. The postcranial skeleton of *Planocephalosaurus robinsonae*. *Palaeontology* 27:575–595.

Fraser, N. C., G. M. Walkden, and V. Stewart. 1985. The first pre-Rhaetic therian mammal. *Nature* 314:161–163.

Fraser, N. C., D. A. Grimaldi, P. E. Olsen, and B. Axsmith. 1996. A Triassic lagerstätte from eastern North America. *Nature* 380:615–619.

Fraser, N. C., K. Padian, G. M. Walkden, and A. L. M. Davis. 2002. Basal dinosauriform remains from Britain and the diagnosis of the Dinosauria. *Palaeontology* 45:79–95.

Frazetta, T. H. 1968. Adaptive problems and possibilities in the temporal fenestration of tetrapod skulls. *J. Morphol.* 125:145–158.

Frey, R. W., S. G. Pemberton, and J. A. Fagerstrom. 1984. Morphological, ethological, and environmental significance of the ichnogenera *Scoyenia* and *Ancorichnus. J. Paleontol.* 58:511–528.

Freytet, P., and J. C. Plaziat. 1982. Continental carbonate sedimentation and pedogenesis—Late Cretaceous and early Tertiary of southern France. *Contrib. Sedimentol.* 12:214.

Fürrer, H. 1995. The Kalkschieferzone (Upper Meride Limestone; Ladinian) near meride (Canton Ticino, southern Switzerland) and the evolution of a Middle Triassic intraplatform basin. *Ecol. Geol. Helvetiae* 88:827–852.

Gall, J. C. 1971. Faunes et paysages du Grès à Voltzia du Nord des Vosges. Essai Paléoécologique sur le Bundsandstein supérieur. *Mém. Serv. Carte Géol. Als. Lorr.* 34:1–318.

————. 1996. Triassic insects of western Europe. *Paleontologia Lombarda* 5. Museo Civico di Storia Naturale di Milano.

Gall, J. C., and L. Grauvogel. 1966. Faune du Buntsandstein I. Pontes d'invertébrés du Buntsandstein supérieur. *Ann. Paléont. Invertébrés* 52:151–161.

Galton, P. M. 1971. The prosauropod dinosaur *Ammosaurus*, the crocodile *Protosuchus*, and their bearing on the age of the Navajo Sandstone of northeastern Arizona. *J. Paleontol.* 45:781–795.

Gans, C., I. Darevskii, and L. P. Tatarinov. 1987. *Sharovipteryx*, a reptilian glider? *Paleobiology* 13:415–426.

Garassino, A., and G. Teruzzi. 1993. A new decapod crustacean assemblage from the Upper Triassic of Lombardy (N. Italy). Paleontologia Lombarda, n.s., 1:1–27.

Gardiner, B. G. 1961. New Rhaetic and Liassic beetles. *Palaeontology* 4:87–89.

————. 1993. Haematothermia: Warm-blooded amniotes. *Cladistics* 9:369–395.

Gauthier, J. A. 1986. Saurischian monophyly and the origin of birds. *Mem. Calif. Acad. Sci.* 8:1–55.

Gauthier, J., and K. de Queiroz. 2001. Feathered dinosaurs, flying dinosaurs, crown dinosaurs, and the name "Aves." In *New perspectives on the origin and early evolution of birds: Proceedings of the international symposium in honor of John H. Ostrom*, ed. J. Gauthier and L. F. Gall, 7–41. New Haven, Conn.: Peabody Museum of Natural History, Yale University.

Gensel, P. 1986. Plant fossils of the Upper Triassic Deep River basin. In *Depositional framework of a Triassic rift basin: The Durham and Sanford sub-basins of the Deep River basin, North Carolina*, ed. P. J. W. Gore, 82–86. Society of Economic Paleontologists and Mineralogists Field Guidebook, Third Annual Mid-Year Meeting, Raleigh, N.C.

Geyer, G., and K.-P. Kelber. 1987. Flügelreste und Lebensspuren von Insekten aus dem Unteren Keuper Mainfrankens. *Neues Jahrbuch Geol. Paläontol. Abhandlungen* 174:331–355.

Goodwin, B. K., K. W. Ramsey, and G. P. Wilkes. 1986. Guidebook to the geology of the Richmond, Farmville, Briery Creek and Roanoke Creek basins, Virginia. Presented at the Virginia Geological Field Conference, 18th Annual Meeting.

Gore, P. J. W. 1988. Paleoecology and sedimentology of a Late Triassic lake, Culpeper basin, Virginia, USA. *Palaeogeogr. Palaeoclimatol. Palaeoecol.* 62:593–608.

Gould, S. J. 1989. *Wonderful life: The Burgess Shale and the nature of history.* Harmondsworth, UK: Penguin.

Gould, R. E., and T. Delevoryas. 1977. The biology of *Glossopteris*: Evidence from petrified seed-bearing and pollen-bearing organs. *Alcheringa* 1:387–399.

Gow, C. E. 1977. New procolophonids from the Triassic *Cynognathus* Zone of South Africa. *Ann. S. Afr. Mus.* 72:109–124.

Grauvogel, L. 1947. Note préliminaire sur la faune du Grès à Voltzia. *C. R. Som. Soc. Geol. France* 66–67.

Grauvogel, L., and D. Laurentiaux. 1952. Un Protodonate du Trias des Vosges. *Ann. Paléont* 38:121–129.

Grauvogel-Stamm, L. 1978. La flore du Grès à Voltzia (Buntsandstein supérieur) des Vosges du Nord (France): Morphologie, anatomie, interpretations phylogénique et paléogéographique. Mémoires Sciences Géologique 50.

Grauvogel-Stamm, L., and K.-P. Kelber. 1996. Plant-insect interactions and coevolution during the Triassic in western Europe. In *Triassic insects of western Europe*, ed. J.-C. Gall, 5–23. Paleontologia Lombarda, n.s., 5.

Gregory, J. T. 1945. Osteology and relationships of *Trilophosaurus*. University of Texas Publication 4401:273–359.

Grimaldi, D., A. Shmakov, and N. Fraser. 2004. Mesozoic thrips and early evolution of the order Thysanoptera (Insecta). *J. Paleontol.* 78:941–952.

Grimaldi, D., Y. Zhang, N. Fraser, and A. Rasnitsyn. 2006. New and revised species of the Mesozoic family Pseudopolycentropodidae (Mecopteroidea), including two mosquito-like species in Burmese amber. *Insect System. Evol.*

Hahn, G., J. C. Lepage, and G. Wouters. 1988. Traversodontiden-Zähne (Cynodontia) aus der Ober-Trias von Gaume (Süd-Belgien). *Bull. Inst. R. Sci. Nat. Belge* 58:171–186.

Hallam, A. 1981. The end-Triassic bivalve extinction event. *Palaeogeogr. Palaeoclimatol. Palaeoecol.* 35:1–44.

———. 1990. The end-Triassic mass extinction event. In *Global catastrophes in earth history: Interdisciplinary conference on impacts, volcanism and mass mortality*, ed. V. L. Sharpton and P. D. Ward, 577–583. Geological Society of America Special Paper 247.

Halstead, L. B., and P. G. Nicoll. 1971. Fossilized caves of Mendip. *Studies Speleol.* 2:93–102.

Hames, W. E., P. R. Renne, and C. Ruppel. 2000. New evidence for geologically instantaneous emplacement of earliest Jurassic Central Atlantic magmatic province basalts on the North American margin. *Geology* 28:859–862.

Harris, J. D., and A. Downs. 2002. A drepanosaurid pectoral girdle from the Ghost Ranch (Whitaker) *Coelophysis* Quarry (Chinle Group, Rock Point Formation, Rhaetian), New Mexico. *J. Vertebr. Paleontol.* 22:70–75.

Harshbarger, J. F., C. A. Repenning, and J. H. Irwin. 1957. Stratigraphy of the uppermost Triassic and Jurassic rocks. U.S. Geological Survey Professional Paper 291:1–74.

Hasiotis, S. T., and R. F. Dubiel. 1995. Termite (Insecta: Isoptera) nest ichnofossils from the Triassic Chinle Formation, Petrified Forest National Park, Arizona. *Ichnos* 4:119–130.

Haubold, H. 1986. Archosaur footprints at the terrestrial Triassic-Jurassic transition. In *The beginning of the Age of Dinosaurs*, ed. K. Padian, 189–201. Cambridge: Cambridge University Press.

Haubold, H., and E. Buffetaut. 1987. Une nouvelle interpretation de *Logisquama insignis*, reptile enigmatique du Trias Superieur d'Asie Centrale. *Compte Rendus Acad. Sci. Paris* 305:65–70.

Hay, O. P. 1930. Second bibliography and catalogue of the fossil Vertebrata of North America. Carnegie Institute of Washington Publication 390:1–186.

Heckert, A. B., and S. G. Lucas. 1996. Revision of the South American aetosaur (Archosauria: Pseudosuchia) record with implications for the absolute age of the Late Triassic Chinle Group, USA. Geological Society of America Abstracts with Program 28:365.

———. 2002. South American occurrences of the Adamanian (Late Triassic: Latest Carnian) index taxon *Stagonolepis* (Archosauria: Aetosauria) and their biochronological significance. *J. Paleontol.* 76:852–863.

Hildebrand, M. 1974. *Analysis of vertebrate structure*. New York: Wiley.

Hitchcock, E. 1858. *Ichnology of New England: A report on the sandstone of the Connecticut Valley, especially its fossil footmarks*. Boston: William White.

Hodych, J. P., and G. R. Dunning. 1992. Did the Manicouagan impact trigger end-of-Triassic mass extinction? *Geology* 20:51–54.

Holser, W. T., and M. Magaritz. 1987. Events near the Permo-Triassic boundary. *Modern Geol.* 11:155–180.

Holt, E. L. 1947. Upright trunks of Neocalamites from the Upper Triassic of western Colorado. *J. Geol.* 55:511–513.

Hopson, J. A. 1969. The origin and adaptive radiation of mammal-like reptiles and nontherian mammals. *Ann. N. Y. Acad. Sci.* 167:199–216.

Hopson, J. A., and H. R. Barghusen. 1986. An analysis of therapsid relationships. In *The ecology and biology of mammal-like reptiles*, ed. N. Hotton III, P. D. MacLean, J. J. Roth, and E. C. Roth, 83–106. Washington, D.C.: Smithsonian Institution Press.

Hopson, J. A., and A. W. Crompton. 1969. Origin of mammals. *Evol. Biol.* 3:15–72.

Hopson, J. A., and J. W. Kitching. 1972. A revised classification of cynodonts (Reptilia; Therapsida). *Palaeontol. Afr.* 14:71–85.

Hubbard, M. D., and E. F. Riek. 1977. New name for a Triassic mayfly from South Africa (Ephemeroptera). *Psyche* 83:260–261.

Huber, P., S. G. Lucas, and A. P. Hunt. 1993. Vertebrate biochronology of the Newark Supergroup Triassic, eastern North America. In *The nonmarine Triassic*, ed. S. G. Lucas and M. Morales, 179–186. New Mexico Museum of Natural History and Science Bulletin 3.

Huene, F. von. 1910a. Über einen echten Rhynchocepalen aus der Trias von Elgin, *Brachyrhinodon taylori*. *Neues Jahrbuch Mineral. Geol. Paläontol.* 1910:29–62.

———. 1910b. Ein primitiver Dinosaurier aus der mittleren Trias von Elgin. *Geol. Paläontol. Abhandlungen*, n.f., 8:315–322.

———. 1914. Beiträge zur Geschichte der Archosaurier. *Geol. Paläontol. Abhandlungen*, n.f., 13:1–53.

———. 1936. *Die fossilen Reptilien des südamerikanischen Gondwanalandes. Ergebnisse der Sauiergrabungen in Südbrasilien 1928/29*. Tübingen: Lieferung Verlag Franz F. Heine.

———. 1939. Die Lebensweise der Rhynchosauriden. *Paläontol. Z.* 21:232–238.

———. 1956. *Palaeontologie und Phylogenie der niederen Tetrapoden*. Jena: Gustav Fischer.

Hungerbühler, A., and H.-D. Sues. 2001. Status and phylogenetic relationships of the Late Triassic phytosaur *Rutiodon carolinensis*. *J. Vertebr. Paleontol.* 21(Suppl. to 3):64A.

Hunt, A. P. 1989. Cranial morphology and ecology among phytosaurs. In *Dawn of the age of dinosaurs in the American Southwest*, ed. S. G. Lucas and A. P. Hunt, 349–354. Albuquerque: New Mexico Museum of Natural History.

Hunt, A. P., and S. G. Lucas. 1990. Re-evaluation of *"Typothorax" meadei*, a late Triassic aetosaur from the United States. *Paläontol. Z.* 64:317–328.

————. 1991. *Rioarribasaurus*, a new name for a Late Triassic dinosaur from New Mexico (USA). *Paläontol. Z.* 65:191–198.

————. 1994. Ornithischian dinosaurs from the Upper Triassic of the United States. In Fraser and Sues, *In the shadow of the dinosaurs*, 227–241.

Huxley, T. H. 1869. On *Hyperodapedon*. *Q. J. Geol. Soc. Lond.* 25:138–157.

————. 1875. On *Stagonolepis robertsoni*, and on the evolution of the Crocodilia. *Q. J. Geol. Soc. Lond.* 31:423–438.

————. 1877. The crocodilian remains found in the Elgin Sandstones, with remarks on the ichnites of Cummingstone. Memoirs of the Geological Survey, UK, Monograph 3:1–52.

————. 1875. On *Stagonolepis robertsoni*, and on the evolution of the Crocodilia. *Q. J. Geol. Soc. Lond.* 31:423–438.

Ivakhnenko, M. F. 1978. [Tailed amphibia from the Triassic and Jurassic of Middle Asia]. *Paleontol. Zhurnal* N3:84–89.

Jacobs, L. L., and P. A. Murry. 1980. The vertebrate community of the Triassic Chinle Formation near St. Johns, Arizona. In *Aspects of vertebrate history*, ed. L. L. Jacobs, 55–71. Flagstaff: Museum of Northern Arizona Press.

Jacobs, W., and M. Renner. 1988. *Biologie und Ökologie der Insekten: Ein Taschenlexikon*. Stuttgart: Gustav Fischer Verlag.

Jadoul, F., F. Berra, and S. Frisia. 1992. Stratigraphic and paleogeographic evolution of a carbonate platform in an extensional tectonic regime: The example of the Dolomia Principale in Lombardy (Italy). *Riv. Ital. Paleontol. Stratigr.* 98:29–44.

Jeans, C. V. 1978. The origin of the Triassic clay assemblages of Europe with special reference to the Keuper Marl and Rhaetic parts of England. *Phil. Trans. R. Soc. Lond. A* 289:549–639.

Jenkins, F. A., Jr. 1970. The Chañares (Argentina) Triassic reptile fauna VII. The postcranial skeleton of the traversodont *Massetognathus pascuali* (Therapsida, Cynodontia). *Breviora* 352:1–28.

————. 1984. A survey of mammalian origins. In *Mammals: Notes for a short course*, ed. P. D. Gingerich and C. E. Badgley. *Univ. Tenn. Studies Geol.* 832–47.

Jenkins, F. A., Jr., and G. E. Goslaw Jr. 1983. The functional anatomy of the shoulder of the Savannah monitor lizard (*Varanus exanthematicus*). *J. Morphol.* 175:195–216.

Jenkins, F. A., Jr., S. M. Gatesy, N. H. Shubin, and W. W. Amaral. 1997. Haramyids and Triassic mammalian evolution. *Nature* 385:715–718.

Johnson, L. A., and M. J. Simms. 1989. The timing and cause of Late Triassic marine invertebrate extinctions: Evidence from scallops and crinoids. In *Mass extinctions: Processes and evidence*, ed. S. K. Donovan, 174–194. New York: Columbia University Press.

Jones, T. D., et al. 2000. Nonavian feathers in a late Triassic archosaur. *Science* 288, 2202–2205.

Kaye, F. T., and K. Padian. 1994. Microvertebrates from the Placerias Quarry: A window on Late Triassic vertebrate diversity in the American Southwest. In Fraser and Sues, *In the shadow of the dinosaurs*, 171–196.

Kelber, K.-P. 1988. Was ist *Equisetites foveolatus*. In *Neue Forschungen zur Erdgeschichte von Crailsheim*, ed. H. Hagdorn, 166–184. Sonderblatt Gesellschaft Naturkunde Württemberg, Stuttgart 1.

Kelber, K.-P., and G. Geyer. 1989. Lebensspuren von Insekten an Pflanzen des Unteren Keupers. *Cour. Forsch.-Inst. Senckenberg* 109:165–174.

Kemp, T. S. 1980. Aspects of the structure and functional anatomy of the Middle Triassic cynodont *Luangwa*. *J. Zool. Lond.* 191:193–239.

————. 1982. *Mammal-like reptiles and the origin of mammals*. London: Academic Press.

————. 1983. The relationships of mammals. *Zool. J. Linnean Soc. Lond.* 77:353–384.

Kent, D. V. 1998. Impacts on earth in the Late Triassic: Discussion. *Nature* 395:126.

Kent, D. V., B. Cornet, B., Witte, W. K., and Schlische, R. W. 1996. High resolution stratigraphy of the Newark rift basin (early Mesozoic, eastern North America). *Geophys. Soc. Am. Bull.* 108:40–77.

Kermack, D. 1984. New prosauropod material from South Wales. *Zool. J. Linnean Soc.* 82:101–117.

Kermack, K. A., F. Mussett, and H. W. Rigney. 1973. The lower jaw of *Morganucodon*. *Zool. J. Linnean Soc.* 53:87–175.

————. 1981. The skull of *Morganucodon*. *Zool. J. Linnean Soc.* 71:1–158.

King, G. M. 1990. *The Dicynodonts: A study in palaeobiology*. London: Chapman and Hall.

King, G. M., and M. A. Cluver. 1991. The aquatic *Lystrosaurus:* An alternative lifestyle. *Historical Biol.* 4:232–341.

Kischlat E.-E., S. G. Lucas, and L. Maciel. 2002. First record of phytosaurs in the Upper Triassic of South America. *J. Vertebr. Paleontol.* 22(Suppl. to 3):74A.

Knoll, A. H., R. K. Bambach, D. E. Canfield, and J. P. Gotzinger. 1996. Comparative Earth history and Late Permian mass extinction. *Science* 272:452–457.

Krebs, B. 1965. *Ticinosuchus ferox* nov. gen. nov. sp.: Ein neuer Pseudosuchier aus der Trias des Monte San Giorgio. *Schweizerische Paläontol. Abhandlungen* 81:1–140.

————. 1976. Pseudosuchia. In *Handbuch der Paläoherpetologie,* ed. O. Kuhn, 13:40–98. Stuttgart: Gustav Fischer Verlag.

Krzemiński, W. E., and E. Krzemińska. 2003. Triassic Diptera: Descriptions, revisions and phylogenetic relations. *Acta Zool. Cracoviensia* 46:153–184.

Krzemiński, W. E., and C. Lombardo. 2001. New fossil ephemeroptera and coleoptera from the Ladinian (Middle Triassic) of Canton Ticino (Switzerland). *Riv. Ital. Paleontol. Stratigr.* 107:69–78.

Krzemiński, W., E. Krzemińska, and F. Papier. 1994. *Grauvogelia arzvilleriana* sp. n.—The oldest Diptera species (Lower/Middle Triassic of France). *Acta Zool. Cracoviensia* 37:95–99.

Kühne, W. G. 1956. *The Liassic therapsid Oligokyphus.* London: Trustees of the British Museum.

Løvtrup, S. 1977. *The phylogeny of Vertebrata.* London: Wiley.

————. 1985. On the classification of the taxon Tetrapoda. *Systematic Zool.* 34:463–470.

Lambiase, J., and M. R. Rodgers. 1988. A model for tectonic control of lacustrine stratigraphic sequences in continental rift basins. Presented at the American Association of Petroleum Geologists Research Conference, Lacustrine exploration: Case studies and modern analogues, Snowbird, Utah.

Langer, M. C., F. Abdala, M. Richter, and M. J. Benton. 1999. A sauropodomorph dinosaur from the Upper Triassic (Carnian) of southern Brazil. *C. R. Acad. Sci. Paris, Sci. Terre Planète* 329:511–517.

Larentiaux-Viera, F. J. Ricour, and D. Laurentiaux. 1952. Un protodonate du Trias de la Dent de Villard (Savoie). *Bull. Soc. Geol. France* 6:319–324.

Larew, H. G. 1992. Fossil galls. In *Biology of insect-induced galls,* ed. J. D. Shorthouse and O. Rohfritsch, 50–59. Oxford: Oxford University Press.

Lefebvre, F., A. Nel, F. Papier, L. Grauvogel-Stamm, and J.-C. Gall. 1998. The first "cicada-like Homoptera" from the Triassic of the Vosges, France. *Palaeontology* 41:1195–1200.

LeTourneau, P. M., and P. E. Olsen. 2003a. *The great rift valleys of Pangea in eastern North America: Tectonics, structure, and volcanism.* Vol. 1. New York: Columbia University Press.

————. 2003b. *The great rift valleys of Pangea in eastern North America: Sedimentology, stratigraphy and paleontology.* Vol. 2. New York: Columbia University Press.

Li, J. 1983. Tooth replacement in a new genus of procolophonid from the Early Triassic of China. *Palaeontology* 26:567–583.

Li, C., O. Rieppel, and M. C. LaBarbera. 2004. A Triassic aquatic protorosaur with an extremely long neck. *Science* 305:1931.

Lillegraven, J. A., Z. Kielan-Jaworowska, and W. A. Clemens. 1979. *Mesozoic mammals: The first two thirds of mammalian history.* Berkeley: University of California Press.

Lis, J., and Z. Wojcik. 1958. Brekcja kostna w lomie I kras kopalny Gliny pod Olkuszem. *Przegl. Geol.* 1958:554–556.

————. 1960. Triasowa brekcja kostna I kras kopalny w kamienolomie Stare Gliny pod Olkuszem. *Kwart Geol.* 4:55–74.

Long, R. A., and P. A. Murry. 1995. Late Triassic (Carnian and Norian) tetrapods from the southwestern United States. New Mexico Museum of Natural History and Science Bulletin 4.

Lozovsky, V. R., and M. A. Shiskin. 1974. [First labyrinthodont find from the Lower Triassic of Mangyshlak]. *Doklady AN SSSR* 214:169–172.

Lucas, S. G. 1993. Vertebrate biochronology of the Triassic of China. In *The nonmarine Triassic,* ed. S. G. Lucas and M. Morales, 301–306. New Mexico Museum of Natural History and Science Bulletin 3.

————. 1997. Upper Triassic Chinle Group, western United States: A nonmarine standard for Late Triassic time. In *Late Palaeozoic and Early Mesozoic circum-Pacific events and their global correlation,* ed. J. M. Dickens, Z. Yang, H. Yin, S. G. Lucas, and S. K. Acharyya, 209–228. Cambridge: Cambridge University Press.

————. 1998. Global Triassic tetrapod biostratigraphy and biochronology. *Palaeogeogr. Palaeoclimatol. Palaeoecol.* 143:347–384.

Lucas, S. G., and A. P. Hunt. 1993. Tetrapod biochronology of the Chinle Group (Upper Triassic), western United States. In *The nonmarine Triassic,* ed. S. G. Lucas and M. Morales, 327–329. New Mexico Museum of Natural History and Science Bulletin 3.

Lucas, S. G., and Z. Luo. 1993. *Adelobasileus* from the Upper Triassic of West Texas: The oldest mammal. *J. Vertebr. Paleontol.* 13:309–334.

Lucas, S. G., A. P. Hunt, and R. A. Long. 1992. The oldest dinosaurs. *Naturwissenschaften* 79:171–172.

Lucas, S. G., A. P. Hunt, and R. Kahle. 1993. Late Triassic vertebrates from the Dockum Formation near Otis Chalk, Howard County, Texas. New Mexico Geological Society Guidebook 44:237–244.

Lucas, S. G., A. B. Heckert, and P. Huber. 1998. *Aetosaurus* (Archosauromorpha) from the Upper Triassic of the Newark Supergroup, eastern United States, and its biochronological significance. *Palaeontology* 41:1215–1230.

Lucas, S. G., A. B. Heckert, N. C. Fraser, and P. Huber. 1999. *Aetosaurus* from the Upper Triassic of Great Britain and its biochronological significance. *Neues Jahrbuch Geol. Paläontol. Monatshefte* 1999:568–576.

Luo, Z. 1994. Sister-group relationships of mammals and transformations of diagnostic mammalian characters. In Fraser and Sues, *In the shadow of the dinosaurs*, 98–128.

Lutrell, G. W. 1989. Stratigraphic nomenclature of the Newark Supergroup of eastern North America. U.S. Geological Survey Bulletin 1572:1–136.

Mader, D. 1990. *Palaeoecology of the flora in the Buntsandstein and Keuper in the Triassic of Middle Europe*, Vol. 1, *Buntsandstein*. Stuttgart: Gustav Fischer Verlag.

Magdefrau. 1953. Neue funde fossile Coniferen im mittleren Keuper von Häfurt (Main). *Geol. Bl. No. Bayern* 3:49–48.

Maier, G. 1997. Tendaguru. In *Encyclopedia of dinosaurs*, ed. P. J. Currie and K. Padian, 725–726. San Diego: Academic Press.

———. 2003. *African dinosaurs unearthed: The Tendaguru expeditions*. Bloomington: Indiana University Press.

Marchal-Papier, F. 1998. Les insects du Buntsandstein des Vosges (NE de la France). Biodiversité et contribution aux modalités de la crise biologique du Permo-Trias. Ph.D. thesis, Université Loius Pasteur de Strasbourg.

Marchal-Papier, F., A. Nel, and L. Grauvogel-Stamm. 2000. Nouveaux Orthoptères (Ensifera, Insecta) du Trias des Vosges (France). *Acta Geol. Hispanica* 35:5–18.

Marsh, O. C. 1896. A new belodont reptile (*Stegomus*) from the Connecticut River Sandstone. *Am. J. Sci.* 2:59–62.

Marshall, J. E. A., and D. I. Whiteside. 1980. Marine influences in the Triassic "uplands." *Nature* 287:627–628.

McElwain J. C., D. J. Beerling, and F. I. Woodward. 1999. Fossil plants and global warming at the Triassic-Jurassic boundary. *Science* 285:1386–1390.

McHone, J. G., and J. H. Puffer. 2003. Flood basalt provinces of the Pangean Atlantic Rift: Regional extent and environmental significance. In LeTourneau and Olsen, *The great rift valleys of Pangea in eastern North America: Tectonics, structure, and volcanism*, 1:141–154.

Mellett, J. S. 1974. Scatological origin of microvertebrate fossil accumulations. *Science* 185:349–350.

Merck, J. W. 1997. A phylogenetic analysis of the euryapsid reptiles. Ph.D. diss., University of Texas at Austin.

Merriam, J. C. 1905. The Thalattosauria, a group of marine reptiles from the Triassic of California. *Mem. Calif. Acad. Sci.* 5:1–38.

Meyer, H. von. 1837. Mitteinlung an Prof. Bronn (*Plateosaurus engelhardti*). *Neues Jahrbuch Mineral. Geol. Paläontol.* 1837:317.

Meyertons, C. T. 1963. Triassic formations of the Danville basin. Virginia Division of Mineral Resources, Report of Investigations, 6:1–65.

Miller, C. N. 1977. Mesozoic conifers. *Bot. Rev.* 43:218–280.

———. 1982. Current status of Paleozoic and Mesozoic conifers. *Rev. Palaeobot. Palynol.* 37:99–114.

Milner, A. R., B. G. Gardiner, N. C. Fraser, and M. A Taylor. 1990. Vertebrates from the Middle Triassic Otter Sandstone Formation of Devon. *Palaeontology* 33:873–892.

Modesto, S. P., and R. J. Damiani. 2003. Taxonomic status of *Thelegnathus browni* Broom, a procolophonid reptile from the South African Triassic. *Annals Carnegie Mus.* 72:53–64.

Nesbitt, S. 2003. A new specimen of *Arizonasaurus* from the Moenkopi Formation (lower Middle Triassic) and its importance to pseudosuchian divergence. *J. Vertebr. Paleontol.* 23(Suppl. to 3):82A.

Newton, E. T. 1894. Reptiles from the Elgin Sandstone—Description of two new genera. *Phil. Trans. R. Soc. Lond.* 185:573–607.

Norman, D. B. 1990. *The illustrated encyclopedia of dinosaurs*. London: Salamander.

Nosotti, S. 1999. New findings of *Tanystropheus longobardicus* (Reptilia, Prolacertiformes) in the

Middle Triassic of Besano (Lombardy, northern Italy). In *Third International Symposium on Lithographic Limestones*, ed. S. Renesto, 20(Suppl.):117–120. Rivista del museo civico di scienze naturali.

Novas, F. E. 1994. New information on the systematics and postcranial skeleton of *Herrerasaurus ischigualastensis* (Theropoda: Herrerasauridae) from the Ischigualasto Formation (Upper Triassic) of Argentina. *J. Vertebr. Paleontol.* 13:400–423.

———. 1996. Dinosaur monophyly. *J. Vertebr. Paleontol.* 16:723–741.

Ochev, V. G. 1993. [To the biogeography of Triassic tetrapods: Questions of the Paleozoic, Mesozoic and Cenozoic stratigraphy]. Saratov, Russia: Saratov University Publishing 7:3–26.

Oldham, T. 1879. *Geological glossary for the use of students.* London: Stanford.

Olsen, P. E. 1977. Paleontology of Triangle Brick Quarry. In *Field guide to the geology of the Durham Triassic basin*, ed. G. L. Bain and B. W. Harvey, 59–61. Raleigh, N.C.: Carolina Geological Society.

———. 1979. A new aquatic eosuchian from the Newark Supergroup (Late Triassic–Early Jurassic) of North Carolina and Virginia. *Postilla* 176:1–14.

———. 1986. A 40-million-year lake record of early Mesozoic orbital climatic forcing. *Science* 234:842–848.

———. 1988. Paleontology and paleoecology of the Newark Supergroup (early Mesozoic, eastern North America). In *Triassic-Jurassic rifting, continental breakup, and the origin of the Atlantic Ocean and passive margins*, ed. W. Manspeizer, 185–230. New York: Elsevier.

———. 1989. Yale Quarry, Kings Bluff, Weehawken, NJ. In *Tectonic, depositional, and paleoecological history of early Mesozoic rift basins, eastern North America*, ed. P. E. Olsen, R. W. Schlische, and P. J. W. Gore, 98–102. American Geophysical Union, Field Trip Guidebook T351.

———. 1999. Giant lava flows, mass extinctions and mantle plumes. *Science* 284:604–605.

Olsen, P. E., and P. M. Galton. 1977. Triassic-Jurassic tetrapod extinctions: Are they real? *Science* 197:983–986.

Olsen, P. E., and J. G. McHone. 2003. The Central Atlantic large igneous province: Introduction. In LeTourneau and Olsen, *The great rift valleys of Pangea in eastern North America: Tectonics, structure, and volcanism*, 1:137–140.

Olsen, P. E., N. H. Shubin, and M. H. Anders. 1987. New Early Jurassic tetrapod assemblages constrain Triassic-Jurassic tetrapod extinction event. *Science* 237:1025–1029.

Olsen, P. E., R. W. Schlische, and P. J. W. Gore. 1989. Tectonic, depositional, and paleoecological history of early Mesozoic rift basins, eastern North America. American Geophysical Union, Field Trip Guidebook T351.

Olsen, P. E., V. Schneider, H.-D. Sues, K. M. Peyer, and J. G. Carter. 2001a. Biotic provinciality of the Late Triassic equatorial humid zone. Geological Society of America Abstracts with Program 33:A27.

Olsen, P. E., H.-D. Sues, and M. A. Norell. 2001b. First record of *Erpetosuchus* (Reptilia: Archosauria) from the Late Triassic of North America. *J. Vert. Paleo.* 20:633–636.

Olsen, P. E., D. V. Kent, H.-D. Sues, C. Koeberl, H. Huber, A. Montanari, E. C. Rainforth, S. J. Powell, M. J. Szajna, and B. W. Hartline. 2002. Ascent of dinosaurs linked to an iridium anomaly at the Triassic-Jurassic boundary. *Science* 296:1305–1307.

Ortlam, D. 1970. *Eocyclotosaurus woschmidt* n. g. n. sp.—Ein neuer Capitosauride aus dem Oberen Buntsandstein des nördlichen Schwarzwaldes. *Neues Jahrbuch Geol. Paläontol. Monatschefte* 1970:568–580.

Osborn, H. F. 1886a. A new mammal from the American Triassic. *Science* 8:540.

———. 1886b. Observations on the Upper Triassic mammals *Dromatherium* and *Microconodon*. *Proc. Acad. Nat. Sci. Philadelphia* 37:359–363.

Owen, R. 1851. Vertebrate air-breathing life in the Old Red Sandstone. *Literary Gazette* 1851:900.

Padian, K. 1983. A functional analysis of flying and walking pterosaurs. *Paleobiology* 9:218–239.

———. 1984. The origin of pterosaurs. In *Third symposium on Mesozoic Terrestrial Ecosystems*, ed. W.-E. Reif and F. Westphal, 163–168. Tübingen: Attempto Verlag.

———. 1986. On the type material of *Coelophysis* Cope (Saurischia: Theropoda), and a new specimen from the Petrified Forest of Arizona (Late Triassic: Chinle Formation). In *The beginning of the Age of Dinosaurs: Faunal change across the Triassic-Jurassic boundary*, ed. K. Padian, 45–60. New York: Cambridge University Press.

———. 1994. What were the tempo and mode of evolutionary change in the Late Triassic to Middle Jurassic? In Fraser and Sues, *In the shadow of the dinosaurs*, 401–407.

———. 1997. Pterosauromorpha. In *Encyclopedia of dinosaurs*, ed. P. J. Currie and K. Padian, 617–618. Los Angeles: Academic Press.

Papier, F., and L. Grauvogel-Stamm. 1995. Les Blattodea du Trias: Le genre *Voltziablatta* n. gen. Du Buntsandstein superieur des Vosges (France). *Palaeontographica Abteilung A* 235:141–162.

Papier, F., and A. Nel. 2001. Les Subioblattidae (Blattodea, Insecta) du Trias d'Asie Centrale. *Paläontol. Z.* 74:533–542.

Papier, F., L. Grauvogel-Stamm, and A. Nel. 1994. *Subioblatta undulata* n. sp., une nouvelle blatte (Subioblattidae Schneider) du Buntsandstein supérieur (Anisien) des Vosges (France): Morphologie, systématique et affinities. *Neues Jahrbuch Geol. Paläontol. Mh.* 1994:277–290.

Papier, F., A. Nel, L. Grauvogel-Stamm, and J.-C. Gall. 1997. La plus ancienne sauterelle Tettigoniidae, Orthoptera (Trias, NE France): Mimétisme ou exaptation? *Paläontol. Z.* 71:71–77.

Parrington, F. R. 1946. On a collection of Rhaetic mammalian teeth. *Proc. Zool. Soc. Lond.* 116:707–728.

Parrish, J. M. 1986. Structure and function of the tarsus in the phytosaurs (Reptilia: Archosauria). In *The beginning of the Age of Dinosaurs: Faunal change across the Triassic-Jurassic boundary*, ed. K. Padian, 35–43. New York: Cambridge University Press.

————. 1987. The origin of crocodilian locomotion. *Paleobiology* 13:396–414.

————. 1993. Phylogeny of the Crocodylotarsi, with reference to archosaurian and crurotarsan monophyly. *J. Vertebr. Paleontol.* 13:287–308.

Parrish, J. M., J. T. Parrish, and A. M. Ziegler. 1986. Permian-Triassic paleogeography and paleoclimatology and implications for therapsid distributions. In *The biology and ecology of mammal-like reptiles*, ed. N. H. Hotton III, P. D. MacLean, J. J. Roth, and E. C. Roth, 109–132. Washington, D.C.: Smithsonian Press.

Parrish, J. T. 1998. *Interpreting Pre-Quaternary climate from the geologic record.* New York: Columbia University Press.

————. 1993. Climate of the supercontinent Pangea. *J. Geol.* 101:215–233.

Peters, D. 2000. A re-examination of four prolacertiforms with implications for pterosaur phylogenies. *Riv. Ital. Paleontol. Stratigr.* 106:293–336.

Peterson, F., and G. N. Piperingos. 1979. Stratigraphic relations of the Navajo Sandstone to Middle Jurassic formations, southern Utah and northern Arizona. U.S. Geological Survey Professional Paper 1035-A:A1–A29.

Peyer, B. 1937. Die Triasfauna der tessiner Kalkalpen, XII: *Macrocnemus bassanii* Nopcs. *Abhandlungen Schweizerische Paläontol. Gesellschaft* 59:1–140.

Pinna, G. 1980. *Drepanosaurus unguicaudatus*, nuova specie di Lepidosaurio del Trias Alpino (Reptilia). *Atti Soc. Ital. Sci. Nat.* 121:181–192.

————. 1984. Osteolgia di *Drepanosaurus unguicaudatus* Lepidosauro triassico del Sottordine Lacertilia. *Mem. Soc. Ital. Sci. Nat.* 24:7–28.

Pough, F. R. 1973. Lizard energetics and diet. *Ecology* 54:837–844.

Rage, J.-C., and Z. Roček. 1989. Redescription of *Triadobatrachus massinoti* (Piveteau, 1936) an anuran amphibian from the early Triassic. *Palaeontogr. Abteilung A* 206:1–16.

Raup, D. M., and J. J. Sepkoski Jr. 1982. Mass extinction in the marine fossil record. *Science* 215:1501–1503.

Rayfield, E. J., R. McDonnell, P. M. Barrett, and K. J. Willis. 2003. Testing Triassic vertebrate biostratigraphy: A GIS approach. In *Proceedings of the GIS Research UK 11th Annual Conference*, ed. J. Wood, 324–327. London: City University.

Read, R. W., and L. J. Hickey. 1972. A revised classification of fossil palm and palm-like leaves. *Taxon* 21:129–137.

Redfield, J. H. 1845. A catalogue of fossil fish of the United States, at present known, with descriptions of those which occur in the New Red Sandstone. Paper read before the 6th annual meeting of American Geologists and Naturalists, April 1845.

Redfield, W. C. 1856. On the relations of fossil fishes of the Liassic and Oolitic periods. *Am. J. Sci.* 22:357–363.

Reig, O. A. 1959. Primeros datos descriptivos sobre nuevos reptiles arcosaurios del Triásico de Ischigualasto. *Rev. Asoc. Geol. Argentina* 13:257–270.

————. 1963. La presencia de dinosaurios saurisquios en los "Estratos de Ischigualasto" (Mesotriásico superior) de las provincias de San Juan y La Rioja (República Argentina). *Ameghiniana* 3:3–20.

Reis, O. M. 1909. *Handlirschia gelasii*, n. sp.: Aus dem Schaumkalk Frankens. *Abhandlungen Bayerische Akademie Wissenschaften II*, Kl. 23, Bd. 3:661–694.

Reisz, R., and Sues, H.-D. 2000. The "feathers" of *Longisquama. Nature* 408, 428.

Renesto, S. 1994. A new prolacertiform reptile from the Late Triassic of northern Italy. *Riv. Ital. Paleontol. Stratigr.* 100:285–306.

————. 1995. Ecology and taphonomy of the reptiles from the Calcare di Zorzino (Norian, Late Triassic, N. Italy). In *Extended Abstracts of the II International Symposium on Lithographic Limestones*, Lleida, Spain, Ediciones de la Universidad Autonoma de Madrid, Cuenca 123–130.

————. 2000. Bird-like head on a chameleon body: New specimens of the enigmatic diapsid reptile *Megalancosaurus* from the Late Triassic of Northern Italy. *Riv. Ital. Paleont. Strat.* 106, 157–180.

Renesto, S., and M. Avanzini. 2002. Skin remains in a juvenile *Macrocnemus bassani* Nopsca (Reptilia, Prolacertiformes) from the Middle Triassic of Northern Italy. *Neues Jahrbuch Geol. Paläontol. Abhandlungen* 224:31–48.

Renesto, S., and N. C. Fraser. 2003. Drepanosaurid (Reptilia: Diapsida) remains from a Late Triassic fissure infilling at Cromhall Quarry (Avon, Great Britain). *J. Vertebr. Paleontol.* 23:703–705.

Renesto, S., and C. Lombardo. 1999. Structure of the tail of a phytosaur (Reptilia, Archosauria) from the Norian (Late Triassic) of Lombardy (Northern Italy). *Riv. Ital. Paleontol. Stratigr.* 105:135–144.

Renesto, S., and A. Paganoni. 1995. A new *Drepanosaurus* (Reptilia, Neodiapsida) from the Upper Triassic of Northern Italy. *Neues Jahrbuch Geol. Paläontol. Abhandlungen* 197:87–99.

Renne, P. R., and A. R. Basu. 1991. Rapid eruption of the Siberian Traps flood basalts at the Permo-Triassic boundary. *Science* 253:176–179.

Retallack, G. J. 1977. Reconstructing Triassic vegetation of eastern Australasia: A new approach for the biostratigraphy of Gondwanaland. *Alcheringa* 1:253–283.

———. 1996. Early Triassic therapsid footprints from the Sydney Basin, Australia. *Alcheringa* 20:301–314.

Retallack, G. J., and W. R. Hammer. 1998. Paleoenvironment of the Triassic therapsid *Lystrosaurus* in the central Transantarctic Mountains, Antarctica. *U.S. Antarctic J.* 31:29–33.

Riek, E. F. 1976. An unusual mayfly (Insecta: Ephemeroptera) from the Triassic of South Africa. *Palaeontol. Afr.* 19:149–151.

Rieppel, O. 1985. Die gattung *Saurichthys* (Pisces, Actinopterygii) aus der mittleren Trias des Monte San Giorgio, Kanton Tessin. *Schweizerische Paläontol. Abhandlungen* 108:1–103.

———. 1989a. A new pachypleurosaur (Reptilia: Sauropterygia) from the Middle Triassic of Monte San Giorgio), Switzerland. *Phil. Trans. R. Soc. Lond. B* 323:1–73.

———. 1989b. The hindlimb of *Macrocnemus bassani* (Nopcsa) (Reptilia, Diapsida): Development and functional anatomy. *J. Vertebr. Paleontol.* 9:373–387.

———. 1992. New species of the genus *Saurichthys* (Pisces: Actinopterygii) from the Middle Triassic of Monte San Giorgio (Switzerland), with comments on the phylogenetic interrelationships of the genus. *Palaeontographica* A221:63–94.

———. 2000. Turtles as diapsid reptiles. *Zool. Scripta* 29:199–212.

Rieppel, O., and R. R. Reisz. 1999. The origin and early evolution of turtles. *Annu. Rev. Ecol. Systematics* 30:1–22.

Riley, H., and S. Stutchbury. 1840. A description of various fossil remains of three distinct saurian animals, recently discovered in the Magnesian Conglomerate near Bristol. *Trans. Geol. Soc. Lond.* 5:349–357.

Rinehart, L. F., A. Heckert, and S. Lucas. 2002. Probability plots, populations and paleobiology: What to do with lagerstätten. *J. Vertebr. Paleontol.* 22(Suppl. to 3):100A.

Robbins, E. I., G. P. Wilkes, and D. A. Textoris. 1988. Coal deposits of the Newark rift system. In *Triassic-Jurassic rifting*, ed. W. Manspeizer, 649–682. Developments in Geotectonics 22.

Roberts, J. K. 1928. The geology of the Virginia Triassic. Virginia Geological Survey, University of Virginia.

Robinson, P. L. 1957. The Mesozoic fissures of the Bristol Channel area and their vertebrate faunas. *Zool. J. Linnean Soc.* 43:260–282.

———. 1962. Gliding lizards from the Upper Keuper of Great Britain. *Proc. Geol. Soc. Lond.* 1601:37–46.

———. 1973. A problematic reptile from the British Upper Trias. *J. Geol. Soc. Lond.* 129:457–479.

Robinson, P. 1971. A problem of faunal replacement on Permo-Triassic continents. *Palaeontology* 14:131–153.

Rogers, R. R., C. C. Swisher III, P. C. Sereno, A. M. Monetta, C. A. Forster, and R. N. Martinez. 1993. The Ischigualasto tetrapod assemblage (Late Triassic, Argentina) and $^{40}Ar/^{39}Ar$ dating of dinosaur origins. *Science* 260:794–797.

Rogers R. R., A. B. Arcucci, and F. Abdala. 2001. Taphonomy of the Chañares Formation tetrapods (Triassic, Argentina): Spectacular preservation in volcanogenic concretions. *J. Vertebr. Paleontol.* 21(Suppl. to 3):94A.

Rohdendorf, B. B. 1961. [The oldest infraorders of Diptera from the Triassic of middle Asia]. *Paleontologicheskij Zhurnal.* 1961(2):90–100.

Romer, A. S. 1966. The Chañares (Argentina) Triassic reptile fauna. I. Introduction. *Breviora* 247:1–14.

———. 1967. The Chañares (Argentina) Triassic reptile fauna. III. Two new gomphodonts, *Massetognathus pascuali* and *M. teruggii*. *Breviora* 264:1–25.

————. 1969. The Chañares (Argentina) Triassic reptile fauna. V. A new chiniquodont cynodont, *Probelesodon lewisis*—Cynodont ancestry. *Breviora* 333:1–24.

————. 1970. The Chañares (Argentina) Triassic reptile fauna. VI. A chiniquodontid cynodont with an incipient squamosal-dentary jaw articulation. *Breviora* 344:1–18.

————. 1971a. The Chañares (Argentina) Triassic reptile fauna. X. Two new but incompletely known long-limbed pseudosuchians. *Breviora* 378:1–10.

————. 1971b. The Chañares (Argentina) Triassic reptile fauna. XI. Two new long-snouted thecodonts, *Chanaresuchus* and *Gualosuchus*. *Breviora* 379:1–22.

————. 1972a. The Chañares (Argentina) Triassic reptile fauna. XII. The postcranial skeleton of the thecodont *Chanaresuchus*. *Breviora* 385:1–21.

————. 1972b. The Chañares (Argentina) Triassic reptile fauna. XIII. An early ornithosuchid pseudosuchian *Gracilisuchus stipanicicorum* gen. et sp. nov. *Breviora* 389:1–24.

————. 1972c. The Chañares (Argentina) Triassic reptile fauna. XIV. *Lewisuchus admixtus*, gen et sp. nov., a further thecodont from the Chañares. *Breviora* 390:1–13.

Romer, A. S., and A. D. Lewis. 1973. The Chañares (Argentina) reptile fauna. XIX. Postcranial materials of the cynodonts *Probelesodon* and *Probainognathus*. *Breviora* 407:1–26.

Roselt, G. 1954. Ein neuer Schachtelhalm aus dem Keuper und Beiträge zur Kenntnis von Neocalamites Brongn. *Geol. Berlin* 3:617–643.

Rowe, T. 1988. Definition, diagnosis and origin of Mammalia. *J. Vertebr. Paleontol.* 8:241–264.

Ruben, J. A. 1998. Gliding adaptations in the Triassic archosaur *Megalancosaurus*. *J. Vertebr. Paleontol.* 18(Suppl. to 3):73A.

Sander, P. M. 1989. The pachypleurosaurids (Reptilia: Nothosauria) from the Middle Triassic of Monte San Giorgio, (Switzerland), with the description of a new species. *Phil. Trans. R. Soc. Lond. B* 325:561–670.

————. 1992. The Norian *Plateosaurus* bonebeds of central Europe and their taphonomy. *Palaeogeogr. Palaeoclimatol. Palaeoecol.* 93:255–299.

Schäfer, P., and E. Fois. 1987. Systematics and evolution of Triassic Bryozoan. *Geol. Palaeontol.* 21:173–225.

Scheuring, B. W. 1978. Mikrofloren aus den Meridekalken des Monte San Giorgio (Kanton Tessin). *Schweizerische Paläontol. Abhandlungen* 100:1–100.

Schlische, R. W., and P. E. Olsen. 1988. Structural evolution of the Newark basin. In *Geology of the Central Newark Basin: 5th Annual Meeting of the New Jersey Geological Association,* ed. J. M. Husch and M. J. Hozik, 43–65. Lawrenceville, N.J.: Rider College.

Schmidt, S. 1984. Paleoecology of nothosaurs. In *Third symposium on Mesozoic terrestrial ecosystems,* ed. W. E. Reif and F. Westphal, 215–218. Tübingen: Attempto Verlag.

————. 1987. Phylogenie der Sauropterygier (Diapsida; Trias-kreide). *Neues Jahrbuch Geol. Paläontol. Abhandlungen* 173:339–375.

Scott, R. A., E. S. Barghoorn, and E. B. Leopold. 1960. How old are the angiosperms? *Am. J. Sci.* 258A:284–299.

Selden, P., and J. C. Gall. 1992. A Triassic mygalomorph spider from the northern Vosges, France. *Palaeontology* 35:211–235.

Selden, P. A., J. M. Anderson, H. M. Anderson, and N. C. Fraser. 1999. Fossil araneomorph spiders from the Triassic of South Africa and Virginia. *J. Arachnol.* 27:401–414.

Sepkoski, J. J., Jr., and D. M. Raup. 1986. Periodicity in marine extinction events. In *Dynamics of extinction,* ed. D. K. Elliott, 3–36. New York: Wiley.

Sereno, P. C. 1991. Basal archosaurs: Phylogenetic relationships and functional implications. Society of Vertebrate Paleontology Memoir 2:1–53.

————. 1994. The pectoral girdle and forelimb of the basal theropod *Herrerasaurus ischigualastensis*. *J. Vertebr. Paleontol.* 13:425–450.

Sereno, P. C., and F. E. Novas. 1992. The complete skull and skeleton of an early dinosaur. *Science* 258:1137–1140.

————. 1994. The skull and neck of the basal theropod *Herrerasaurus ischigualastensis*. *J. Vertebr. Paleontol.* 13:451–476.

Sereno, P. C., and R. Wild. 1992. *Procompsognathus:* Theropod, "thecodont" or both? *J. Vertebr. Paleontol.* 12:435–458.

Sereno, P. C., C. A. Forster, R. R. Rogers, and A. M. Monetta. 1993. Primitive dinosaur skeleton from Argentina and the early evolution of Dinosauria. *Nature* 361:64–66.

Sharov, A. G. 1966. [Unique discoveries of reptiles from Mesozoic beds of Central Asia.] *Byulleten Moskovskogo Obshchestva Ispytatelei Prirody, Otdel Geologicheskii* 61:145–146.

————. 1968. [Phylogeny of the Orthopteroidea.] *Trudy Paleontol. Inst. Akad. Nauk SSSR* 118:1–218.

————. 1970. [Unusual reptile from the Lower Triassic of Fergana.] *Paleontol. Z.* 1970:127–131.

Shcherbakov, D. E., E. D. Lukashevich, and V. A. Blagoderov. 1995. Triassic *Diptera* and initial radiation of the order. *Int. J. Dipterological Res.* 6:75–115.

Shiskin, M. A., V. G. Ochev, V. R. Lozovskii, and I. V. Novikov. 2000. Tetrapod biostratigraphy of the Triassic of Eastern Europe. In *The Age of Dinosaurs in Russian and Mongolia*, ed. M. J. Benton, M. A. Shiskin, D. M. Unwin, and E. N. Kurochkin, 120–139. New York: Cambridge University Press.

Shubin, N. H., and H.-D. Sues. 1991. Biogeography of early Mesozoic continental tetrapods: Patterns and implications. *Paleobiology* 17:214–230.

Shubin, N. H., A. W. Crompton, H.-D. Sues, and P. E. Olsen. 1991. New fossil evidence on the sister-group of mammals and early Mesozoic faunal distribution. *Science* 251:1063–1065.

Sigogneau-Russell, D. 1989. Haramiyidae (Mammalia, Allotheria) en provenance du Trias supérieur de Lorraine (France). *Palaeontographica* A206:137–198.

Sigogneau-Russell, D., and G. Hahn. 1994. Late Triassic microvertebrates from central Europe. In Fraser and Sues, *In the shadow of the dinosaurs*, 197–213.

Sill, W. D. 1974. The anatomy of *Saurosuchus galilei* and the relationship of the rauisuchid thecodonts. *Bull. Mus. Compar. Zool.* 146:317–362.

Simms, M. J. 1990. Triassic palaeokarst in Britain. *Cave Sci.* 17:93–101.

Simms, M. J., and A. H. Ruffell. 1990. Climatic and biotic change in the Later Triassic. *J. Geol. Soc. Lond.* 147:321–327.

Simms, M. J., A. H. Ruffell, and A. L. A. Johnson. 1994. Biotic and climatic changes in the Carnian (Triassic) of Europe and adjacent areas. In Fraser and Sues, *In the shadow of the dinosaurs*, 352–365.

Simpson, G. G. 1926a. Are *Dromatherium* and *Microconodon* mammals? *Science* 63:548–549.

———. 1926b. Mesozoic Mammalia. V *Dromatherium* and *Microconodon*. *Am. J. Sci.* 12:87–108.

Small, B. J. 1985. The Triassic thecodontian reptile *Desmatosuchus*: Osteology and relationships. Master's thesis, Texas Tech University, Lubbock.

Small, B. J., and A. Downs. 2002. An unusual Late Triassic archosauriform from Ghost Ranch, New Mexico. *J. Vertebr. Paleontol.* 22(Suppl. to 3):108.

Smith, A. B. 1990. Echinoid evolution from the Triassic to the Lower Liassic. *Cahiers de l'Université Catholique de Lyon, Série Sciences*, 3:79–117.

Smith, A. G., D. G. Smith, and B. M. Funnell. 1994. *Atlas of Mesozoic and Cenozoic coastlines*. Cambridge: Cambridge University Press.

Snyder, R. C. 1954. The anatomy and function of the pelvic girdle and hindlimb in lizard locomotion. *Am. J. Anat.* 95:1–46.

Southwood, T. R. E. 1973. The insect/plant relationship: An evolutionary perspective. In *Insect/plant relationships*, ed. H. F. Van Emden, 3–30. Symposium of the Royal Entomological Society London 6. London: Blackwell.

Spray, J. G., S. P. Kelley, and D. B. Rowley. 1998. Evidence for a Late Triassic multiple impact event on Earth. *Nature* 392:171–173.

Stanley, G. D. 1988. The history of early Mesozoic reef communities. *Palaios* 3:170–183.

Sues, H.-D. 1991. Venom-conducting teeth in a Triassic reptile. *Nature* 351:141–143.

———. 2003. An unusual new archosauromorph reptile from the Upper Triassic Wolfville Formation of Nova Scotia. *Can. J. Earth Sci.* 40:635–649.

Sues, H.-D., and D. Baird. 1993. A skull of a sphenodontian lepidosaur from the New Haven Arkose (Upper Triassic) of Connecticut. *J. Vertebr. Paleontol.* 13:370–372.

———. 1998. Procolophonidae (Reptilia: Parareptilia) from the Upper Triassic Wolfville Formation of Nova Scotia. Canada. *J. Vertebr. Paleontol.* 18:525–532.

Sues, H.-D., and P. E. Olsen. 1990. Triassic vertebrates of Gondwanan aspect from the Richmond basin of Virginia. *Science* 249:1020–1023.

Sues, H.-D., P. E. Olsen, and P. A. Kroehler. 1994. Small tetrapods from the Upper Triassic of the Richmond basin (Newark Supergroup), Virginia. In Fraser and Sues, *In the shadow of the dinosaurs*, 161–170.

Sues, H.-D., P. E. Olsen, and J. G. Carter. 1999. A Late Triassic traversodont cynodont from the Newark Supergroup of North Carolina. *J. Vertebr. Paleontol.* 19:351–354.

Sues, H.-D., P. E. Olsen, D. M. Scott, and P. S. Spencer. 2000. Cranial osteology of *Hypsognathius fenneri*, a latest Triassic procolophonid reptile from the Newark Supergroup of eastern North America. *J. Vertebr. Paleontol.* 20:275–284.

Sullivan, R., and S. G. Lucas. 1999. *Eucoelophysis baldwini*, a new theropod dinosaur from the Upper Triassic of New Mexico, and the status of the original types of *Coelophysis*. *J. Vertebr. Paleontol.* 19:81–90.

Sun, G., Q. Ji, D. L. Dilcher, Q. Zheng, K. C. Nixon, and X. Wang. 2002. Archaefructaceae, a new basal angiosperm family. *Science* 296:899–904.

Tanner, L. H., J. F. Hubert, B. P. Coffey, and D. P. McInerney. 2001. Stability of atmospheric CO$_2$ levels across the Triassic/Jurassic boundary. *Nature* 411:675–677.

Tanner, L. H., S. G. Lucas, and M. G. Chapman. 2004. Assessing the record and causes of Late Triassic extinctions. *Earth Sci. Rev.* 65:103–139.

Tarlo, L. B. 1959. A new Middle Triassic reptile fauna from fissures in the Middle Devonian limestones of Poland. *Proc. Geol. Soc. Lond.* 1568:63–64.

Tatarinov, L. P. 1980. [Toward a prehistory of mammals.] In [*Paleontology and Stratigraphy, 26th International Geological Congress*], ed. B. S. Sokolov, 103–114. Moscow: Nauka.

————. 1994. Terrestrial vertebrates from the Triassic of the USSR with comments on the morphology of some reptiles. In *Evolution, ecology and biogeography of the Triassic reptiles*, ed. J.-M. Mazin and G. Pinna, 165–170. Paleontologia Lombarda, n.s., 2.

Taylor, T. N., and E. L. Taylor. 1993. *The biology and evolution of fossil plants.* Englewood Cliffs, N.J.: Prentice-Hall.

Thayer, P. 1970. Stratigraphy and geology of Dan River Triassic basin, North Carolina. *Southeastern Geol.* 12:1–31.

Thomas, H. H., and N. Bancroft. 1913. On the cuticles of some recent and fossil cycadean fronds. *Trans. Linnean Soc.* 8:155–204.

Tidwell, W. D., A. D. Simper, and G. F. Thayn. 1977. Additional information concerning the controversial Triassic plant: *Sanmiguelia. Palaeontographica* 163B:143–151.

Traverse, A. 1986. Palynology of the Deep River Basin, North Carolina. In *Depositional framework of a Triassic rift basin: The Durham and Sanford sub-basins of the Deep River Basin, North Carolina*, ed. P. J. W. Gore, 66–71. Society of Economic Paleontologists and Mineralogists Meeting Field Trip 3 Guidebook.

————. 1987. Pollen and spores date origin of rift basins from Texas to Nova Scotia as early Late Triassic. *Science* 236:1469–1472.

Tschanz, K. 1989. *Lariosaurus buzzii* n. sp. from the Middle Triassic of Monte San Giorgio (Switzerland), with comments on the classification of nothosaurs. *Palaeontographica* A208:153–179.

Tucker, M. E. 1977. The marginal Triassic deposits of South Wales: Continental facies and paleogeography. *Geol. J.* 12:169–199.

————. 1978. Triassic lacustrine sediments from South Wales: Shore-zone clastics, evaporates and carbonates. In *Modern and ancient lake sediments*, ed. A. Matter and M. E. Tucker, 205–244. Oxford: Blackwell.

Tucker, M. E., and M. J. Benton. 1982. Triassic environments, climates, and reptile evolution. *Palaeogeogr. Palaeoclimatol. Palaeoecol.* 40:361–379.

Unwin, D. M., Alifanov, V. R., and Benton, M. J. 2000. Enigmatic small reptiles from the Middle-Late Triassic of Kirgizstan. In *The Age of Dinosaurs in Russia and Mongolia*, ed. M. J. Benton, M. A. Shiskin, D. M. Unwin, and E. N. Kurochkin, 177–186. New York: Cambridge University Press.

Van Amerom, H. W. J. 1966. *Phagophytichnus elowskii* nov. ichnogen and nov. ichnosp., eine Missbildung infolge von Insektenfrassaus dem spanischen Stephanien (Provinz Leon). *Leidse Geol. Meded. Leiden* 38:181–184.

Van Houten, F. B. 1982. Redbeds. In *McGraw-Hill encyclopedia of science and technology*, 5:441–442. New York: McGraw-Hill.

Veevers, J. J. 1989. Middle/Late Triassic (230 ± 5 Ma) singularity in the stratigraphic and magmatic history of the Pangaean heat anomaly. *Geology* 17:784–787.

Vishnyakova, V. N. 1998. Cockroaches (Insecta, Blattodea) from the Triassic Madygen Locality, Central Asia. *Paleontol. J.* 32:505–512.

Walkden, G. M., and N. C. Fraser. 1993. Late Triassic fissure sediments and vertebrate faunas: Environmental change and faunal succession at Cromhall, South West Britain. *Modern Geol.* 18:511–535.

Walkden, G. M., J. Parker, and S. Kelley. 2002. A Late Triassic impact ejecta layer in southwestern Britain. *Science* 298:2185–2188.

Walker, A. D. 1961. Triassic reptiles from the Elgin area: *Stagonolepis, Dasygnathus* and their allies. *Phil. Trans. R. Soc. Lond. B* 244:103–204.

————. 1964. Triassic reptiles from the Elgin area: *Ornithosuchus* and the origin of carnosaurs. *Phil. Trans. R. Soc. Lond. B* 248:53–134.

Wang, Z. S., E. T. Rasbury, G. N. Hanson, and W. J. Meyers. 1998. Using the U-Pb system of calcretes to date the time of sedimentation of clastic sedimentary rocks. *Geochim. Cosmochim. Acta* 62:2823–2835.

Ward, L. F. 1900. The older Mesozoic. In *Status of the Mesozoic floras of the United States*, 213–748. U. S. Geological Survey 20th Annual Report, part 2.

Warrington, G. 1970. The stratigraphy and palaeontology of the "Keuper" series in the central Midlands of England. *Q. J. Geol. Soc.* 126:183–223.

Watson, D. M. S. 1912. The skeleton of *Lystrosaurus*. *Records Albany Mus.* 2:287–295.

Weber, R. 1968. Die fossile Flora der Rhät-Lias-Übergangs-schicten von Bayreuth (Oberfranken) unter besonderer Berücksichtigung der Coenologie. *Erlanger Geol. Abhandlungen* 72:1–73.

Weems, R. E. 1979. A large parasuchian (phytosaur) from the Upper Triassic portion of the Culpeper Basin of Virginia. *Proc. Biol. Soc. Wash.* 92:682–688.

————. 1987. A Late Triassic footprint fauna from the Culpeper basin Northern Virginia (USA). *Trans. Am. Phil. Soc.* 77:1–79.

————. 1992a. A re-evaluation of the taxonomy of Newark Supergroup saurischian dinosaur tracks, using extensive statistical data from a recently exposed tracksite near Culpeper, Virginia. In *Proceedings of the 26th Forum on the Geology of Industrial Minerals*, ed. P. C. Sweet, 113–127. Virginia Division of Mineral Resources Publication 119.

————. 1992b. The "terminal Triassic catastrophic extinction event" in perspective: A review of Carboniferous through Early Jurassic terrestrial vertebrate extinction patterns. *Palaeogeogr. Palaeoclimatol. Palaeoecol.* 94:1–29.

Weishampel, D. B., and D. B. Norman. 1989. Vertebrate herbivory in the Mesozoic: Jaws, plants and evolutionary metrics. Geological Society of America Special Paper 238:87–100.

Whalley, P. 1986. Insects from the Italian Upper Trias. *Riv. Mus. Civico. Sci. Nat. Bergamo* 10:51–90.

Wheeler, W. H., and D. A. Textoris. 1978. Triassic limestone and chert of playa origin in North Carolina. *J. Sediment. Petrol.* 48:765–776.

Whiteside, D. I. 1986. The head skeleton of the Rhaetian sphenodontid *Diphydontosaurus avonis* gen. et sp. nov., and the modernizing of a living fossil. *Phil. Trans. R. Soc. Lond. B* 312:379–430.

Whittington, H. B. 1975. The enigmatic animal *Opabininia regalis*, Middle Cambrian, Burgess Shale, British Columbia, Canada. *Phil. Trans. R. Soc. Lond. B* 271:1–43.

Wible, J. R. 1991. Origin of Mammalia: The craniodental evidence re-examined. *J. Vertebr. Paleontol.* 11:1–28.

Wible, J. R., and J. A. Hopson. 1993. Basicranial evidence for early mammal phylogeny. In *Mammal phylogeny*, ed. F. S. Szalay, M. J. Novacek, and M. C. McKenna, 45–62. New York: Springer-Verlag.

Wild, R. 1973. Die Triasfauna der Tessiner Kalkalpen. XXIII. *Tanystropheus longobardicus* (Bassani) (Neue Ergebnisse). *Abhandlungen Schweizerischen Paläontol. Gesellschaft* 95:1–162.

————. 1978. Die Flugsaurier (Reptilia, Pterosauria) aus der Oberen Trias von Cene bei Bergamo, Italien. *Boll. Soc. Paleont. It.* 17:176–256.

————. 1984. Flugsaurier aus der Obertrias von Italien. *Naturwissenschaften* 71:1–11.

Wills, L. J. 1907. On some fossiliferous Keuper rocks at Bromsgrove, Worcestershire. *Geol. Mag.* 4:28–34.

————. 1908. Note on the fossils from the Lower Keuper of Bromsgrove. Report of the British association for the Advancement of Science 1908 (1907), 312–313.

————. 1950. *The palaeogeography of the Midlands.* 2nd ed. Liverpool University Press.

Witmer, L. M. 1997. The evolution of the antorbital cavity of archosaurs: A study in soft-tissue reconstruction in the fossil record with an analysis of the function of pneumaticity. Society of Vertebrate Paleontology Memoir 3:i–v, 1–73. Supplement to *J. Vertebr. Paleontol.* 17(1).

————. 2001. Nostril position in dinosaurs and other vertebrates and its significance for nasal function. *Science* 293:850–853.

Woodward, A. S. 1907. On a new dinosaurian reptile (*Scleromochlus taylori* gen. et sp. nov.) from the Trias of Lossiemouth, Elgin. *Q. Jl. Geol. Soc. Lond.* 63:140–146.

Wu, X.-C. 1994. Late Triassic–Early Jurassic sphenodontians from China and the phylogeny of the Sphenodontia. In Fraser and Sues, *In the shadow of the dinosaurs*, 38–69.

Yemane, K. 1993. Contribution of Late Permian palaeogeography in maintaining a temperate climate in Gondwana. *Nature* 361:51–54.

Zangerl, A. 1935. Die Triasfauna de Tessiner Kalkalpen. IX. *Pachypleurosaurus edwardsi, Cornalia* sp. Osteologie-Variationsbreite. *Abh. Schweiz. Palaeont. Ges.* 56:1–80.

Zardoya, R., and A. Meyer. 1998. Complete mitochondrial genome suggests diapsid affinities of tyrtles. *Proc. Natl. Acad. Sci. U S A* 95:14226–14231.

Zeigler, K. E. 2003. Taphonomic analysis of the Snyder Quarry, Chinle Group, Chama basin, New Mexico. In *Paleontology and geology of the Upper Triassic (Revueltian) Snyder Quarry, New Mexico*, ed. K. E. Zeigler, A. B. Heckert, and S. G. Lucas, 49–62. New Mexico Museum of Natural History and Science Bulletin 24.

Zeigler, K. E., S. G. Lucas, and A. B. Heckert. 2001. Variation in the Late Triassic Canjilon Quarry (Upper Chinle Group, New Mexico) phytosaur skulls: Sexual dimorphism in phytosaurs. *J. Vertebr. Paleontol.* 21(Suppl. to 3):117A.

Zeigler, K. E., A. B. Heckert, and S. G. Lucas. 2003. Paleontology and geology of the Upper Triassic

(Revueltian) Snyder Quarry, New Mexico. New Mexico Museum of Natural History and Science Bulletin 24.

Ziegler, A. M., and W. S. McKerrow. 1975. Silurian marine red beds. *Am. J. Sci.* 275:31–56.

Ziegler, A. M., J. M. Parrish, J. Yao, E. D. Gyllenhaal, D. B. Rowley, J. T. Parrish, S. Nie, A. Bekker, and M. L. Hulver. 1993. Early Mesozoic phytogeography and climate. *Phil. Trans. R. Soc. Lond. B.*

Zorn, H. 1971. Die triasfauna der tessiner Kalkalpen. XXI. Paläontologische, stratigraphische und sedimentologishe Untersuchungen des Salvatoredolomits (Mitteltrias) der Tessiner Kalkalpen. *Abhandlungen Schweizer. Paläotol. Fesellsch.* 81:1–140.

index

Italicized page numbers refer to illustrations.

NICHOLAS FRASER is Curator of Vertebrate Paleontology and Director of Research and Collections at the Virginia Museum of Natural History. He has published widely on Triassic tetrapods, and is interested in environmental changes that were instrumental in the rise of the dinosaurs.

DOUGLAS HENDERSON is an artist who combines his interest in dinosaurs and Earth history with traditional natural history illustration. His work has appeared in several national and international touring exhibits as well as numerous magazines and books, including the children's books *Living with Dinosaurs*, *Dinosaur Tree*, *Dinosaur Ghosts*, and, more recently, *Asteroid Impact*.